Library of
Davidson College
VOID

Invertebrate – microbial interactions

Invertebrate – microbial interactions

*JOINT SYMPOSIUM OF
THE BRITISH MYCOLOGICAL SOCIETY
AND THE BRITISH ECOLOGICAL SOCIETY
HELD AT THE UNIVERSITY OF EXETER
SEPTEMBER 1982*

EDITED BY
J.M. ANDERSON, A.D.M. RAYNER &
D.W.H. WALTON

CAMBRIDGE UNIVERSITY PRESS
*CAMBRIDGE
LONDON NEW YORK NEW ROCHELLE
MELBOURNE SYDNEY*

Published by the Press Syndicate of the University of Cambridge
The Pitt Building, Trumpington Street, Cambridge CB2 1RP
32 East 57th Street, New York, NY 10022, USA
296 Beaconsfield Parade, Middle Park, Melbourne 3206, Australia

© British Mycological Society 1984

First published 1984

Printed in Great Britain by The Pitman Press, Bath

Library of Congress catalogue card number: 83–14416

British Library cataloguing in publication data

Invertebrate–microbial interactions. – (British
   Mycological Society symposium; 6)
   1. Invertebrates – Diseases – Congresses
I. Anderson, J.M.   II. Rayner, A.D.M.
   III. Walton, D.W.H.   IV. British Mycological Society   V. British Ecological Society
   592'.02   SF997.5.I5

ISBN 0 521 25395 0

Contents

	List of contributors	vii
	Preface	ix
1	Myxomycetes, microorganisms and animals: a model of diversity in animal – microbial interactions M. F. Madelin	1
2	Soil nutrient transformations in the rhizosphere via animal – microbial interactions D. C. Coleman, R. E. Ingham, J. F. McClellan and J. A. Trofymow	35
3	Interactions between microorganisms and soil invertebrates in nutrient flux pathways of forest ecosystems J. M. Anderson and P. Ineson	59
4	Animal – microbial interactions in wood decomposition M. J. Swift and Lynne Boddy	89
5	The evolution of insect – fungus relationships in the primary invasion of forest timber P. Dowding	133
6	The role of ingested enzymes in the digestive processes of insects M. M. Martin	155
7	Biochemical aspects of symbiosis between termites and their intestinal microbiota J. A. Breznak	173
8	The arthropod gut as an environment for microorganisms D. E. Bignell	205

9	Physiological aspects of destructive pathogenesis in insects by fungi: a speculative review *A. K. Charnley*	229
10	The transmission of Dutch elm disease: a study of the processes involved *Joan F. Webber and C. M. Brasier*	271
11	The interrelationships between microbial entomopathogens and insect hosts: a system study approach with particular reference to the entomophthorales and the eastern spruce budworm *D. F. Perry and G. H. Whitfield*	307
	Index of specific names	333
	Subject index	341

Contributors

J. M. Anderson, *Wolfson Ecology Laboratory, Department of Biological Sciences, University of Exeter, Exeter EX4 4PS, UK.*

D. E. Bignell, *Department of Zoology, Westfield College, University of London, Hampstead, London NW3 7ST, UK.*

Lynne Boddy, *Department of Microbiology, University College, Newport Road, Cardiff CF2 1TA, Wales, UK.*

C. M. Brasier, *Forest Research Station, Alice Holt Lodge, Wrecclesham, Farnham, Surrey GU10 4LH, UK.*

J. A. Breznak, *Department of Microbiology and Public Health, Michigan State University, East Lansing, Michigan 48824 – 1101, USA.*

A. K. Charnley, *School of Biological Sciences, University of Bath, Claverton Down, Bath BA2 7AY, Avon, UK.*

D. C. Coleman, *Colorado State University, Fort Collins, Colorado 80521, USA.*

P. Dowding, *Botany School, Trinity College, Dublin 2, Eire.*

P. Ineson, *Wolfson Ecology Laboratory, Department of Biological Sciences, University of Exeter, Exeter EX4 4PS, UK.*

R. E. Ingham, *Colorado State University, Fort Collins, Colorado 80521, USA.*

J. F. McClellan, *Colorado State University, Fort Collins, Colorado 80521, USA.*

M. F. Madelin, *Department of Botany, University of Bristol, Bristol BS8 1UG, UK.*

M. M. Martin, *Division of Biological Sciences, University of Michigan, Ann Arbor, Michigan 48109, USA.*

D. F. Perry, *Environment Canada, Canadian Forestry Service, Forest Pest Management Institute, Sault Ste Marie, Ontario P6A 5M7, Canada.*

M. J. Swift, *Department of Botany, University of Zimbabwe, PO Box MP.167, Mount Pleasant, Harare, Zimbabwe.*

J. A. Trofymow, *Colorado State University, Fort Collins, Colorado 80521, USA.*

Joan F. Webber, *Forest Research Station, Alice Holt Lodge, Wrecclesham, Farnham, Surrey GU10 4LH, UK.*

G. H. Whitfield, *Agriculture Canada Research Station, Lethbridge, Alberta T1J 4B1, Canada.*

Preface

The scientific search for cause and effect, for order and explanation amongst the complexity and diversity of natural ecosystems, is nowhere more elusive than amongst the lower orders of animals and plants. Invertebrates and microorganisms dominate the world biota in terms of diversity with some 2 million species of invertebrates, including about 1 million species of arthropods, more than 100 000 species of fungi and a virtually inestimable number of bacterial 'species'.

Amongst these organisms there are an infinite number of possibilities for two-way, three-way, and even more complex interactions. The spectrum of bipartite relationships ranges from obligate, mutualistic associations between species at one extreme through commensalism to passive non-specific relationships at the other extreme. The more complex tripartite interactions involving higher plants, microbial pathogens and their invertebrate vectors, and those between vertebrates, pathogens and vectors are still only elementary examples of the potential multiplicity of relationships within natural systems.

It must be evident then that animal–microbial interactions are a dominant feature of the real world and affect most aspects of our health, wealth and natural environment. And yet our knowledge of the functional importance of these interactions is still severely limited despite the interest which they have attracted from biologists over many decades. Leaving aside the medical and agricultural entomological literature our detailed knowledge of invertebrate–microbial interactions has a very skewed distribution on the spectrum of relationships described above. Among the numerous symposia and review volumes there is a predominance of studies on termites (particularly the role of gut protozoa in the nutrition of lower termites, fungus-growing termites or leaf-cutting ants), and an extensive literature on fungal associations with wood-boring Coleoptera and Hymenoptera, nematophagous fungi and insect pathogens. In some cases the stimulus for research has clearly been an economic one: for example, the literature on the lower termites completely outweighs their importance in natural ecosystems of the humid tropics, where processes of decomposition and nutrient recycling are often dominated by the higher termites which comprise about 80% of the described termite species.

In other cases the stimulus for research may have been the deceptive simplicity of the system, for example the *Termitomyces*/Macrotermitinae and *Attamyces*/Attini associations, or the fact that wood-decaying organisms offer a well defined system which is amenable to study and manipulation in the field and laboratory. The same features of large, defined resource units and their durability over long periods of time also have evolutionary implications for some of the specialized interactions found between insects and fungi attacking trees and dead wood. However, the emphasis on these discrete systems defined by a particular resource or organism has led to a neglect of more complex systems which are less defined and less readily studied but which are more typical of natural communities. For example, studies of plant–herbivore interactions, which have come into vogue over the last decade, have revealed the functional importance of grazing for nutrient turnover in grasslands, the influence of animal feeding activities on the growth and reproduction of plants, the importance of food quality for animals and, in particular, the complex evolutionary interplay between plant chemical defences and insect populations. There has, however, been no serious attempt to use these concepts in decomposer systems although there are functional analogues in soil animal/fungal/bacterial associations. For the most part mycologists have tended to view fungal growth in the absence of bacteria and animals; bacteriologists have largely ignored the animal habitat of soil bacteria, as well as the effects of animals on microbial microsite activities and the synergistic impact of mixed bacteria, protozoa and nematode populations on nutrient mobilization and growth dynamics; while the zoologists have mainly ignored the very principles which have led to a greater understanding of herbivore systems.

Our intention in bringing together the participants for this meeting was to provide a forum for the presentation and discussion of the interdisciplinary ideas essential to studies of interactions. The keynote papers in this volume are more specialist than general but do provide

Preface

examples of various levels of complexity of interactions and specificity of experimental approach. To an extent they reflect the preoccupation of research with the subjects mentioned above, yet within this framework are exciting new ideas for extending work in the future on a much broader front.

The first paper by Madelin on myxomycetes reviews a relatively recent field of research which is still at an early descriptive stage. Myxomycetes are ubiquitous organisms found in most soils in numbers suggesting an ecological role comparable to many fungal and bacterial groups. But in addition the unique fungal/bacterial/protozoan features of myxomycetes result in a very wide spectrum of interactions with other organisms, which not only serve as a model for the range of animal–microbial interactions but may also greatly increase the functional significance of the group in soil processes. Developments of successful culture methods could provide useful material for laboratory investigations of the higher levels of interaction complexity.

Coleman *et al.* and Anderson and Ineson are working with complex systems defined in only the broadest terms by the grassland rhizosphere and forest leaf litter respectively – both diffuse habitats which are difficult to delimit under natural conditions. The biota is highly complex and the systems approach, using microcosms, is necessary to provide a feasible experimental unit which is functionally defined. From a functional point of view it is currently impossible, and probably unnecessary, to consider all the component organisms in decomposer systems in the same way that the impact of grazing animals on nutrient turnover in grasslands can be abstracted from all the component plant and animal species in the community.

The following two papers, by Swift and Boddy, and Dowding, are studies of animal–microbial interactions in woody resources. Here it is possible to investigate the interactions far more specifically in terms of the component organisms, but the functional attributes still have to be treated holistically. Both Martin and Breznak consider the mechanisms of interactions in very specific terms, the former through bacterially mediated processes in termite nutrition. Breznak's study is effectively another systems approach in which the animal is used as a test-tube or model system for experimentation. This is the highest level of resolution of the mechanisms of symbiosis and offers an insight into possible new experimental approaches to studying the role of gut microorganisms in mediating chemical processes through gnotobiotic systems. Further development of gnotobiotic systems requires knowledge of the gut as an environment for microorganisms, including sites for physical attachment which could prevent 'washout'. Bignell's paper provides essential background for this approach and is complemented by Charnley's study of the mechanisms of attack on insect cuticle by microbial pathogens. Both studies have a similar approach towards the resolution of invertebrate–microbial interactions but more is known of cuticle breakdown because of the economic implications for biological control.

The two final papers illustrate the high level of understanding which has been developed in systems where there are few organisms, defined resources and economic interests promoting funding and research. The Dutch elm disease story (Webber and Brasier) is both elegant in detail and interesting with respect to the new insights into the nature of the association which these authors provide. Finally, Perry and Whitfield show how detailed knowledge of the interactions between spruce budworms and bacterial pathogens can be modelled to provide a tool for forestry management.

While these papers comprised the core of the meeting the offered papers provided a rich variety of supporting detail and controversy which stimulated much fruitful discussion.

We would like to express our thanks to all participants who made the meeting such a success. We are also grateful to the British Ecological Society and the British Mycological Society for providing funds for overseas speakers, the University of Exeter for excellent facilities, and finally to Professor John Webster as our host.

September 1983

J. M. Anderson
A. D. M. Rayner
D. W. H. Walton

1
Myxomycetes, microorganisms and animals: a model of diversity in animal–microbial interactions

M. F. MADELIN
Department of Botany, University of Bristol, Bristol BS8 1UG, UK

Key words: trophic interactions; *Mycetozoa*; coprophily; food chains; phagocytosis; invertebrate predation; Myxomycete ecology; Myxomycete nutrition

The ecological significance of myxomycetes and their interactions with other organisms

About 430 species of myxomycetes have been described of which 322 species in 52 genera are known to occur in Britain (Ing, 1980, 1982). Though many are cosmopolitan, the great majority inhabit the temperate regions, while curiously few appear to occur in tropical rain forests (Sanderson, 1922a; Farr, 1976). However, the latter conclusion is based on a dearth of fructifying material rather than on experimental isolation. Indeed, because field studies so far have almost all been concerned with the occurrence of sporangia and not with feeding stages, very little is known about the nature and magnitude of the natural roles of myxomycetes in the soils, litters, decaying wood and other vegetable materials in which their swarm cells, myxamoebae and plasmodia, feed and grow. The fact that sporangia usually form after the plasmodium has migrated away from the site of its main feeding activity means that their abundance on a given substratum is seldom related to the preceding abundance of the vegetative phase in that particular part of the substratum, or indeed in the substratum generally. For this reason, information on the actual ecological roles of myxomycetes is not abundant despite the wealth of floristic and taxonomic data. Ecological information tends to be scattered in the literature and often is incidental to other sorts of study. Some relevant information derives from experimental studies, but most available physiological and genetical information on myxomycetes derives from work on only a small number of relatively easily cultured species such as *Physarum polycephalum* and *Didymium iridis*. The former is probably an example of just one ecological type of myxomycete among many, namely the fungivore. It is chiefly because it is a tractable model organism that it has been so thoroughly studied. It seems unlikely that myxomycetes as a whole will prove to be any more uniform than the fungi when their ecology and associated characteristics have been elucidated, so one must guard against generalizing too much from the example of *Physarum*. Indeed many common and conspicuous species have resisted all efforts to make them grow and fruit under laboratory conditions, even in gross culture with other microorganisms, and only a few attempts to grow myxomycetes free of bacteria have been successful. This contrasts with the readiness of many true fungi to grow axenically in comparatively simple nutrient media, and probably reflects the natural dependence of most myxomycetes on ingestion of associated microorganisms for food. This dependence in turn provides the most obvious basis for interactions between myxomycetes and other organisms.

Predation upon food-microorganisms is not the only interaction with other organisms exhibited by myxomycetes. They themselves fall victim to other microorganisms and microfauna, and other indirect relationships occur which are of ecological significance. Fig. 1.1 presents the principal features of the trophic interactions which are undoubtedly those of most ecological significance. Unfortunately there appear to be no objective quantitative data by which nutrient or energy flows through the different food chains involving myxomycetes can at present be assessed. Intuitively one suspects that the major ones are inputs from bacteria and fungi and outputs through arthropods, but these are little more than guesses in the absence of actual measurements of biomass and of material and energy fluxes. In this laboratory we are currently investigating the numbers of 'plasmodium-forming units' in soils. These are units analogous to the 'colony-producing units' of soil mycologists and bacteriologists. We have found that myxomycetes are present in nearly every soil which we have examined (mostly British, but from all continents except South America and Antarctica) at usually 10 to 1000 g dry wt^{-1}, but occasionally even up to 10 000 g dwt^{-1}. If one assumes the least possible size for a plasmodium-forming unit, namely that of a single vegetative cell or spore of which the volume is of the order of 1000 times that of a bacterium, these numbers represent a biomass equivalent to that of between 10^4 and 10^7 bacteria per gram dry weight of soil. In view of this sometimes substantial level of abundance of myxomycetes, their virtual ubiquity, and their many interactions with other organisms, it seems likely that their general role in terrestrial ecology has been greatly underestimated in the past.

One reason for the breadth of their range of interactions with other microorganisms and animals is their remarkable combination of 'plant', 'microbial' and 'animal' features. To this extent they compose a model of different animal–microbial interactions against which to view these phenomena in general. They are however part of a still largely unexplored field where direct observations still play an important part in the development of understanding.

Because myxomycetes have several discrete stages in their life cycle, and in particular have more than one feeding stage, the interactions of these individual stages with different sorts of associated microorganisms and animals will be considered separately. Attention will first be paid to the feeding of the unicellular and plasmodial vegetative stages on other organisms, and then to the reverse aspect, namely the feeding of microorganisms and invertebrate animals upon the several distinct

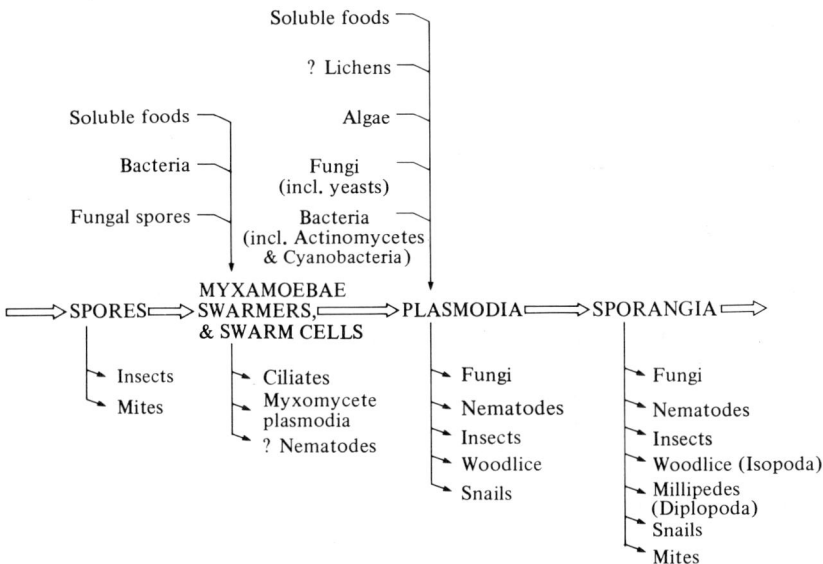

Fig. 1.1 Summary of trophic interactions between the different life cycle stages of myxomycetes and other microorganisms, and fauna. Sources of food for myxomycetes are above the horizontal arrows which represents the myxomycete life cycle; consumers of myxomycetes are below the line. 'Soluble foods' would include products of extracellular digestion and autolysis of other organisms, as well as inorganic nutrients.

stages in the myxomycete life cycle. Some physiological and ecological phenomena which may be interpreted as consequences of trophic interactions will then be considered. Finally indirect interactions of animals and myxomycetes which do not involve predation will be reviewed.

Feeding of myxomycetes on other organisms
Feeding of swarm cells and myxamoebae on bacteria

The germinating spore releases one or sometimes more haploid myxamoebae or flagellate cells. The latter are typically unequally biflagellate at the anterior end, while myxamoebae have no flagella. Myxamoebae and flagellate cells inter-convert, sometimes under the control of availability of free water, while at other times influenced by nutritional factors (Goodman, 1972). Further, flagellate cells can change in behaviour and appearance between being comma-shaped, actively swimming or 'hopping' cells – sometimes distinguished as *swarmers* – and amoeboid cells which creep over firm surfaces referred to as *swarm cells*. Often however the latter term is used for all flagellate cells of myxomycetes (de Bary, 1887; Indira, 1969).

Before 1890 when Lister proved that swarm cells ingest and digest bacteria, it was generally held that they obtained all of their food in solution. Since then many more observations of feeding on bacteria have been made. In culturing haploid cells of myxomycetes, especially of the species most commonly used in research laboratories, two-membered (or monoxenic) culture with *Klebsiella* (*Aerobacter*) *aerogenes* or *Escherichia coli* is commonly employed. It is general experience that little or no multiplication occurs on plain or nutrient agar media unless suitable food bacteria are present (e.g. Parker, 1946; Carlile, 1974). It is evident that these cells depend largely on phagotrophy, even though osmotrophic (absorptive) nutrition possibly makes a concurrent contribution.

Axenic growth of myxamoebae has been achieved with only a few myxomycetes. One early procedure was to add formalin-killed bacteria (*Aerobacter aerogenes* or *Escherichia coli*) to corn meal agar (Henney & Henney, 1968). However, cultivation of amoebae with no bacteria present was apparently achieved first by Ross (1964), who grew myxamoebae of *Badhamia obovata* (*B. curtisii*) on a complex agar medium, and at a slower rate in a complex chemically undefined nutrient *solution*. Subsequently Haskins (1970), Goodman (1972), and Henney, Asgari & Henney (1974) respectively grew *Echinostelium minutum*, *Physarum polycephalum* and *P. flavicomum* in complex undefined liquid media, while McCullough & Dee (1976) eventually succeeded in culturing myxamoebae of *P. polycephalum* in chemically defined liquid media. However, these achievements do not necessarily mean that wholly absorptive nutrition is possible in nature. The successful cultivation of *P. polycephalum* in defined liquid medium was later shown to be due to mutations (McCullough, Dee & Foxon, 1978). It was thought possible that the mutations were of a regulatory nature, allowing the expression of genes normally manifested only in the more readily culturable plasmodia, rather than that they enabled the amoebae to synthesize a nutrient formerly supplied by bacteria. Further evidence against wholly osmotrophic nutrition of myxamoebal populations in nature lies in the observation that myxamoebae of *Echinostelium minutum* grown axenically failed to give rise to plasmodia, but did so when re-exposed to living bacteria or yeast (Haskins, 1970).

Lister (1890) recorded that the capture and ingestion of bacilli always took place at the posterior end of amoeboid swarm cells where pseudopodia, often in the form of extremely delicate threads, extended out into the aqueous medium. He saw individual swarm cells of *Stemonitis fusca* with as many as six or seven large bacilli attached to these pseudopodia. The captured bacteria were drawn towards the body of the swarm cell where an extension of the cell folded over each bacillus and absorbed it into the interior. He also observed long posterior pseudopodia which captured bacteria in *Trichia fragilis* and *Didymium difforme*.

Indira (1969) similarly described the intake of food particles by elongate amoeboid swarm cells of *Stemonitis herbatica* as occurring by pseudopodial ingestion, but in addition reported that pseudopodia were occasionally seen to form and close, as if ingesting food, at a definite region slightly below the flagellar (anterior) end. Evidently there is some variation in the site of ingestion in amoeboid swarm cells, although the posterior site seems to be the most important. Gilbert (1928*b*), in an extensive study of feeding by swarm cells, reported that only in the phase of undulatory creeping movements was ingestion of bacteria seen and never in swarmers displaying free-swimming rotatory movement. Indira (1969) however reported that free-swimming swarmers of *Stemonitis herbatica* captured and ingested bacteria by a novel method. When one passed near a bacterial colony there were vigorous lashing movements of its flagella which seemed not only to aid detection of bacteria but also to sweep them close to the body of the swarmer. When a bacterium had reached the curve of the comma-shaped cell, just below the flagellar end, the swarmer bent, trapping the bacterium in the bend. It then revolved rapidly for a while. On halting it relaxed and the bacterium was then seen to be enclosed in a food vacuole. Thus bacteria seemed to be ingested by these swarmers at a definite region of the body, not by means of pseudopodia, but by means of movements of the swarmer as a whole. Whether phagocytosis of bacteria of the type seen in *S. herbatica* is common during the free-swimming phase must remain in doubt until other species have been closely and critically studied.

Indira (1969) suggested that at least in the swarmers of *Stemonitis herbatica* the flagella were partly sensory in function, serving to detect the presence of food particles. Smart (1938) similarly referred to creeping swarm cells using the flagellum as if it were an antenna. Chemotaxis of non-flagellate myxamoebae towards live bacteria over distances exceeding 600 μm across surfaces has been reported by Konijn & Koevenig (1971). The principal attractant was apparently of molecular weight 200–400, but minor attractants were also present. An artificial mixture of amino acids was also effective, and L-tryptophan was the most active of single amino acids. Cyclic AMP was without positive effect, although it attracts amoebae of some cellular slime moulds and is secreted by bacteria.

There are indications that swarm cells have specific surface sites for attachment of prey, in that captured particles of powdered carmine apparently adhere to the tips of pseudopodia much less tenaciously than do bacteria (Lister, 1890). However, more information is available concerning phagocytosis by cellular slime mould amoebae (dictyostelids) than for myxamoebae. Those of *Dictyostelium discoideum* added to dense suspensions of *Escherichia coli* captured bacteria at a decreasing rate for about 60 min, when each amoeba had taken, on average, 67 bacteria, suggesting that there were only a limited number of binding sites to which bacteria could attach (Glynn, 1981). A half-saturation constant (K_m) for phagocytosis of bacteria was determined at just over 5×10^7 bacteria ml^{-1}. Similar quantitative studies on myxomycete amoebae and swarm cells would be both physiologically and ecologically interesting.

Little appears to be known about the range of bacteria on which myxamoebae and swarm cells can feed. Laboratory cultivation frequently employs *Escherichia coli*, *Klebsiella* (*Aerobacter*) *aerogenes* or *Enterobacter aerogenes*, while Gilbert (1928b) reported that swarm cells *in vitro* ingested and seemed to thrive on cells of *Nocardia* (*Actinomyces*) *asteroides*, a soil-inhabiting actinomycete pathogenic to man and animals.

The fate of bacteria ingested by swarm cells appears to be swift and total digestion. Lister (1890) observed that digestion was evident in as little as 15 min, and after one hour nothing was left but a faint residiuum. He was rarely able to see discharge of the residuum of bacilli from swarm cells. Feest and Madelin (unpublished) have noted the disappearance of ingested bacteria in swarm cells of *Didymium squamulosum* in about 15 minutes, and sometimes in even as little as 6 minutes.

Feeding of swarm cells and myxamoebae on eukaryotic microbes

In natural habitats swarm cells and myxamoebae will almost inevitably be closely associated with microbes other than bacteria. Their capacity to consume eukaryotic microorganisms will doubtless be limited by the relative sizes of predator and prey, but it is known that many fungal spores and unicellular algae are ingested (Gilbert, 1928*a*; Lister, 1890). On the other hand there is no record of myxomycete swarm cells ingesting each other, whether of the same or different species, even though the considerably larger zygotes and plasmodia of myxomycetes have often been seen to ingest microcysts and ungerminated myxomycete spores (Gilbert, 1928*b*). Indira (1969) observed amoeboid swarm cells of *Stemonitis herbatica* trying to ingest myxomycete spore cases and cysts but failing.

Gilbert (1928*a*) reported that during their phase of more or less amoeboid creeping movement the swarm cells of *Dictydiaethalium plumbeum* ingested fungus spores in essentially the same way as they caught bacteria, via a tenuous pseudopodium put out from the posterior part of the swarm cell. Myxamoebae on the other hand ingested spores and bacteria in the same manner as did plasmodia, that is by extending a part of the body and surrounding or engulfing the prey. Spores of different fungal species vary in their susceptibility to ingestion and digestion, apparently for different reasons, and four classes were distinguished by Gilbert (1928*a, b*; Table 1.1). Swarm cells feeding on the readily ingested and digested spores became very large and multiplied rapidly. They were often packed with spores which, at division of the swarm cell, were apportioned between the sister cells. The second category comprised ingestible spores which were sooner or later egested without apparent change, though no tests of the viability of these spores were reported. Spores over $8 \mu m$ in diameter or exceeding about

Table 1.1. *The susceptibility of fungal spores to ingestion and digestion by swarm cells of about 20 species of myxomycetes*

I.	Spores easily ingested and digested Sporangiospores: *Circinella simplex, C. spinosa, Helicostylum piriforme, Mucor javanicus, M. ramannianus* Ascospores: *Cryptovalsa sparsa* (6–7 × 1–2 µm), *Eutypella scoparia* (4 × 1 µm), *Mollisia* sp. (7–8 × 3–4 µm) Basidiospores: *Daedalea quercina, Hydnum septentrionale* Conidia and sprout cells: *Acrostalagmus fragrans, Candida brevis, C. tropicalis* (= *Monilia candida*), *Diaporthe oxyspora, Endomycopsis burtonii* (= *Dematium chodatii*), *Endothia parasitica, Isaria felina* (3 × 2 µm), *Oospora humi, Pullularia nigrans, Thyronectria denigrata, Trichoderma lignorum, Trichosporon cutaneum* (= *Oidium cutaneum*), *Verticillium tenerum* (= *Acrostalagmus cinnabarinus*), yeasts (unidentified)
II.	Spores ingested in quantities, but seem not to be digested Basidiospores: *Cortinarius semisanguineus*
III.	Spores too large to be taken in Sporangiospores: *Rhizopus oryzae* Ascospores: *Chaetomium cochliodes* (≥9 × 7 µm), *C. globosum* (≥9 × 7 µm), *Diaporthe oxyspora, Endothia parasitica* (≥9–10 × 5 µm), *Thyronectria denigrata* (≥9–10 × 5 µm) Basidiospores: *Amanita phalloides* (9–12 × 8–9 µm), *Flammulina velutipes*[a] (= *Collybia velutipes*) (7–9 × 3–4 µm), *Ganoderma applanatum* (= *Fomes applanatus*)
IV.	Spores apparently unfavourable and practically never taken in Conidia: *Aspergillus niger, A.* spp., *Penicillium camembertii, P. roquefortii, P.* spp., *Stysanus stemonitis*

[a] Larger swarm cells occasionally took in spores.
Data from Gilbert, 1928*b*.

one-fifth of the volume of the swarm cell were too large to be ingested. The rather large swarm cells of *Dictydiaethalium plumbeum* are 10–20 µm long by 5–10 µm wide, so that a fungus spore around 7 × 3 µm is close to the limiting size for ingestion. Fungus spores of the fourth category were not ingested and even seemed to repel the swarm cells. Repeated freezing of such spores in liquid air or washing in alcohol rendered the spores ingestible but not digestible, for they were later thrown out unchanged. Gilbert (1928*b*) speculated that the treatment with alcohol removed some substance, possibly a layer of air or grease, and that the composition of the wall or substances adhering to it were factors determining ingestibility. More recent work has revealed that many spores of fungi have thin but discrete surface layers, including rodlet layers, of various compositions (Cole & Pope, 1981). It is not impossible that grazing of fungus spores by phagotrophic microbes including myxomycetes has exerted sufficient selection pressure for specific features which protect against ingestion to have arisen in some true fungi, particularly among small-spored species inhabiting litter and soil. Gilbert (1928b) found that small fungus spores with a more or less moist or mucilaginous coat were preferred to the dry or perhaps greasy spores of *Penicillium roquefortii*.

Gilbert (1928b) recognized several intergrading categories into which myxomycetes fell according to the length of time that their swarm cells remained in the free-swimming rotating state before entering that of undulatory creeping movement. Since he believed that the swarm cell with rotating movement did not ingest food, the relative times spent in each state should affect the capacity for consuming spores. Whether similar relative durations of the states would prevail under field conditions as in the laboratory is unknown. In *Lycogala epidendrum*, *Physarum viride* and *Stemonitis fusca* the swarm cells in culture were nearly always in the rotatory state and so were rarely seen to ingest fungus spores. A few species, including *Arcyria denudata*, *Arcyria incarnata* and *Hemitrichia clavata*, had swarm cells which in culture ingested fungus spores to only a limited degree, and seldom became gorged with them. Other myxomycetes which assumed the creeping or undulatory form early after germination and which readily ingested spores included *Badhamia lilacina*, *B. magna*, *Comatricha typhoides*, *Dictydiaethalium plumbeum*, *Didymium nigripes* var. *xanthopus*, *Hemitrichia vesparium*, *Leocarpus fragilis*, *Stemonitis splendens* var. *flaccida* and *Trichia floriformis*. Because the fungus spores which Gilbert found capable of being ingested and digested were for the most part from species that occur in moist woodland situations where myxomycetes are abundant, he thought it possible that they may form a considerable part of the natural food of the swarm cells.

During the digestion of fungus spores, which predictably takes considerably longer than for bacteria, swarm cells move sluggishly and often withdraw their flagella for a time (Gilbert, 1928b). After a number of hours spores ingested by the swarm cells of *Dictydiaethalium plumbeum* become smaller, more indistinct, and finally are digested with the exception of certain portions such as oil globules, which are then expelled (Gilbert, 1928a).

Gilbert (1928a) found that myxomycete swarm cells showed no obvious migration towards easily digested fungus spores. On the other hand some spores ordinarily not ingested seemed to repel the swarm cells. Inorganic matter and indigestible spores were occasionally engulfed and soon cast out again, suggesting that the swarm cells were unable to differentiate between metabolically useful and useless materials. However this is only partly true because swarm cells were capable of discriminating between digestible and indigestible spores in a mixture, the former being ingested in large quantities, the latter rarely.

One is led to conclude, chiefly from these observations of Gilbert, that the ingestion of fungus spores by myxomycete swarm cells is a complex process involving recognition and possible rejection at more than one step. Initial surface adhesion, the size of the spore, and intravacuolar recognition appear to be three important determinants of the use of fungus spores as food, but the whole process requires modern study, especially because of its potential ecological and even practical significance.

Feeding of plasmodia on bacteria

Most plasmodia live in environments where bacteria are likely to be plentiful. There seems to be little doubt that these latter constitute an important source of food, if not the major one. Nevertheless, a capacity for osmotrophic nutrition also exists, since the plasmodia of a number of species have been cultured axenically on or in media with dissolved nutrients; however, its contribution to nourishment of plasmodia in natural habitats is uncertain.

Plasmodia brought into culture from natural sources are possibly without exception closely associated with bacteria which may prove difficult to eliminate. Sobels (1950), for example, isolated six species of bacteria (as well as certain yeasts) from plasmodia of *Badhamia utricularis*. An early view (Skupienski, 1920) that species-specific myxomycete–bacterium symbioses existed is no longer held. Cohen (1941) found no general correlation between tribes and genera of bacteria and the ability of plasmodia of *Badhamia foliicola* and *B. magna* to feed on them in two-membered culture; nevertheless the slimy capsules formed by several species of bacterium markedly inhibited feeding, and loss of slime with age increased susceptibility to attack. Acid-fast bacteria of the genera *Mycobacterium*, *Proactinomyces* and *Actinomyces* supported fair to excellent growth of the plasmodia, and colonies of other actinomycetes and *Bacillus mycoides* were consumed, observations which are noteworthy in view of the significance of actinomycetes in soils. On the other hand another actinomycete, *Nocardia* sp., was not utilized by plasmodia of an aquatic form of *Didymium difforme*, even though it fed upon *Bacillus subtilis, Sarcina lutea, Escherichia coli* and *Pseudomonas fluorescens* (Gray & Lanning, 1978).

Cyanobacteria (blue–green algae) may also serve as food (Indira, 1968). For example the plasmodium of *Physarum gyrosum* can sometimes spread over a crust of cyanobacteria at the soil surface, suggesting that the latter organism could furnish a source of food. Indira also collected fructifications of *Physarum cinereum* on a crust of cyanobacteria on the wall of a house, while Skulberg (1958) found *Didymium nigripes* living in an aquatic habitat with many algae, and successfully cultured its plasmodia upon mannite agar to which the cyanobacterium *Oscillatoria rubescens* was added at weekly intervals.

Although the axenic culture of certain plasmodia is readily achieved, growth of others is much better when live bacteria are present, and may be slow and reluctant when freed from bacteria, with fruiting inhibited (Hok, 1954; Lazo, 1961). In respect of its vigorous axenic growth and fruiting *Physarum polycephalum* is exceptional.

The mechanism of feeding of plasmodia upon bacteria may vary with circumstance. Cohen (1941) reported that plasmodia *in vitro* crawled over dry colonies of bacteria as a fan and displayed 'central feeding', while plasmodia which met slimy colonies of bacteria banked up against the edge of the colony without covering it and conducted 'peripheral feeding'. The ingestion of food particles takes place as the advancing margin of a plasmodium flows around and encloses each particle which then lies within a vacuole (Camp, 1937; Gray & Alexopoulos, 1968). However, Benedict (1965) reported that the delicate aphanoplasmodium of *Stemonitis fusca* produced lateral pseudopodia which engulfed bacteria.

Plasmodia of a number of species display chemotaxis, and reactions to bacteria have been described by Konijn & Koevenig (1971), who found that plasmodia of *Didymium iridis* and *Physarum pusillum* were attracted to *Escherichia coli* before engulfing them. Cyclic AMP, which some bacteria produce, did not attract the plasmodia, but a mixture of amino acids did. Watanabe (1932) used tactic behaviour towards bacterial streaks as a criterion of the ability of plasmodia to feed on bacteria, but Cohen (1941) found that taxis and feeding were not necessarily linked.

If present in excessive numbers bacteria can be deleterious towards some plasmodia. Howard (1931) reported that the growth of bacteria in gross cultures of *Physarum polycephalum* on pieces of fungus fruit body was detrimental to the plasmodium, which actually moved away from bacterial colonies.

Feeding of plasmodia on yeasts

Naturally occurring associations between yeasts and plasmodia have been demonstrated by isolating *Cryptococcus laurentii* var. *laurentii* (*Torulopsis laurentii*) from plasmodial tracks of *Licea flexuosa*; *Rhodotorula minuta* and another yeast from tracks of *Badhamia utricularis*; and *Candida scottii* from an unidentified myxomycete. Further, *Licea flexuosa* grew well in culture with *C. laurentii* var. *laurentii* and *Saccharomyces cerevisiae*, and *Badhamia utricularis* with *C. laurentii* var. *laurentii* and *R. minuta* (Sobels, 1950). Sobels & Cohen (1953) used washed cells of either *S. cerevisiae*, *Saccharomyces ellipsoideus* or *Torula aclotiana* (*Rhodotorula mucilaginosa*) when transferring myxomycete plasmodia from gross to two-membered culture, while many other workers also have used selected yeast species to grow myxomycete plasmodia. Rather unexpectedly, Gray & Lanning (1978) reported that an aquatic form of *Didymium difforme*, capable of completing the whole of its life cycle while submerged, could not feed on *Saccharomyces cerevisiae*.

Feeding of plasmodia on fungal hyphae

Vegetative mycelium is consumed by plasmodia of numerous myxomycetes, attack upon individual hyphae apparently occurring by extraplasmodial digestion rather than ingestion. This feature may be related to the shape of intact fungal hyphae, which is inappropriate for ingestion. A vivid account of hyphal lysis is given by Lister (1888), who described how hyphae of an unidentified mycelium melted away 'like sugar in boiling water' as the plasmodium of *Badhamia utricularis* encroached upon them. A similar account was given by Howard & Currie (1932a) and interpreted as evidence for chemical destruction of hyphae. More recently, Stirling, Cook & Pope (1979) reported that *Physarum polycephalum*, which in nature is fungivorous, produces chitinases, while Knowles & Carlile (1978) recorded that growth of the plasmodium of this species is supported by N-acetyl-D-glucosamine, D-glucosamine hydrochloride, glucose and mannose – all possible products of hydrolysis of fungal walls. Although Howard & Currie (1932a) observed that hyaline hyphae were attacked more readily than were coloured ones, Sobels (1950) found that the plasmodium of *Badhamia utricularis* readily digested the dematiaceous aerial mycelium

Table 1.2. *The species of myxomycetes shown by Howard & Currie (1932b) to consume* in vitro *the vegetative mycelia of a wide variety of fungi*

Arcyria occidentalis	Hemitrichia vesparium	Physarum tenerum
Badhamia magna	Leocarpus fragilis	Physarum virescens
Badhamia rubiginosa	Lindbladia effusa	Physarum viride
Badhamia utricularis	Lycogala epidendrum	Stemonitis fusca
Brefeldia maxima	Physarum cinereum	Trichia decipiens
Fuligo septica	Physarum flavicomum	Trichia persimilis
Hemitrichia clavata	Physarum polycephalum	Trichia scabra

of *Cladosporium herbarum*. Howard & Currie (1932*a*) also observed that the thickness of hyphal walls determined their susceptibility to digestion, thin-walled hyphae from the hymenium of a fruit body dissolving away within a few seconds of contact by the margin of a plasmodium.

Fungi whose vegetative mycelia have been seen to be consumed by plasmodia include unidentified species of *Aspergillus* and *Penicillium*, *Aspergillus glaucus*, *Sterigmatocystis* sp., *Stysanus* sp. and various other 'moulds' (Hilton, 1914; Skupienski, 1928). Howard (1931) also reported that the plasmodium of *Physarum polycephalum* completely utilized vegetative hyphae of several Basidiomycotina, including *Cyathus stercoreus*, *Exidia glandulosa*, *Lenzites betulinus*, *Nidularia pulvinata*, *Tremella mesenterica* and *Tremella* sp., but utilized *Merulius americanus* less well and failed to touch various Mucorales and the basidiomycete *Guepinia spathularia*. Hyphae of *Penicillium* and *Alternaria* spp. were utilized, but the spores left untouched.

Extending this work to a wider range of organisms, Howard & Currie (1932*b*) found that plasmodia of 21 species (see Table 1.2) *in vitro* digested the vegetative mycelia of a wide variety of fungi responsible for the decay of wood and plant debris. Individual myxomycetes varied in their ability to consume mycelia of different species of fungus, and individual fungi (of which 49 chiefly wood-inhabiting species were tested) varied in their susceptibility to attack by different species of myxomycete. Further, the behaviour of the plasmodia in digesting fungus mycelia was influenced by the medium on which the fungus was growing. It thus appears likely that under diverse natural conditions the pattern of myxomycete–fungus interactions might vary considerably, but there is little available information on interactions in nature. Elliott & Elliott (1920) kept a large newly fallen oak branch under observation for eight years and noted that it was not until the seventh year that

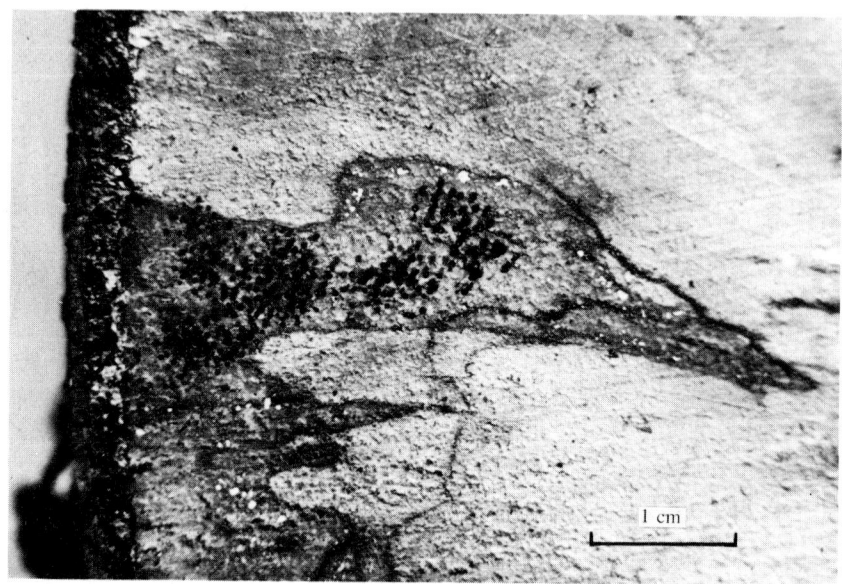

Fig. 1.2. *Comatricha nigra* sporulating at the cut surface of a section from a felled beech trunk, decayed extensively by *Stereum hirsutum*. The myxomycete appeared after a few weeks' incubation in corresponding positions on cut surfaces at least 0.5 m apart, so presumably occupied a single decay column. (Photograph by courtesy of Drs A. D. M. Rayner and D. Coates.)

fructifications of *Physarum nutans* appeared, followed by *Stemonitis fusca* in the eighth year. The distribution of *P. nutans* on the log correlated with that of fructifications of the Discomycetes *Bulgaria polymorpha* and *Coryne sarcoides* which, having originally been abundant, disappeared after four and five years respectively. This suggests that the mycelia of the two Discomycetes were absorbed and destroyed during the plasmodial stage of the myxomycete, which subsequently took about three years to fruit. Sobels (1950) similarly observed that myxomycetes followed attack by wood-rotting fungi in fallen dead trees, with no trace of the fungi subsequently being found. *Comatricha nigra* has been found occupying a decay column several metres long in a recently felled beech trunk decayed by *Stereum hirsutum*, and *Badhamia utricularis* has been found associated with mycelium of *Xylaria hypoxylon* in incubated cut beech lengths (A. D. M. Rayner & D. Coates, personal communication; Figs. 1.2 and 1.3). There is thus evidence for an important role for myxomycetes in the decay of dead

Fig. 1.3. Plasmodium of *Badhamia utricularis* (arrows) feeding on mycelial mats of different individuals of *Xylaria hypoxylon* produced, after incubation in polythene bags, from decay columns in cut beech logs. The myxomycete occurred in corresponding positions on several slices cut serially from the logs and incubated separately. (Photograph by courtesy of Drs A. D. M. Rayner and D. Coates.)

timber, and despite the general lateness of the appearance of their fructifications, they may be vegetatively active from a comparatively early stage. Indeed Ostrofsky & Shigo (1981) have found myxomycete cysts within discoloured wood of living red maple (*Acer rubrum*).

Feeding of plasmodia on fungus fruit bodies

The mere presence of a plasmodium or sporangia on a fruit body is not proof that the slime mould has actually fed upon it, since a plasmodium may migrate laterally from the substratum on which it has been feeding before it fruits. Thus although there are many records of myxomycetes on fruit bodies (both for temperate and tropical zones), and in many cases the physical relationship and the conditions of the plasmodium and fruit body are such that it may reasonably be presumed that the myxomycete is fungivorous (Fig. 1.4), this is certainly not

Fig. 1.4. Plasmodium of *Badhamia utricularis* advancing over, and feeding upon, a basidiocarp of *Phlebia merismoides* on the surface of a much decayed birch log. The extent of the plasmodium can be discerned by its shiny surface; the uninvaded surface of the basidiomycete is matt. Conspicuous strands link this plasmodial fan to others elsewhere on the extensive basidiocarp. (Photograph by M. F. Madelin and T. Colborn.)

always the case. Lister (1918) reported *Trichamphora pezizoidea* on the gelatinous lobes of *Auricularia mesenterica* but was unclear whether it obtained food from the fungus or the underlying wood. Sunhede (1973) found *Reticularia lycoperdon* on an insect-damaged fruit body of *Fomes fomentarius* growing on *Betula*, but it almost certainly had emerged from small cavities and pores in the decaying wood. Records of myxomycetes on agarics are uncommon. Eliasson (1981) discovered sporangia of *Leocarpus fragilis* on the cap of the agaric *Rozites caperata*, and remarked that since the fleshy fruit bodies of agarics are ephemeral there is seldom time for myxomycetes to fructify on them. Similarly, extensive vegetative growth on agarics is rather unlikely even though it can be effected in the laboratory.

Experimental studies on the feeding of plasmodia on basidiocarps appear to have begun with Lister's (1888) report that *Coriolus versico-*

lor, Bjerkandera adusta, and *Stereum hirsutum* were especially suitable as food for the plasmodium of *Badhamia utricularis*. He presented slices of various fruit bodies, including agarics, to cultures of this plasmodium and found a range of susceptibilities from complete digestion, through partial digestion, to little effect. *Amanita rubescens* was consumed slowly but a deposit of undigested spores was left, related perhaps to the different composition of hyphal and spore walls. Sanderson (1922b) fed *Physarum viride* var. *rigidum* on fruit bodies of *Schizophyllum commune* and *Hirneola hispida*, but found it died when transferred onto the Pyrenomycetes *Daldinia concentrica, Nummularia pithodes* and *Ustulina zonata*. Howard (1931) obtained the plasmodium of *Physarum polycephalum* from pilei of *Pleurotus cornucopiae* and thereafter grew it upon pilei of various agarics, Polyporaceae and Thelephoraceae. Howard & Currie (1932a) extended the list of myxomycetes known to have genuinely fungivorous plasmodia beyond the aforementioned three species to include *Badhamia foliicola, B. magna, B. rubiginosa, Brefeldia maxima, Hemitrichia clavata, Leocarpus fragilis, Lycogala epidendrum, Physarum flavicomum, P. tenerum, P. virescens*, and some (but not all) strains of *Fuligo septica, Trichia decipiens*, and *Lamproderma* sp. They concluded that some species, such as *Badhamia utricularis* and *Physarum polycephalum*, were very general parasites of fungi.

Some fruit bodies are apparently not attacked by plasmodia. This may be due partly to lack of durability, as with agarics (see above), but the texture of some fruit bodies (such as the hard stromatic ones of *Daldinia concentrica, Nummularia pithodes*, and *Ustulina zonata* which are not attacked by *Physarum viride* var. *rigidum* (Sanderson, 1922b)) might also confer protection. The possibility that there could be secondary metabolites in fruit bodies which deter plasmodia also requires investigation.

Feeding of plasmodia on fungal spores

In some fungi the vegetative mycelium or fruit body may be readily consumed by plasmodia yet the spores may either not be ingested, or, if they are, they may be egested without apparent change. However, effects on the viability of such spores have seldom been investigated. Howard (1931) reported that the conidia of *Alternaria* and *Penicillium* spp. were left untouched by *Physarum polycephalum*, and at least those of *Alternaria* remained demonstrably viable. Pull and Madelin (unpublished) found that conidia of *Alternaria brassicicola*, ingested by *Badhamia utricularis* and later egested showing little structural change, were mostly non-viable. Lister (1888) noted that the plasmodium of *B. utricularis* became deep brown when creeping over a *Coniophora* sp. on wood owing to undigested brown spores held in suspension in the cytoplasm. In the same myxomycete, Elliott (1916) observed darkening after feeding on *Coprinus micaceus*. Fed on *Amanita rubescens*, it left a deposit of undigested spores (Lister, 1888).

Feeding of plasmodia on algae

Algal cells often occur in the natural habitats of plasmodia, such as surfaces of soil or decaying logs or below water, and it is likely that they serve there as food. *Lamproderma scintillans* and *Didymium aquatile* have both been recorded creeping along stream beds (Lister, 1918; Gottsberger & Nannenga-Bremekamp, 1971), and *Didymium difforme* can pass all stages of its life cycle under water (Ward, 1886). However, these records did not indicate the food on which the submerged plasmodia had fed. Parker (1946) grew plasmodia of two *Physarum* species (probably *P. gyrosum* and *P. nutans*) beneath water in dishes where they fed upon a green alga, *Chlorococcum* sp., which grew upon the bottoms and sides of the vessels, and Gray & Lanning (1978) fed plasmodia of an aquatic strain of *Didymium difforme* on *Chlorella*. Lazo (1961) attempted two-membered culture of the plasmodia of several myxomycetes with several different unicellular algae, the cultures being incubated beneath fluorescent lights. He found that bacterium-free plasmodia of *Physarum didermoides* and *Fuligo cinerea* incorporated cells of *Chlorella protothecoides*, *C. xanthella* and *C. ellipsoidea*, but that plasmodia of *Physarum polycephalum*, *P. gyrosum* and *Fuligo septica* did not. Six other species of *Chlorella* and one of *Euglena* were not incorporated. The algal cells apparently remained alive within the plasmodia, which turned green but lost this colour if kept in darkness. The alga–plasmodium associations grew much better on oatmeal agar plates than did axenic plasmodia and were also more tolerant of acid conditions, a feature which might have ecological significance. Zabka & Lazo (1962) demonstrated that radio-labelled phosphorus was transferred from cells of *Chlorella xantha* to plasmodia of *Fuligo cinerea* and *vice versa*, but not to the substrate. Whether the transfer from the alga to the plasmodium occurred from living or dead algal cells was not determined, but the association appeared to be one from which both organisms benefited.

It is thus evident that not only can certain simple algae serve in nature as food for plasmodia of at least a number of species of myxomycete, but that specific mutualistic symbioses may also exist, a potentially important possibility which requires further investigation.

Feeding of plasmodia on lichens

There are numerous records of myxomycetes fruiting on lichens but no proof that any plasmodia actually *feed* upon the latter. *Colloderma oculatum* in a Swedish spruce forest was found on *Cladonia* spp. on decaying wood, but it was possible that the latter was the primary substratum and that the plasmodium left the wood before fructifying (Eliasson, 1981). *Listerella paradoxa* has mostly been found on the podetia of *Cladonia* spp., but can also occur on other substrata (Eliasson & Gilert, 1982). Ing (1967b) reported that a grey plasmodium of *Diacheopsis insessa* developed on lichens on damp-chambered bark, and fruited two days later, but again it is not certain whether it actually consumed lichen material.

Feeding of plasmodia on protozoa and on other myxomycetes

Once a plasmodium is generated in culture it usually consumes any residual myxamoebae, swarm cells, microcysts and ungerminated spores (Gilbert, 1928b; Indira, 1969; Carlile, 1974; Ashworth & Dee, 1975; Blackwell & Gilbertson, 1980b). Presumably, the same process happens in natural environments, but its quantitative significance is quite unknown. If myxamoebae and swarm cells can be ingested by plasmodia, protozoa might well also be consumed, but there seems to be no information about this.

The mechanism by which the aphanoplasmodium of *Stemonitis herbatica* captures and ingests swarm cells of the same species is described by Indira (1969). When a swarm cell is near, the plasmodium extends a filiform pseudopodium which attaches itself to the swarm cell and then retracts, drawing the prey close to the body of the plasmodium where more pseudopodia close over the swarm cell and engulf it in a food vacuole. Curiously, undigested swarm cells were sometimes egested.

Feeding of other organisms on myxomycetes
Predation upon myxamoebae and swarm cells

Lister (1901) observed that paramecia present in a gross culture of *Badhamia utricularis* sometimes devoured nearly all of the swarm cells and microcysts of the slime mould. Since Anderson *et al.* (1978) observed the nematode *Mesodiplogaster lheritieri* to consume the protozoan *Acanthamoeba polyphaga*, myxomycete amoebae and swarm cells might be expected also to be vulnerable to ingestion by eelworms, but no direct data are available.

True fungi living on the plasmodia of myxomycetes

It is generally thought that myxomycetes in their plasmodial phase are not colonized by fungi (Hawksworth, 1981). However mycelia, for example of *Mucor mucedo*, are sometimes seen growing on plasmodia in cultures on agar or solid substrata (Ing, 1967a). Such mycelia are sometimes consumed by the plasmodia, but whether they themselves ever benefit from the association is unclear. It is possible that some of the numerous fungi which sporulate upon sporangia of myxomycetes actually infect during the plasmodial stage. A yeast-like fungus that does benefit from its association with plasmodia has recently been described (Madelin & Feest, 1982). *Dipodascus macrosporus* was first isolated from the slime trails of migrating plasmodia of *Badhamia utricularis* in gross culture on corn meal agar where it was seen growing as minute sporulating colonies. It was evident that the yeast did not merely survive passage through the plasmodium, but in the process actually multiplied in food vacuoles. It was egested, often in large numbers, when strands of the posterior network of the plasmodium ruptured, allowing vacuoles to spill their contents directly upon the surface of the agar medium; egestion also occurred when vacuoles occasionally discharged yeast cells directly into fenestrations in the marginal fan. Different strains of the fungus differed in their virulence. The vigour of the parasite within a plasmodium depended on the particular strain of *B. utricularis* and on whether the supporting medium was plain or corn meal agar. Under appropriate conditions the parasite could multiply to the point at which the plasmodium became packed with *Dipodascus* cells and died. Infection occurred simply by ingestion of the yeast. On one occasion the yeast was apparently introduced into an unparasitized plasmodium on pieces of hymenomycetous fruit body fed to the plasmodium in the laboratory. A search for the yeast in natural habitats was therefore conducted (Platt and Madelin, unpublished). The centrifuged washings of bark samples (approximately 100 cm^2) from standing and fallen birch trees (*Betula verrucosa*) were placed in the paths of uninfected plasmodia of *B. utricularis* in culture.

Table 1.3. *Known species of myxomyceticolous fungi*

Acremonium fungicola (*Nectria violacea*)	*Stilbella ovalispora* (?*Nectria hirsuta*)
Acremonium sp. (*Nectria candicans*)	*Stilbella tomentosa* (*Byssostilbe stilbigera*)
Aphanocladium album	*Verticillium catenulatum*
Gliocladium sp. (*Nectria sporangiicola*)	*Verticillium lindauianum*
Sesquicillium microsporum	*Verticillium rexianum* (*Nectria myxomyceticola*)
Stilbella orbicularis	(*Rhynchonectria longispora*)

Names of teleomorphs, where known, are in parentheses.
After Ing, 1976.

In one out of 21 samples the plasmodium became infected with *Dipodascus*. A second yeast has also been discovered which inhabits *B. utricularis* plasmodia in a similar relationship (Platt and Madelin, unpublished). As long ago as 1888 Lister observed the presence of a small sprouting fungus which he said always accompanied the plasmodium of *B. utricularis*, and which might have been one of these yeasts.

True fungi living on the sporangia of myxomycetes
Sporangia of myxomycetes in the field are often covered with mycelial growths which usually colour them white. These generally flourish on what is effectively an already terminated phase in the life cycle after the majority of spores have been liberated, but sometimes they lead to the production of abnormal spores and capillitium by the myxomycete (Ing, 1967a; Hawksworth, 1981). Similarly, ubiquitous saprotrophic fungi from several genera (*Aspergillus*, *Dendryphiella*, *Penicillium*, *Scopulariopsis*, *Sporothrix* and *Trichoderma*) are occasionally found on myxomycete sporangia after these have started to decay (Ing, 1974; Hawksworth, 1981). *Mucor mucedo*, a saprotroph, grows over plasmodia in culture and damp chambers, and also grows on *Trichia decipiens*, *T. floriformis*, *Comatricha nigra* and *Badhamia utricularis* in the field (Ing, 1967a). A nematode-trapping species of *Harposporium* occurs on *Diderma* sp. (Ing, 1967a), while a species of *Arthrobotrys*, another genus with species which trap nematodes, sporulated upon the sporangium of a *Didymium* sp. in gross culture (Feest and Madelin, unpublished). Rather complex nutritional interactions in natural environments are thus suggested.

Of particular interest are fungus species which are more or less myxomyceticolous. Usually these are the imperfect states of species of *Nectria* (Table 1.3; Samuels, 1973). Ing (1976) notes that they are often similar to, or even identical with, species which parasitize insects, and suggests that the common factor in both hosts is presumably chitin. The chemistry of the walls of myxomycetes, however, merits study by modern methods. Recent analyses of the walls of microcysts and microsclerotia of *Physarum flavicomum* reveal lipids, proteins, and polysaccharides composed of neutral sugars and D-galactosamine, but not chitin (Henney & Chu, 1977).

Invertebrates feeding upon plasmodia

Most records of invertebrates consuming plasmodia relate to insects, but predation by other invertebrates is known. Woodlice (Isopoda) and snails have been seen to eat plasmodia as well as the sporangia of myxomycetes (Ing, 1967a; Eliasson, 1981). Eliasson (1981) suggests that ingestion by snails is probably the reason why plasmodia under regular observation sometimes disappear unexpectedly. In view of the abundance of woodlice and snails, the feeding of these animals on phaneroplasmodia might well prove to be quantitatively significant in natural food webs, though there is no evidence of feeding specialization.

Insects recorded as feeding upon plasmodia are mostly Diptera and Coleoptera. While some may be general saprozoic feeders others appear to be more specifically associated with myxomycetes. Buxton (1954) bred a number of dipterous species (*Lonchaea vaginalis, Leptocera fontinalis, Bradysia* sp., *Scatopse fuscipes, Drapetis nigritella* and *Hylemyia cilicrura*) from plasmodia of *Fuligo septica* or from the plasmodium-containing substratum. None were clearly specific to myxomycetes. On the other hand, none of the latter four species was bred from the more mature growths of *Fuligo*, including its aethalia, so the association appeared to be with only the young plasmodium. Another dipteran, *Nemopodia nitidula*, was bred from an unidentified plasmodium.

A specific association occurs between the dipteran, *Epicypta testata*, and plasmodia of *Tubifera ferruginosa* and *Reticularia lycoperdon* (Sellier & Chassain, 1976). The adult fungus gnat deposits its eggs in the plasmodium, where they soon hatch. The larva secretes a sheath which probably prevents it from adhering to the myxomycete cytoplasm. Metamorphosis of the larva and fruiting of the myxomycete coincide by an unexplained mechanism. The adult emerging from the sheath has to traverse the substance of the myxomycete to escape, and in so doing becomes laden with spores which adhere to its integument and to the fine hairs on the appendages and wings. The adult insect thus disperses the spores.

Coleoptera associated with plasmodia all appear to be small beetles of the genus *Anisotoma* (family Leiodidae). Sunhede (1973) frequently observed *Anisotoma humeralis* on plasmodia of *Reticularia lycoperdon* in Sweden, while Wheeler (1980) reported a new species, *Anisotoma plasmodiophaga*, from Central America, the larvae and adults of which fed on the plasmodium of an unidentified myxomycete beneath bark on a log. As described below, species of *Anisotoma* are most commonly encountered in association with sporangia. However, circumstantial evidence indicates that plasmodia may be at least as common a food source (Lawrence & Newton, 1980).

Records of other sorts of insect associated with plasmodia are few, but suggest the existence of widespread eating of plasmodia. Eliasson (1981) has observed ants (Hymenoptera) carrying away pieces of the plasmodium of *Reticularia jurana*, and Chassain (1973) reported that slowly moving insects, such as the collembolan *Achorutes muscorum*, may actually become trapped when a plasmodium fruits.

Invertebrates feeding on sporangia

Many of the earlier records of invertebrates associated with myxomycete sporangia are given by Ing (1967a). Most records relate to insects, but they also include woodlice (Isopoda) and millipedes (Diplopoda), and nematodes too have been seen in damp colonies of sporangia (Ing, 1967a); Keller & Smith (1978) have also recorded ingestion by mites. The insects include Collembola, Hymenoptera, Lepidoptera, Diptera and Coleoptera. Small sporangia as well as large aethalia may be consumed, and the predator may be a larva or an adult. The depredations of insects may reduce even the large aethalia of *Fuligo septica* to a mere shell (Sanderson, 1922a). Unlike those of true fungi, myxomycete fructifications consist mainly of spores. In general they represent a spatially and temporally limited food source (Lawrence & Newton, 1980), and monophagy or oligophagy may not therefore be a successful strategy in insects which feed on them. Nevertheless myxomyceticolous Coleoptera appear to prefer spores of slime moulds to those of true fungi (Lawrence & Newton, 1980). Ing (1967a) reported that collembolans, both nymphal and adult, feed widely on the small myxomycetes which grow on bark in moist chambers, and suggests that bark-inhabiting collembolans might habitually use slime moulds as food. He also has seen them on rotten wood eating sporangia of *Comatricha typhoides* and *Cribraria piriformis*.

Table 1.4. *Records of beetles associated with sporangia of myxomycetes*[a]

HYDROPHILIDAE
Megasternum obscurum

HISTERIDAE
Bacanius rhombophorus

LEIODIDAE
Agathidium angulare
A. arcticum
A. badium
A. brevisternum
A. californicum
A. confusum
A. contiguum
A. dentatum
A. estriatum
A. exiguum
A. mollinum
A. oniscoides
A. pulchrum
A. rhinoceros
A. rotundatum
A. rotundulum
A. seminulum
A. sphaerula
A. sp.
Amphicyllis globiformis
A. globus
Anisotoma axillaris
A. basalis
A. blanchardi
A. castanea
A. confusa
A. discolor
A. errans
A. geminata
A. glabra
A. humeralis
A. nevadensis
A. obsoleta
A. orbicularis

SCAPHIDIIDAE
Baeocera charybda
B. nana
B. picea
B. spp.

STAPHYLINIDAE
Atheta amicula
A. aterrima
A. fungiicola
A. inoptata
A. oblita
Cypha longicornis
Oxytelus tetracarinatus
Philonthus fimetarius
Quedius cruentus
Q. mesomelinus
Xantholinus punctulatus

CLAMBIDAE
Clambus nigriclavis
C. spp.

EUCINETIDAE
Eucinetus morio

PELTIDAE
Thymalus limbatus

SPHINDIDAE
Aspidiphorus lareyniei
A. orbiculatus
Eurysphindus hirtus
E. spp.
Odontosphindus clavicornis
O. denticollis
O. grandis
Protosphindus chilensis
Sphindus americanus
S. crassulus
S. dubius
S. trinifer
S. spp.

ENDOMYCHIDAE
Sphaerosoma piliferum
Symbiotes latus

LATHRIDIIDAE
Enicmus cordatus
E. fungicola
E. hirtus
E. minutus
E. rugosus
E. tenuicornis
E. testaceus

contd.

       Lathridius consimilis
       L. nodifer
       Revelieria californica

       CIIDAE
       Cis boleti

[a] Based on records in Ing (1967a), Russell (1979) and Lawrence & Newton (1980).

  Among dipterous insects which feed on myxomycetes, three species of Mycetophilidae (*Mycetophilus vittipes*, *Platurocypta punctum*, and *P. testata*, associated respectively with *Arcyria incarnata*, *Lycogala epidendrum*, and what was probably *Reticularia lycoperdon*) are not known to breed other than on myxomycete fructifications (Buxton, 1954). Other Diptera are associated with myxomycetes but perhaps not specifically so. For example, the fungivorous larvae of the mycetophilid genus *Phronia*, of which there are 86 known species, generally graze on slime moulds but also on other fungi growing on the surfaces of decaying logs (Gagné, 1975). Indeed, most dipterous larvae found on myxomycetes are probably saprophagous. They include *Bradysia* sp., *Brittenia fraxinicola*, *Lonchaea vaginalis*, *Drosophila repleta*, *Leptocera fontinalis*, *Hylemyia cilicrura*, and a Cecidomyid (Buxton, 1954).
  Table 1.4 lists Coleoptera recorded as feeding on sporangia of myxomycetes. Species of *Anisotoma* are probably the beetles most commonly and abundantly associated with slime moulds, at least in the holarctic region (Wheeler, 1979). Members of the family Sphindidae have only been recorded in the spore masses of myxomycetes. Not unexpectedly, larvae of species specializing in feeding on myxomycete spores possess feeding systems adapted to cope with very small food particles (Lawrence & Newton, 1980).
  Damage to sporangia by insects is difficult to quantify but is probably substantial, at least under some circumstances. Sanderson (1922a) suggested that the apparent scarcity of myxomycetes in the virgin forest of Malaya was in fact more apparent then real, and may have been largely the result of insect life. Ants abounded, and vigorously devoured the sporangia of some myxomycetes, while larvae of various Coleoptera, Lepidoptera, and Diptera also played a part.

Some consequences of trophic interactions
Possible influence of bacteria on germination of myxomycete spores

In view of the importance of bacteria in the nutrition of swarm cells and myxamoebae it would not be surprising if the presence of bacteria influenced the germination of myxomycete spores. Germination in the laboratory is very variable (Gray & Alexopoulos, 1968), which suggests that unresolved factors may be operating. Pinoy (1907) suggested that bacteria play a part in the process, perhaps by softening the spore wall, but the rapid germination of many spores in water has led to scepticism about such a role. Nevertheless it seems plausible that spores might well respond to other chemical signals from bacteria. Germination of spores of the cellular slime mould *Dictyostelium discoideum*, another organism that feeds on bacteria, is triggered by a specific substance secreted by *Aerobacter aerogenes* (Hashimoto, Tanaka & Yamada, 1976). To show that germination of myxomycete spores is truly independent of bacteria necessitates precautions, since spores from field collections are frequently contaminated with them. Even spores from the central portion of a spore mass in a fructification from which the wall has been removed may carry bacteria (Parker, 1946).

Attraction responses

The chemotactic responses of some myxomycete plasmodia to substances in certain fungus fruit bodies may be indicative of their specialization towards feeding on these structures. Emoto (1932) reported that plasmodia of *Physarum viride* and *P. rigidum* were attracted by extracts of the fruit bodies of 11 different Hymenomycetes, while Madelin, Audus & Knowles (1975) reported a thermostable extractive in fruit bodies and mycelium of *Stereum hirsutum* which attracted plasmodia of *Badhamia utricularis*. Its presence in the mycelium as well as the fruit bodies may be significant in the light of circumstantial evidence that plasmodia in nature may consume the mycelia of wood-rotting fungi (see above). The *Stereum* attractant was volatile and effective at very low concentrations, suggesting that it acts differently from the sugars and peptone which Carlile (1970) found to cause chemotaxis of *Physarum polycephalum*. Insofar as the latter materials may be products of the digestion of fruit bodies which takes place in intimate contact with the plasmodium, a lesser degree of chemotactic sensitivity may suffice for effective location of absorbable nutrients.

Defence against predation

Nothing seems to be known about means by which myxamoebae and swarm cells may escape or reduce predation, apart from those inherent in their being motile, but features exist which may contribute to the protection of plasmodia, sporangia, and spores. The seemingly vulnerable plasmodia are invested in a coat of slime. In at least *Physarum polycephalum* this is a polysaccharide for which no microbial degradative enzyme has yet been found, even in the producing organism (Sauer, 1982). Such a layer might help hold some potential predators at bay.

Sobels (1950) found that variously prepared extracts of plasmodia of strains of *Badhamia utricularis, Fuligo septica, Fuligo* sp. and *Physarum confertum*, exercised various degrees of inhibition and lysis toward the yeasts *Cryptococcus (Torulopsis) neoformans, Cryptococcus laurentii* var. *laurentii (Torulopsis laurentii), Saccharomyces cerevisiae, Saccharomyces cerevisiae* var. *ellipsoideus*, and the bacteria *Staphylococcus aureus* and *Escherichia coli*. Locquin (1948) extracted antibiotics with fungistatic and bacteriostatic properties (or just the former) from plasmodia of seven strains of *Fuligo septica, Badhamia utricularis*, and *Trichamphora pezizoidea. In vivo* secretion of these antibiotics has not been demonstrated, so much uncertainty remains about the natural roles, if any, of these substances.

Antimicrobial activities have also been demonstrated in sporangia and spores. Locquin (1948) and Locquin & Prévot (1948) extracted substances with antibacterial activity from sporangia of *Fuligo septica, Lycogala epidendrum, L. flavo-fuscum, Mucilago spongiosa, Reticularia lycoperdon, Stemonitis fusca* and *Tubifera ferruginosa*, and reported that the antibiotic from the aethalium of *L. flavo-fuscum*, which was active against bacteria and *Mucor* sp., appeared to be localized in the spores, which constitute the bulk of the aethalium.

There is an indication that, besides antibiotics, substances with insect-deterrent properties might be present in some sporangia, for Sanderson (1922*a*) reported that insects which quickly destroyed sporangia of many species in Malayan forests left sporangia of *Physarella oblonga* severely alone.

Additional dispersal mechanisms

Although wind is the most obvious mechanism for dispersal of myxomycete spores, predators on myxomycete sporangia may become heavily contaminated with spores, both externally and internally, and so may also serve as vectors. Eliasson (1981) observed small unidentified beetles, powdered with spore dust, on fructifications of *Arcyria incarnata*, *Enerthenema papillatum*, and *Stemonitis axifera*, and suggested that insects are probably the most important agent of dispersal for some myxomycetes, carrying spores to new localities and bringing them into cavities and insect tunnels in wood. Keller & Smith (1978) found that the mite *Tyrophagus putrescentiae* ingested spores of *Stemonitis flavogenita* and *Didymium* sp., and that intact spores of the latter which eventually appeared in the faecal pellets were germinable. Woodlice too might act as agents of dispersal: *Androniscus dentiger* introduced viable spores of *Didymium iridis*, either in or on its body, into the sterile contents of a jar intended for the establishment of a colony of woodlice, and spores of myxomycetes, apparently unaffected by digestion, occur within the digestive glands of *Androniscus* and *Trichoniscus* from beech litter (Ing, 1967a).

Indirect interactions between myxomycetes and animals
Coprophilous myxomycetes

It has long been known that dung of herbivorous and omnivorous mammals may serve as a substratum for myxomycete sporangia whether in the field or after incubation in damp chambers (Jahn, 1916; Lister, 1918). Eighty species in 23 genera have been reported on dung and some species which are exclusively or chiefly coprophilous are known (Eliasson & Lundqvist, 1979). The sporangia always develop late, after three weeks or more, but this does not necessarily imply that their trophic activity is insignificant at earlier stages. Dung also may have special importance as a substratum and habitat for myxomycetes in particular circumstances. Blackwell & Gilbertson (1980a) list records of myxomycetes from desert habitats around the world. They report 33 species from the Sonoran Desert of Arizona, and observe that in that situation the dung of herbivorous animals is abundant and important as a substratum for myxomycetes.

Whether there exist any truly endocoprophilous myxomycetes, that is, species whose spores survive passage through the animal's gut, is not known. Eliasson & Lundqvist (1979) remarked that no obvious adaptations to the endocoprophilous habit, for example ones related to the dispersal of spores to the surrounding vegetation to be eaten by vectors, have been observed. They noted that the very few investigators who had examined the fungus flora of gut and stomach contents taken from butchered animals had not reported myxomycetes, although this might be due to inappropriate techniques. Murray, Feest and Madelin (unpublished) have found that large numbers of viable myxomycete cells (in excess of 1000 'plasmodium-forming units' per gram dry weight of contents) were present in the anterior, middle and posterior regions of the gut of cast-producing earthworms, surviving passage through the alimentary tract and eventually multiplying rapidly within the egested cast.

It therefore appears that animal excrement is a natural food source within which a considerable number of myxomycetes flourish and contribute to decay. However, dung is nutritionally and ecologically very complex (Webster, 1970), and the role of myxomycetes in its decomposition is virtually unknown.

Indirect benefits from insect activities

There are records of myxomycetes benefiting indirectly from insect activities in ways that may be naturally important. The plasmodia of *Reticularia jurana, Amaurochaete atra, A. tubulina,* and *Symphytocarpus flaccidus* developing in insect tunnels in wood may pour out onto the wood surface through insect tunnels and fissures to fruit (Eliasson, 1977, 1981). Sunhede (1973) observed the same for *Reticularia lycoperdon*. A somewhat different role played by boring insects is described by Sanderson (1922a), who twice observed that *Physarum reniforme* appeared in Malayan forests on rubber trees (*Hevea brasiliensis*) just below a diseased portion of the cortex where fluid was exuding from holes made by boring beetles (*Xyleborus* sp.). The colony of sporangia followed the line of the stream of fluid.

Acknowledgements. I am indebted to Mr Alan Feest for much stimulating discussion on myxomycetes during the past two years and for commenting on the manuscript. I am also grateful to the librarians of the University of Bristol, particularly Mrs S. Pettit and Mr J. A. H. Brooks, for their unfailing helpfulness.

References

Anderson, R. V., Elliott, E. T., McLellan, J. F., Coleman, D. C., Cole, C. V. & Hunt, H. W. (1978). Trophic interactions in soils as they affect energy and nutrient dynamics. III. Biotic interactions of bacteria, amoebae, and nematodes. *Microbial Ecology*, **4**, 361–71.

Ashworth, J. M. & Dee, J. (1975). *The Biology of Slime Moulds*. London: Edward Arnold.

Bary, A. de (1887). *Comparative morphology and biology of the fungi, Mycetozoa and bacteria.* (English translation.) London: Clarendon Press.

Benedict, W. G. (1965). Plasmodial activity in *Stemonitis fusca* Roth. *Canadian Journal of Botany*, **43**, 355–9.

Blackwell, M. & Gilbertson, R. L. (1980a). Sonoran Desert Myxomycetes. *Mycotaxon*, **11**, 139–49.

Blackwell, M. & Gilbertson, R. L. (1980b). *Didymium eremophilum*: a new myxomycete from the Sonoran Desert. *Mycologia*, **72**, 791–7.

Buxton, P. A. (1954). British diptera associated with fungi. 2. Diptera bred from Myxomycetes. *Proceedings of the Royal Entomological Society of London, A*, **29**, 163–71.

Camp, W. G. (1937). The structure and activities of the myxomycete plasmodia. *Bulletin of the Torrey Botanical Club*, **64**, 307–35.

Carlile, M. J. (1970). Nutrition and chemotaxis in the myxomycete *Physarum polycephalum*: the effect of carbohydrate on the plasmodium. *Journal of General Microbiology*, **63**, 221–6.

Carlile, M. J. (1974). The myxomycete *Physarum nudum*: life cycle and pure culture of plasmodia. *Transactions of the British Mycological Society*, **62**, 213–15.

Chassain, M. (1973). Capture d'un insecte collembole par deux myxomycetes. *Documents Mycologiques*, **8**, 37–8.

Cohen, A. L. (1941). Nutrition of the Myxomycetes. II. Relations between plasmodia, bacteria, and substrate in two-membered culture. *Botanical Gazette*, **103**, 205–24.

Cole, G. T. & Pope, L. M. (1981). Surface wall components of *Aspergillus niger* conidia. In *The Fungal Spore: Morphogenetic Controls*, ed. G. Turian & H. R. Hohl, pp. 195–215. London: Academic Press.

Eliasson, U. (1977). Recent advances in the taxonomy of myxomycetes. *Botaniska Notiser*, **130**, 483–92.

Eliasson, U. (1981). Pattern of occurrence of myxomycetes in a spruce forest in south Sweden. *Holarctic Ecology*, **4**, 20–31.

Eliasson, U. & Gilert, E. (1982). A SEM-study of *Listerella paradoxa* (Myxomycetes). *Nordic Journal of Botany*, **2**, 249–55.

Eliasson, U. & Lundqvist, N. (1979). Fimicolous myxomycetes. *Botaniska Notiser*, **134**, 551–68.

Elliott, W. T. (1916). Some observations upon the assimilation of fungi by *Badhamia utricularis* Berk. *Transactions of the British Mycological Society*, **5**, 410–13.

Elliott, W. T. & Elliott, J. S. (1920). The sequence of fungi and mycetozoa. *Journal of Botany*, **58**, 273–4.

Emoto, Y. (1932). Uber die Chemotaxis der Myxomyceten-Plasmodien. *Proceedings of the Imperial Academy of Japan*, **8**, 460–3.

Farr, M. L. (1976). Flora Neotropica. *Monograph No. 16, New York Botanical Garden*.

Gagné, R. J. (1975). A revision of the Nearctic species of the genus *Phronia* (Diptera: Mycetophilidae). *Transactions of the American Entomological Society*, **101**, 227–318.

Gilbert, F. A. (1928a). Feeding habits of the swarm cells of the myxomycete *Dictydiaethalium plumbeum*. *American Journal of Botany*, **15**, 123–31.

Gilbert, F. A. (1928b). Observation on the feeding habits of the swarm cells of myxomycetes. *American Journal of Botany*, **15**, 473–84.

Glynn, P. J. (1981). A quantitative study of the phagocytosis of *Escherichia coli* by myxamoebae of the slime mould *Dictyostelium discoideum*. *Cytobios*, **30**, 153–66.

References

Goodman, E. M. (1972). Axenic culture of myxamoebae of the myxomycete *Physarum polycephalum*. *Journal of Bacteriology*, **111**, 242–7.
Gottsberger, G. & Nannenga-Bremekamp, N. E. (1971). A new species of *Didymium* from Brazil. *Proceedings of the Koninklijke Nederlandse Akademie van Wetenschappen, Series C, Biological and Medical Sciences*, **74**, 264–8.
Gray, T. R. G. & Lanning, S. (1978). Culture of an aquatic form of *Didymium difforme*. *Transactions of the British Mycological Society*, **70**, 289–91.
Gray, W. D. & Alexopoulos, C. J. (1968). *Biology of the Myxomycetes*. New York: Ronald Press.
Hashimoto, Y., Tanaka, Y. & Yamada, T. (1976). Spore germination promoter of *Dictyostelium discoideum* excreted by *Aerobacter aerogenes*. *Journal of Cell Science*, **21**, 261–71.
Haskins, E. F. (1970). Axenic culture of the myxomycete *Echinostelium minutum*. *Canadian Journal of Botany*, **48**, 663–4.
Hawksworth, D. L. (1981). A survey of the fungicolous conidial fungi. In *Biology of Conidial Fungi*, vol. 1, ed. G. T. Cole & B. Kendrick, pp. 171–244. New York: Academic Press.
Henney, H. R., Asgari, M. & Henney, M. R. (1974). Growth of the haploid and diploid phases of *Physarum flavicomum* in the same partially defined media. *Canadian Journal of Microbiology*, **20**, 967–70.
Henney, H. R. & Chu, P. (1977). Chemical analyses of cell walls from microcysts and microsclerotia of *Physarum flavicomum*; comparison to slime coat from microplasmodia. *Experimental Mycology*, **1**, 83–9.
Henney, M. R. & Henney, H. R. (1968). The mating-type systems of the myxomycetes *Physarum rigidum* and *P. flavicomum*. *Journal of General Microbiology*, **53**, 321–32.
Hilton, A. E. (1914). Notes on the cultivation of plasmodia of *Badhamia utricularis*. *Journal of the Queckett Microscopical Club, ser. 2*, **12**, 381–4.
Hok, K. A. (1954). Studies of the nutrition of myxomycete plasmodia. *American Journal of Botany*, **41**, 792–9.
Howard, F. L. (1931). The life history of *Physarum polycephalum*. *American Journal of Botany*, **18**, 116–33.
Howard, F. L. & Currie, M. E. (1932a). Parasitism of myxomycete plasmodia on sporophores of Hymenomycetes. *Journal of the Arnold Arboretum*, **13**, 270–84.
Howard, F. L. & Currie, M. E. (1932b). Parasitism of myxomycete plasmodia on fungus mycelia. *Journal of the Arnold Arboretum*, **13**, 438–46.
Indira, P. U. (1968). Some slime moulds from Southern India. IX. *Journal of the Indian Botanical Society*, **47**, 330–41.
Indira, P. U. (1969). The life cycle of *Stemonitis herbatica*. *Transactions of the British Mycological Society*, **53**, 25–38.
Ing, B. (1967a). Myxomycetes as food for other organisms. *Proceedings, South London Entomological and Natural History Society*, 1967, 18–23.
Ing, B. (1967b). Notes on myxomycetes. II. *Transactions of the British Mycological Society*, **50**, 555–62.
Ing, B. (1974). Mouldy myxomycetes. *Bulletin of the British Mycological Society*, **8**, 25–30.
Ing, B. (1976). The natural history of Slapton Ley Nature Reserve. XI. Myxomycetes (slime moulds). *Field Studies*, **4**, 441–55.
Ing, B. (1980). A revised census catalogue of British Myxomycetes – Part 1. *Bulletin of the British Mycological Society*, **14**, 97–111.
Ing, B. (1982). A revised census catalogue of British Myxomycetes – Part 2. *Bulletin of the British Mycological Society*, **16**, 26–35.

Jahn, E. (1916). Coprophilie bei Myxomyceten. *Verhandlung des Botanischen Vereins der Provinz Brandenburg*, **57**, 207-8.
Keller, H. W. & Smith, D. M. (1978). Dissemination of myxomycete spores through the feeding activities (ingestion-defecation) of an acarid mite. *Mycologia*, **70**, 1239-41.
Knowles, D. J. C. & Carlile, M. J. (1978). Growth and migration of plasmodia of the myxomycete *Physarum polycephalum*: the effect of carbohydrate including agar. *Journal of General Microbiology*, **108**, 9-15.
Konijn, T. M. & Koevenig, J. L. (1971). Chemotaxis in myxomycetes or true slime moulds. *Mycologia*, **63**, 901-6.
Lawrence, J. F. & Newton, A. F. (1980). Coleoptera associated with the fruiting bodies of slime moulds (Myxomycetes). *The Coleopterists Bulletin*, **34**, 129-43.
Lazo, W. R. (1961). Growth of green algae with myxomycete plasmodia. *American Midland Naturalist*, **65**, 381-3.
Lister, A. (1888). Notes on the plasmodium of *Badhamia utricularis* and *Brefeldia maxima*. *Annals of Botany*, **2**, 1-24.
Lister, A. (1890). Notes on the ingestion of food-material by the swarm-cells of Mycetozoa. *Journal of the Linnean Society (Botany)*, **25**, 435-41.
Lister, A. (1901). On the cultivation of Mycetozoa from spores. *Journal of Botany*, **39**, 5-8.
Lister, G. (1918). The Mycetozoa: a short history of their study in Britain; an account of their habits generally; and a list of species recorded from Essex. *Essex Field Club Special Memoirs*, **6**, 1-54.
Locquin, M. (1948). Culture des Myxomycètes et production de substances antibiotiques par ces champignons. *Comptes rendus hebdomadaire des séances de l'Académie des sciences*, **227**, 149-50.
Locquin, M. & Prévot, A. R. (1948). Etudes de quelques antibiotiques produits par les myxomycètes. *Annales de l'Institut Pasteur*, **75**, 8-13.
McCullough, C. H. R. & Dee, J. (1976). Defined and semi-defined media for the growth of amoebae of *Physarum polycephalum*. *Journal of General Microbiology*, **95**, 151-8.
McCullough, C. H. R., Dee, J. & Foxon, J. L. (1978). Genetic factors determining the growth of *Physarum polycephalum* amoebae in axenic medium. *Journal of General Microbiology*, **106**, 297-306.
Madelin, M. F., Audus, F. & Knowles, D. (1975). Attraction of plasmodia of the myxomycete, *Badhamia utricularis*, by extracts of the Basidiomycete, *Stereum hirsutum*. *Journal of General Microbiology*, **89**, 229-34.
Madelin, M. F. & Feest, A. (1982). *Dipodascus macrosporus* sp. nov. (Hemiascomycetes), associated with plasmodia of *Badhamia utricularis*. *Transactions of the British Mycological Society*, **79**, 331-5.
Ostrofsky, A. & Shigo, A. L. (1981). A myxomycete isolated from discoloured wood of living red maple. *Mycologia*, **73**, 997-1000.
Parker, H. (1946). Studies in the nutrition of some aquatic myxomycetes. *Journal of Elisha Mitchell Scientific Society*, **62**, 231-47.
Pinoy, E. (1907). Rôle des bactéries dans le dévelopment de certains Myxomycètes. *Annales de l'Institut Pasteur*, **21**, 686-700.
Ross, I. K. (1964). Pure culture of some Myxomycetes. *Bulletin of the Torrey Botanical Club*, **91**, 23-31.
Russell, L. K. (1979). Coleoptera associated with slime molds (Mycetozoa) in Oregon and California, USA. (Leiodidae, Sphindidae, Lathridiidae). *Pan-Pacific Entomologist*, **55**, 1-9.
Samuels, G. J. (1973). The myxomyceticolous species of *Netria*. *Mycologia*, **65**, 401-20.

Sanderson, A. R. (1922a). Notes on Malayan Mycetozoa. *Transactions of the British Mycological Society*, **7**, 239–56.
Sanderson, A. R. (1922b). On the parasitic habit of the plasmodium of *Physarum viride* var. *rigidum* Lister. *Transactions of the British Mycological Society*, **7**, 299–300.
Sauer, H. W. (1982). *Developmental Biology of Physarum*. Cambridge University Press.
Sellier, R. & Chassain, M. (1976). Observations sur le mode de dissemination de spores d'un champignon myxomycete par un insect diptère: Mycetophilide. *Bulletin de la Société des Sciences Naturelles de l'Ouest de la France*, **74**, 81–5.
Skulberg, O. M. (1958). Notiz über *Didymium nigripes* (Myxomycetes) aus einer Abwasserbiozönose. *Schweizerische Zeitschrift für Hydrologie*, **20**, 210–17.
Skupienski, F. X. (1920). *Recherches sur le cycle évolutif des certains myxomycètes*. Paris: Imprimerie M. Flinikowski.
Skupienski, F. X. (1928). Badania bio-cytologiczne nad *Didymium* difforme Duby-Cześć pierwsza. *Acta Societatis botanicorum Poloniae*, **5**, 255–336.
Smart, R. F. (1938). The reactions of the swarm cells of myxomycetes to nutrient materials. *Mycologia*, **30**, 254–84.
Sobels, J. C. (1950). *Nutritions de quelques myxomycètes en cultures pures et associées et leurs proprietés antibiotiques*. Gouda, Netherlands: N.V. Drukkerij v/h Koch & Knuttel.
Sobels, J. C. & Cohen, A. C. (1953). The isolation and culture of opsimorphic organisms. II. Notes on the isolation, purification and maintenance of myxomycete plasmodia. *Annals of the New York Academy of Sciences*, **56**, 944–8.
Stirling, J. L., Cook, G. A. & Pope, A. M. S. (1979). Chitin and its degradation. In *Fungal Walls and Hyphal Growth*, ed. J. H. Burnett & A. P. J. Trinci, pp. 169–88. Cambridge University Press.
Sunhede, S. (1973). Studies in the myxomycetes. I. On the growth and ecology of the myxomycete *Reticularia lycoperdon* Bull. *Svensk botanisk tidskrift*, **67**, 172–6.
Ward, H. M. (1886). The morphology and physiology of an aquatic myxomycete. *Studies from the Biological Laboratories of Owens College*, **1**, 64–86.
Watanabe, A. (1932). Uber die Bedeutung der Nährbakterien für die Entwicklung der Myxomyceten-Plasmodien. *Botanical Magazine (Tokyo)*, **46**, 247–55.
Webster, J. (1970). Coprophilous fungi. *Transactions of the British Mycological Society*, **54**, 161–80.
Wheeler, Q. D. (1979). Slime mould beetles of the genus *Anisotoma* (Leiodidae): classification and evolution. *Systematic Entomology*, **4**, 251–309.
Wheeler, Q. D. (1980). Studies on neotropical slime mold/beetle relationships, Part I: Natural history and description of a new species of *Anisotoma* from Panama (Coleoptera: Leiodidae). *Proceedings of the Entomological Society of Washington*, **82**, 493–8.
Zabka, G. G. & Lazo, W. R. (1962). Reciprocal transfer of materials between algal cells and myxomycete plasmodia in intimate association. *American Journal of Botany*, **49**, 146–8.

2
Soil nutrient transformations in the rhizosphere via animal–microbial interactions

D. C. COLEMAN, R. E. INGHAM,
J. F. McCLELLAN, and J. A. TROFYMOW
Colorado State University, Fort Collins, Colorado 80523, USA

Key words: microcosms; nitrogen mineralization; grazing; heterotrophic interactions; nutrient mobilization

Introduction

One of the major areas of interest in production ecology is in the realm of interactions and interfaces, both within and between ecosystems (Marshall, 1976; Coleman, Reid & Cole, 1983). Soils represent a particularly complex system of interactions. Thus, soil may be considered as a mosaic of various combinations of sand, silt, and clay, arranged in genetic horizons with their associated organic matter fractions and biota. As in many other systems, there are localized concentrations of organisms at microsites (aggregates or soil particles in general) as well as at larger sized interfaces such as that of leaf litter in soil and also between root and soil (Tisdall & Oades, 1982).

In a conceptual model (Fig. 2.1), there are discrete locations for existence, growth and activity of various organisms. Complex nutrient changes occur in which transformations from labile to stable, organic to inorganic may be principally controlled by microbial or faunal activities (Fig. 2.2). This paper will examine nutrient transformations, i.e. the initial uptake and immobilization by bacteria, fungi, and actinomycetes, and the subsequent release of the materials by other activities (mineralization) which are, to a considerable extent, mediated by the soil micro- and mesofauna.

Our objectives in this paper are threefold: (1) to examine principal animal–microbial–plant processes in ecosystems; (2) to present recent results of experimental microcosm studies; (3) to discuss impacts of these processes on community and ecosystem function.

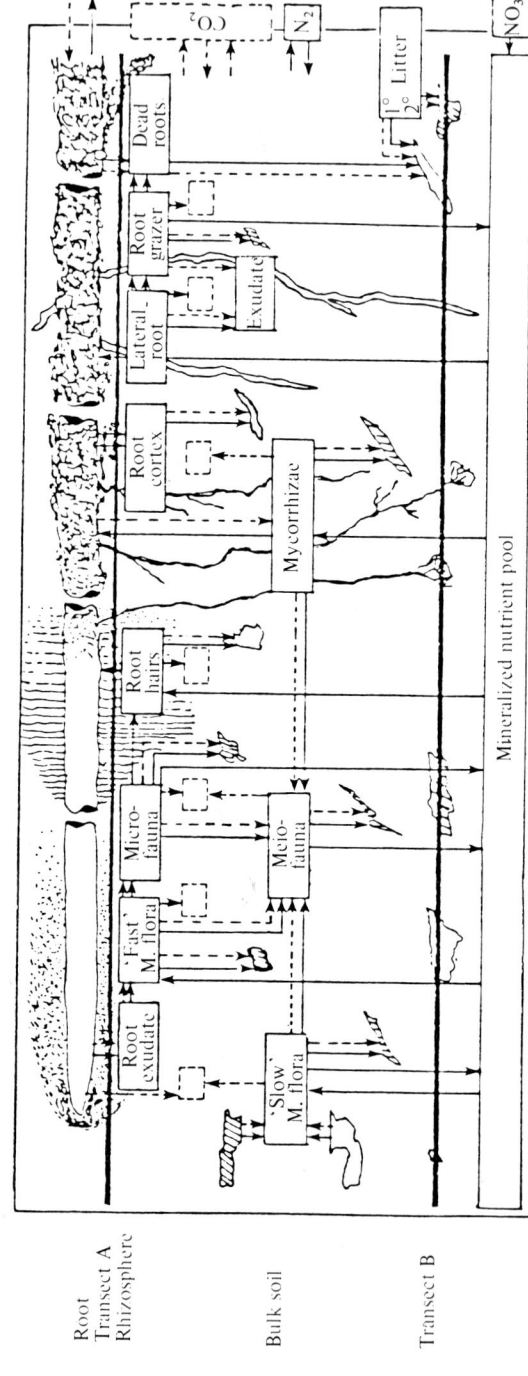

Fig. 2.1. Conceptual model of root in soil. Transect A (rhizosphere) shows structural features and interactions; transect B depicts the bulk (root-free) soil. (From Trofymow & Coleman, 1982.)

Historical perspective

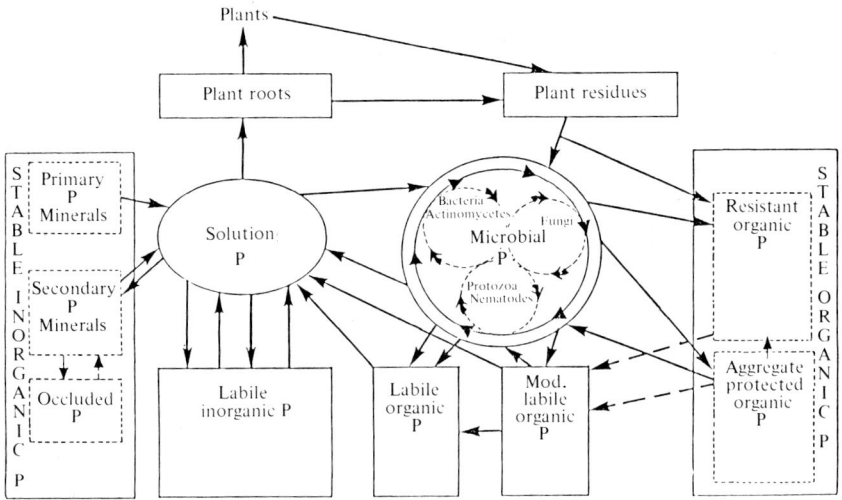

Fig. 2.2. Flows of stable and labile forms of inorganic and organic phosphorus in soil. Notice major flows into and out of solution P, regulated by microbial and faunal activity. (From Stewart & McKercher, 1982.)

Historical perspective
Microbial–faunal interactions

The history of investigations about the soil microflora and fauna contains much information on feeding, and distribution and abundance of microarthropods (Bornebusch, 1930; Jacot, 1940; Witkamp, 1960; Petersen, 1982), of protozoa (Cutler, Crump & Sandon, 1922; Stout, 1980), and of nematodes (Sohlenius, 1980; Yeates & Coleman, 1982). These studies emphasized feeding behaviour and followed the subsequent rise and fall of animal predator and microbial prey populations. In addition, the effects of fauna have been determined on total energy flow in populations and in communities, reviewed by Macfadyen (1968, 1969) and Luxton (1982). Energy dissipation during decomposition by the fauna is very minor compared to that contributed by the microflora. However, faunal effects on nutrient transformations and cycling within a community may be considerably greater than their contribution to simple energetics (Coleman, Reid & Cole, 1983; Clarholm, 1981). Hanlon (1981) found that rates of fungal respiration increased, peaked and then declined with increasing levels of collembolan grazing. Thus the effects of grazers appear to depend both upon their numbers and their activity. We will examine several aspects of direct effects such as feeding, which ruptures and lyses cells, and phoresy, the transport of various diaspores or propagules of microflora, as well as considering the more indirect effects of possible stimulation or altering of hormone production by various microorganisms.

Rhizodeposition studies

Historical reviews of rhizosphere chemistry and biology are ably presented by Darbyshire & Greaves (1973), Foster (1981), and Foster & Martin (1981). In considering the processes involving organic compounds in the rhizosphere in an ecological context, the term rhizodeposition (Shamoot, McDonald & Bartholomew, 1968) will be used.

Inputs to soil from the growing, maturing and senescing root have been placed in five categories (Fig. 2.3) (Rovira, Foster & Martin, 1979): (1) exudates, which are compounds of low molecular weight (MW), leaking from all cells into intercellular spaces or the soil; (2) secretions, low MW compounds released as the results of metabolic processes, not lost passively, as are exudates; (3) plant mucilages from a variety of sources, namely: the root cap, hydrolysates of primary cell wall polysaccharide, mucilages from epidermal cells with only primary cell walls (including root hairs), and mucilages from microbial degradation of outer multilamellate primary cell walls of old, dead epidermal cells; (4) mucigels, the heterogeneous array of polysaccharides including secondarily produced microbial polysaccharides (Jenny & Grossenbacher, 1963); (5) lysates, arising from autolysis of older epidermal cells when the plasmalemma fails. These will in turn provide a source of organic matter (OM) for microbial production and synthesis.

Estimates of rhizodeposition range from 30–50% (Coleman, 1976) to 60% (Lynch & Panting, 1980) of the total input of fibrous roots in a growing season. A detailed review of root mucilaginous substances and microorganisms associated with them (Chaboud & Rougier, 1981) also considers the major impacts which roots have on fixation of nitrogen by symbiotic and heterotrophic microbes, and on gaseous denitrification losses.

Rhizosphere organisms and plant growth factors

Several scientists in Europe and the USSR have suggested that bacteria, or bacteria and protozoa in combination, may produce hormones (IAA, gibberellins and/or kinetins) which markedly affect plant growth responses. Principal among these workers are Nikolyuk &

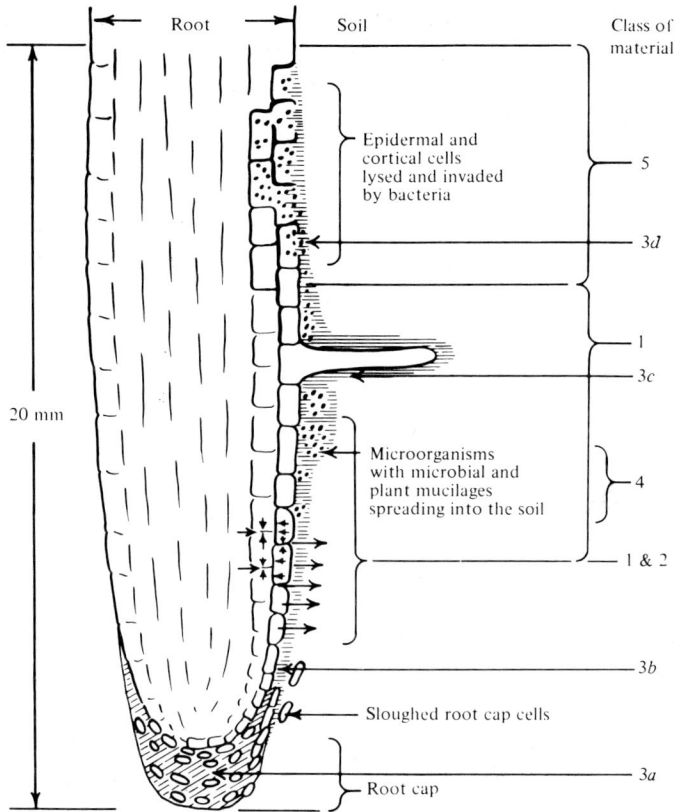

Fig. 2.3. Root growing in soil types of material entering soil: 1, exudates; 2, secretions from metabolic processes; 3, plant mucilages from various sources; 4, mucigels – a product of root–microbial components; 5, lysates from older root cells. (From Rovira, Foster & Martin, 1979.)

Geltzer (1972), who presented interesting data on field plots of cotton and flax in Uzbekistan SSR, where they measured growth increases of 10–30% in dry matter yields. The subject is reviewed in considerable detail by Brown (1975, 1976).

Another factor of considerable recent interest is the work on plant growth-promoting rhizo-microbes. Kloepper, Leong, Teintze & Schroth (1980a, b) and Schroth & Hancock (1982) have investigated fungal–bacterial interactions in several California field crops. They have suggested that certain plant growth-promoting bacteria have a marked inhibitory effect on other microbes, particularly pathogenic fungi. The postulated mechanisms of competition are complex, and centre on production of siderophores which compete for ferric iron in the rhizosphere. The subject is reviewed in more detail by Coleman *et al.* (1983).

Conceptual models of root–rhizosphere nutrient cycling

There have been two major pathways of decomposition and nutrient cycling postulated to occur in the root–rhizosphere region, and these have been termed 'fast' and slow' cycles (Anderson, Coleman & Cole, 1981). The fast cycle represents the release of low molecular weight carbon compounds such as amino acids and various mono- and oligosaccharides through root exudation or sloughing of cells. The more labile materials predominate in the rhizosphere region (Fig. 2.4) at or near the root tip. In comparison, the slow cycle is concerned with decomposition of less labile organic compounds such as those in cell walls or from partially decomposed root material or other sources of plant debris. This would include such substances as cellulose and lignin from plant cell walls, and polymeric nitrogen (chitin) present in arthropod exoskeletons, nematode eggs, and walls of many fungal hyphae. In both pathways, the microbes, with or without grazers, release some of the bound minerals such as nitrogen and phosphorus to the soil solution where they can be taken up by plants and microbes, and passed through other food chains. Mycorrhizae facilitate transport of certain nutrients, i.e. phosphorus, and perhaps nitrogen (Bowen & Smith, 1981), directly into the plant root.

Recent studies of nutrient cycling using microcosms

To investigate these phenomena further, we have looked at the effects on nutrient cycling of varying known components of soil microflora and micro- or mesofauna in the presence or absence of growing plants.

Nutrient cycling in the absence of plants

In early studies (Coleman *et al.*, 1977; Coleman *et al.*, 1978; Cole, Elliott, Hunt & Coleman, 1978) we examined microbial and faunal interactions in soil incubations without plants. These had known amounts of simulated root exudate added, and received gnotobiotic assemblages of bacteria (*Pseudomonas* sp.) and bacterial feeders. The experiments led us to some early conclusions about animal–microbial interactions. Both an amoeboid protozoan (*Acanthamoeba* sp.) and a bacterial feeding nematode (*Mesodiplogaster lheritieri*) increased system activity as measured by phosphorus and nitrogen mineralization (Woods *et al.*, 1982) as well as by carbon dioxide output in soil microcosms with a simple sugar (glucose) as the fast cycle carbon source. A more detailed analysis of the amoebal activity showed significant increase in release of inorganic P (bicarbonate-extractable phosphorus) with amoebae feed-

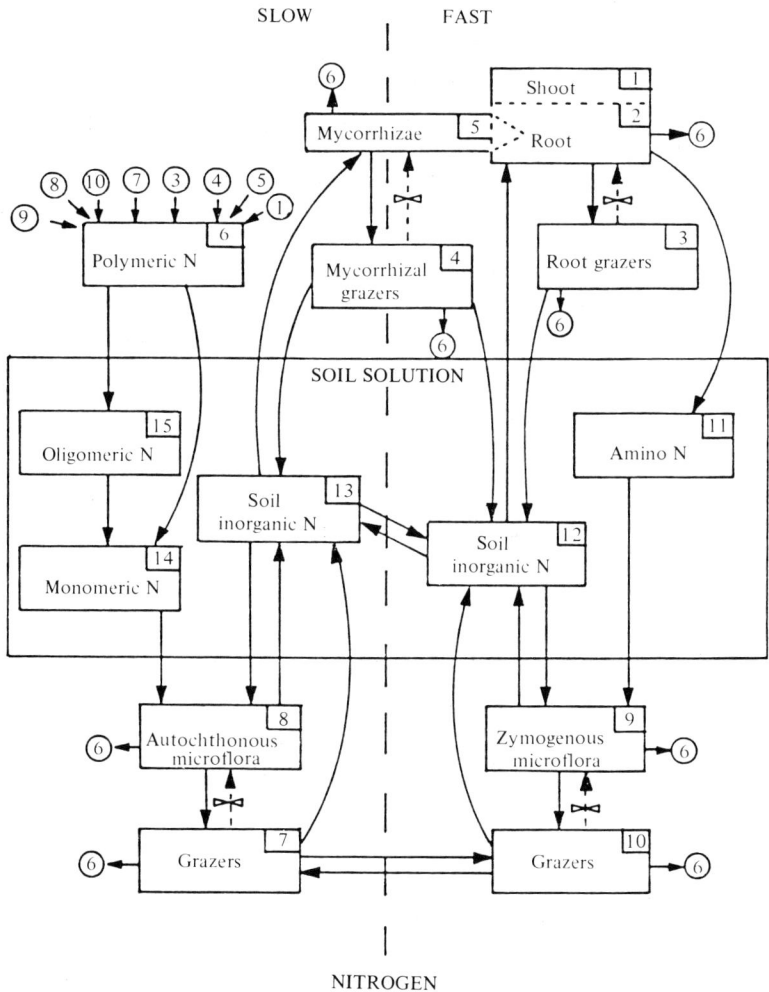

Fig. 2.4. Conceptual model of nitrogen flow in soil systems showing labile components (right-hand side) and non-labile components (left-hand side). Decomposition products enter the soil solution for uptake by microbes (including mycorrhizae) and plant roots.

ing on the bacteria, in contrast to the ungrazed controls (Fig. 2.5). When the complexity of the system was increased by the interaction between omnivorous nematodes and amoebae, respiration was significantly greater in the more complex system in spite of decreased bacterial numbers (Fig. 2.6) (Coleman *et al.*, 1978). Calculations of production efficiency (production of grazer tissue)/(microbial tissue consumed) of the two grazers showed that the amoebae were four times more efficient at producing new body tissue than an equivalent biomass of bacterial-feeding nematodes. The subsequent loss of unlysed cells from the nematode guts may have had a significant feedback on early bacterial growth (days 2–5 of a 24-day incubation). This process is considered further later in this paper. The addition of ^{14}C-labelled glucose (Anderson, Elliott, Coleman & Cole, 1981) showed that the microbial grazing activity by the nematodes increased substrate utilization and nitrogen and phosphorus mineralization (Fig. 2.7). By returning nutrients to the soil solution, particularly in the first few days of the

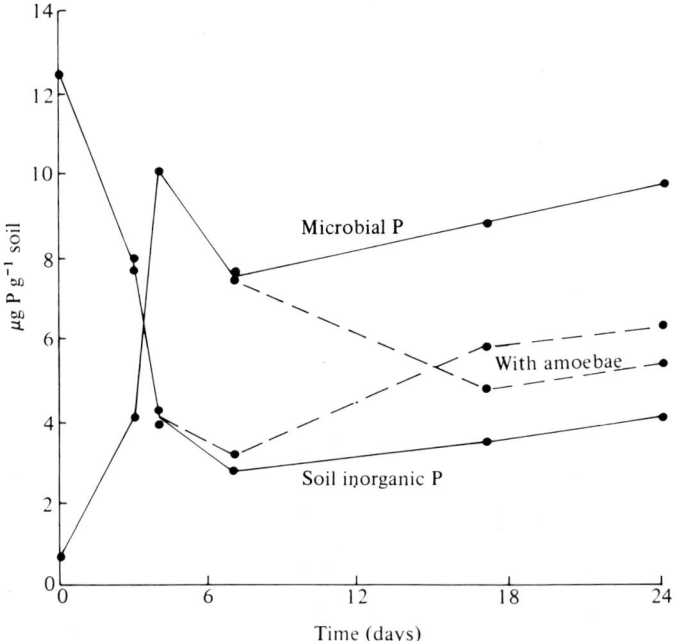

Fig. 2.5. Mineralization of microbial P in gnotobiotic microcosms with (– – – –) and without (————) amoebae (*Acanthamoeba polyphaga*). Rapid immobilization of soil inorganic P by day 7 is significantly reduced by amoebae feeding on bacteria (*Pseudomonas* sp.). (From Cole *et al.*, 1978.)

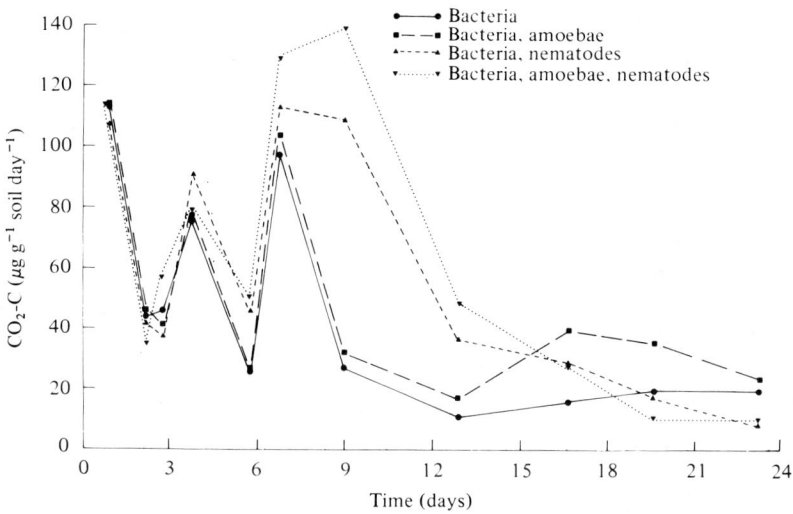

Fig. 2.6. Respiration in glucose-amended microcosms (200 μg glucose C g^{-1} soil added on days 0, 3, and 6). CO_2 output with bacteria and amoebae or nematodes is increased, with most marked enhancement between days 7 and 13 with two grazers in combination. (From Coleman et al., 1978.)

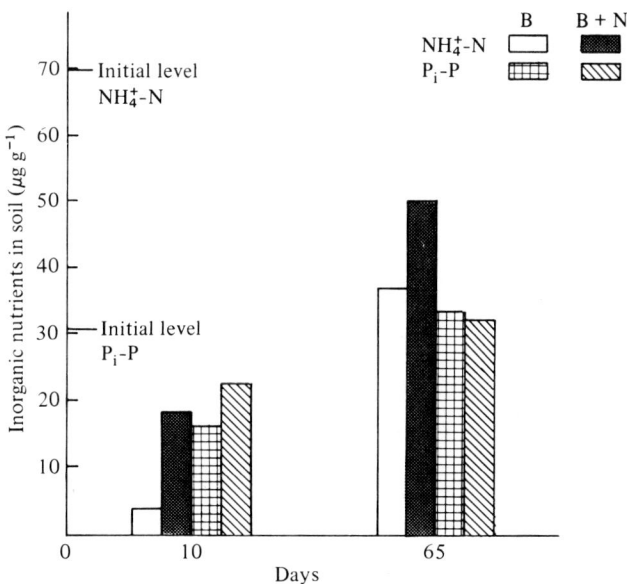

Fig. 2.7. Immobilization by bacteria alone, and subsequent return of NH_4^+-N and P_i-P in grazed systems (Day 10). Mineralization as per cent of ungrazed was greater on Day 10 than Day 65. B, bacteria; N, nematodes. (From Anderson et al., 1981a.)

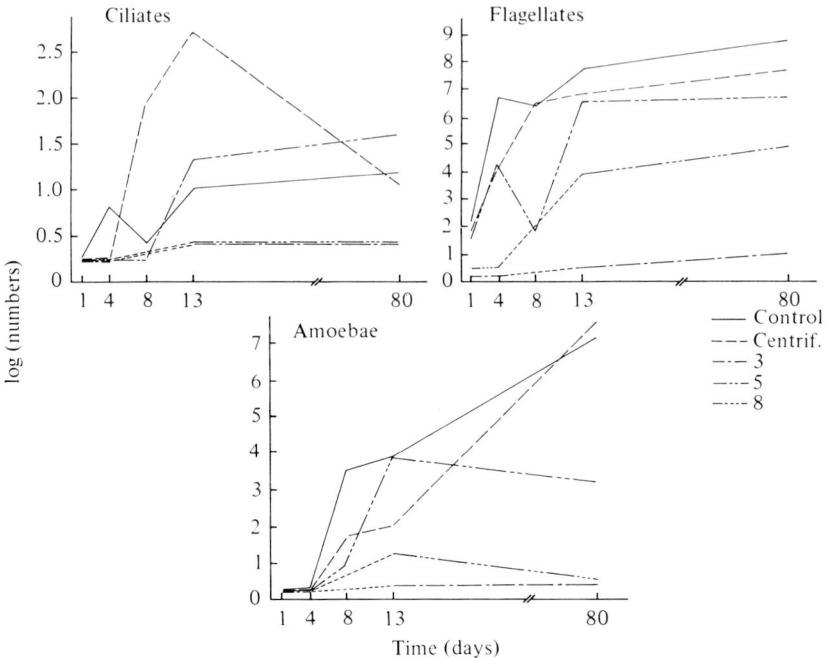

Fig. 2.8. Growth responses of ciliates, flagellates, and amoebae in soil microcosms receiving centrifuged, or 3, 5, and 8 μm filtrates of soil suspensions, or a complete mixture (control) of soil microbiota.

experiment, the microbivorous fauna maintained a somewhat higher metabolic activity, although total microbial biomass was less when grazed.

In other studies entire microbial and protozoan communities were manipulated via a selective filtration process using Nuclepore filters (McClellan, Frey, Campion & Coleman 1981). Changes were followed in respiration and nitrogen and phosphorus dynamics. Communities with several types of protozoa (ciliates, flagellates and amoebae, Fig. 2.8) and bacteria and fungi showed greater carbon and nitrogen mineralization (Figs. 2.9 and 2.10) than in those treatments which had only flagellates and a few amoebae. There was a corresponding decrease in bacterial populations (Fig. 2.11). The greatest increase in CO_2 was associated with the 8 μm filtration treatment, which also showed the largest number of grazing organisms but relatively fewer bacteria than the 3 μm and 5 μm filtrations. The 8 μm filtration also showed significantly more inorganic nitrogen as NH_4^+. In the 3 μm and 5 μm

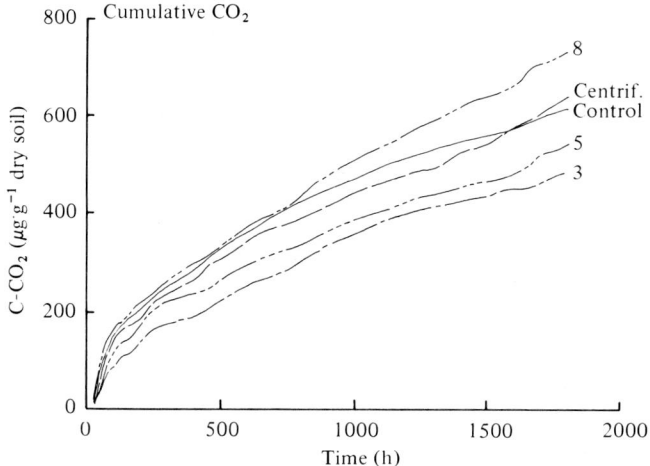

Fig. 2.9. Cumulative CO_2 output of microcosms with varying numbers of protozoa over the 80-day experiment of Fig. 2.8. Note the greater CO_2 output from the 8 μm treatment, with more amoebae.

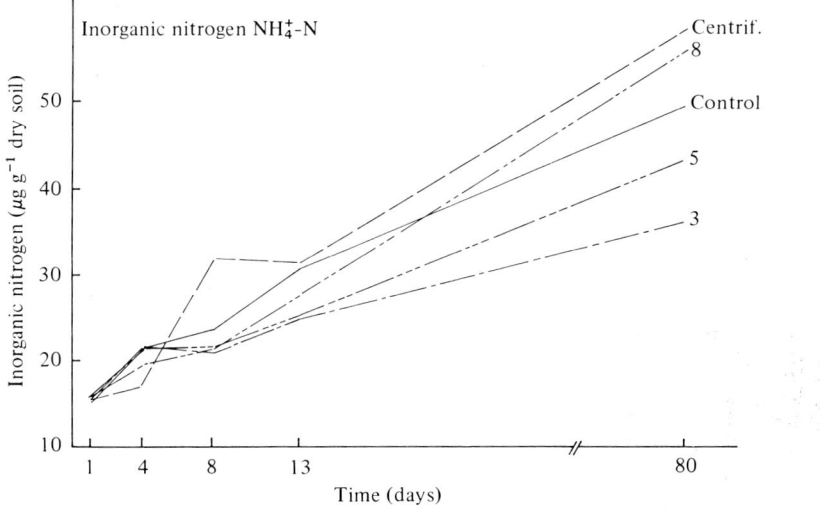

Fig. 2.10. Mineralization of NH_4^+-N in microcosms, with significantly more produced by day 80 of experiment in Fig. 2.8, 8 μm treatment.

filtrations with fewer grazers and relatively more bacteria the mineralization of nitrogen was commensurately decreased. Thus a greater species complexity shows greater turnover of nutrients than in simple incubations with one or two species.

In recent studies using more refractory substrates such as cellulose and chitin, mixed faunal systems often increased rates of decomposition and mineralization of nitrogen and phosphorus. Inoculation of high numbers of fungal-feeding nematodes into soil microcosms containing cellulose and a single fungal species decreased respiration rates, while soil inoculated with lower initial numbers had higher rates of respiration (Fig. 2.12) (Trofymow & Coleman, 1982). Bacteriophagic nematodes in cellulose and/or chitin-amended systems increased both rates of respiration and bacterial populations over those systems containing bacteria alone (Gould et al., 1981; Trofymow & Coleman, 1982).

The method of feeding of the animals may govern the overall response. Bacterial feeders with a holophagic feeding mode release nutrients and, under certain conditions, eliminate considerable quantities of unharmed bacteria during periods of excess consumption (Smerda, Jensen & Anderson, 1971). In contrast, the stylet piercing of hyphae by a fungal-feeding nematode or other suctorial feeder immobilizes nutrients by leaving large portions of the fungi inactive or with empty cell walls. It remains to be seen whether fungal activity as measured by certain metabolic indicators such as fluorescein diacetate (Söderstrom,

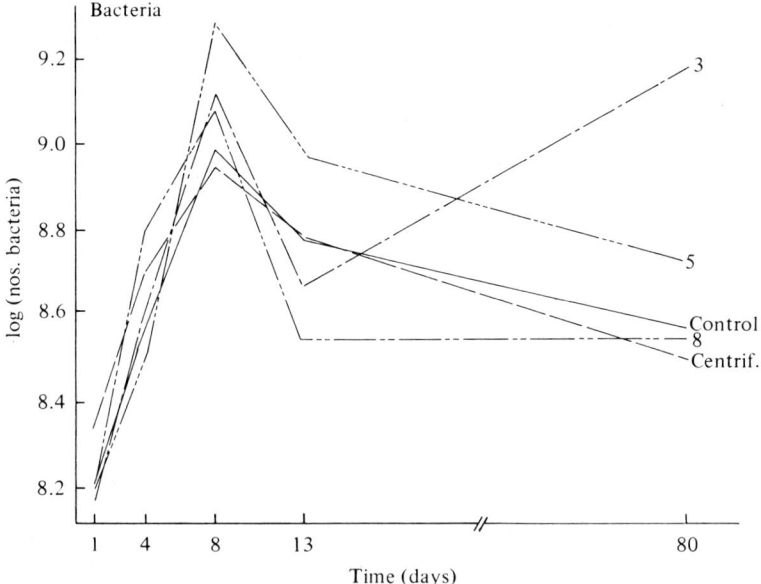

Fig. 2.11. Bacterial population dynamics in soil microcosms, showing reduction in numbers in 8 μm treatment, similar to that in control treatment (see Fig. 2.8), by 80 days.

○——○ Fungus alone (*Fusarium oxysporum*) (800 μg cellulose C g^{-1} soil)
●——● Fungus and nematodes (*Aphelenchus avenae* 36 g^{-1})
○---○ Fungus alone (314 μg cellulose C g^{-1} soil)
●---● Fungus and nematodes (4 g^{-1}) (314 μg cellulose C g^{-1} soil)

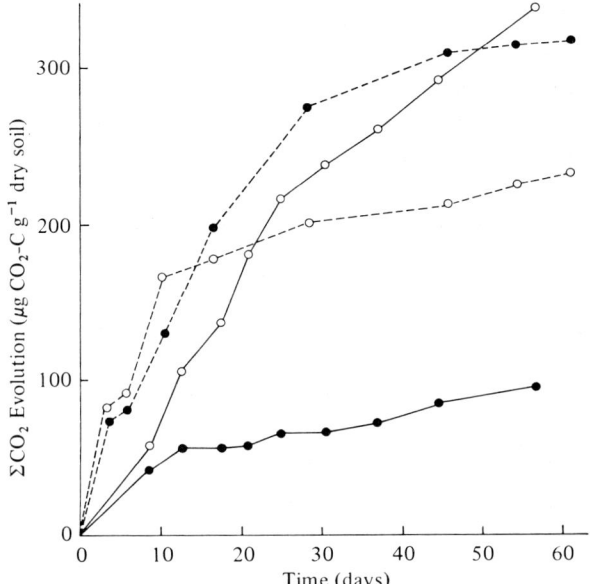

Fig. 2.12. CO_2 evolution from soil amended with cellulose and two levels of fungal-feeding nematodes. (Modified from Trofymow & Coleman, 1982.)

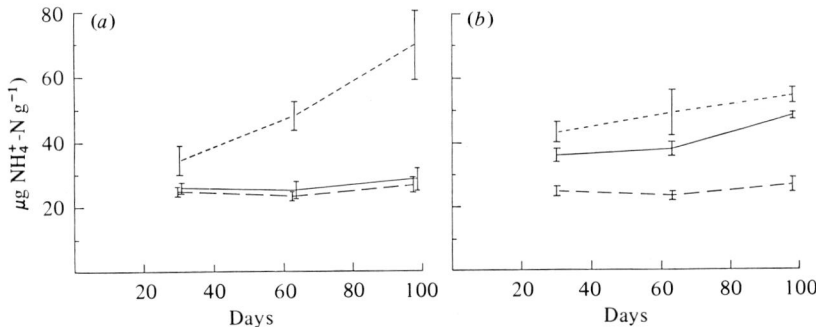

Fig. 2.13. Soil NH_4^+-N (mean and 95% confidence intervals) in soil amended with cellulose and chitin.
(*a*) Bacterial treatments: ———— uninoculated control; ———— bacteria alone (*Flavobacterium* sp.); ------- bacteria and bacterial-feeding nematode (*Pelodera* sp.). (*b*) Fungal treatments: ———— uninoculated control; ———— fungus alone (*Fusarium oxysporum*); ------- fungus and fungal-feeding nematode (*Aphelenchus avenae*). (From Trofymow & Coleman, 1982.)

1977; Ingham & Klein, 1982) would be a useful tracer to further identify the selective or more general feeding modes of some of the fungal feeders.

Nitrogen mineralization from a resistant organic nitrogen source such as chitin can be enhanced in systems containing microbial-grazing nematodes. Fig. 2.13 illustrates nitrogen mineralization in bacterial and fungal systems amended with cellulose and chitin. The additions of grazers in both systems increased levels of mineral nitrogen.

Increased rates of nutrient mobilization from organic matter via animal–microbial interactions may be especially important for plants growing in climatically stochastic environments such as shortgrass prairie and deserts. Since root extension and plant growth are heavily controlled by moisture, high rates of nutrient mineralization during moisture events would be critical for the plant.

Nutrient cycling in the presence of plants

Additional studies have been conducted showing the effects of animal–microbial mediated nutrient transformations in the presence of plants. Using seedlings of blue grama [*Bouteloua gracilis* (HBK) Griffiths] growing in microcosms, there was an increased ammonium nitrogen uptake in incubations containing bacteria and amoebae. Microcosms which received moderate (90 micrograms) or high (200 microgram) levels of ammonium nitrogen per gram of soil had significantly elevated plant tissue nitrogen levels when compared with those with plants and bacteria alone (Elliott, Coleman & Cole, 1979). The plant tissue nitrogen was twice that of controls (1.2 v. 2.4%) which had no amoebae grazing on bacteria. The experiment ran for approximately 45 days under less than optimal light conditions (c. 200–300 μE), and during this rather limited period there was no difference in net dry weight.

Additional studies growing blue grama (Ingham, Trofymow, Ingham & Coleman, unpublished) in larger microcosms (Trofymow, Gurnsey & Coleman, 1980) with a longer time span and higher light intensity have shown considerable increases in plant dry weight as well as nutrient uptake by the plant growing in the presence of various microbe-feeding nematodes. These experiments contained either bacterial- or fungal-feeding nematodes, and included purified chitin as a particulate organic nitrogen substrate added to soil low in inorganic nitrogen.

The bacteria and fungus inoculated into the soil were known chitin decomposers. Although plants growing in soil with bacteria grew more rapidly than plants in sterile soil, plants in microcosms inoculated with bacterial-feeding nematodes grew faster and had a higher nitrogen content than sterile plants or plants with bacteria alone (Fig. 2.14). Plants in soil with fungus and bacteria also had enhanced growth and nitrogen content over plants with only bacteria, but addition of fungal-feeding nematodes to the system had no significant effect. In a similar experiment in unamended soils (Ingham, 1981), the most vigorous plant responses were found in the most complex biological milieu, i.e. bacteria, fungi, bacterial-feeding nematodes and fungal-feeding nematodes (Fig. 2.15).

Discussion

Present options on the importance of the soil microfauna in enhancing decomposition are not unanimous. Bååth *et al.* (1978, 1981) measured growth of pine seedlings in soil and organic matter from field plantations of Scots pine (*Pinus silvestris*). The soils were acid and rather poor in nutrient with a carbon to nitrogen ratio of 44 : 1 in the humic material. With various members of the soil fauna (microarthropods, enchytraeids and protozoa) present, there was some enhancement of decomposition rate, but no significant increase in plant growth or in nutrient concentrations. They concluded that the fauna were not very important for nutrient return in the field. We suggest however that there might have been significant effects of mycorrhizae on nutrient uptake, as shown by Melin & Nilsson (1952) for ammonium uptake by ectomycorrhizae, but apparently this was not followed by the Swedish group.

Ingham (1981) found a considerable concentration of microbe feeding nematodes in the rhizosphere portion of the total soil, which one would expect to be areas of greater nutrient availability and turnover, particularly for high-energy carbon-containing compounds. At times he found as much as 20–40% of the total soil nematode populations in his microcosms in only 3% of the total soil, i.e. rhizosphere soil (Fig. 2.16). It therefore seems important in rhizosphere soil studies, especially with apparently suboptimal C : N ratios, to determine whether localized microsites with more favourable carbon to nitrogen ratios exist, and ascertain if only a small fraction of the total root system is needed to take up the majority of the nutrients required by the plant during the growing season. This sort of attention to spatial heterogeneity at the soil microsite level could lead to new insights in soil ecology. Sampling of bulk soil without distinguishing microsite regions, such as the rhizosphere, may dilute important interactions to an insignificant level.

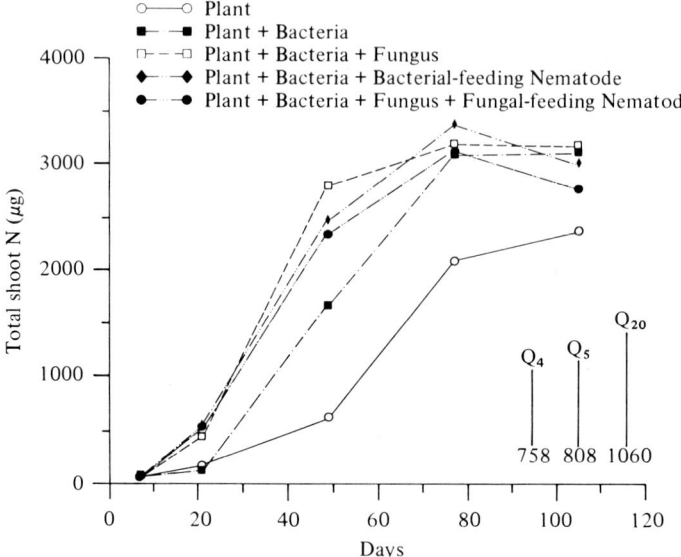

Fig. 2.14. Total shoot nitrogen in *Bouteloua gracilis* grown in chitin amended soil with different biological treatments. (From Ingham, Trofymow, Ingham & Coleman, unpublished.)

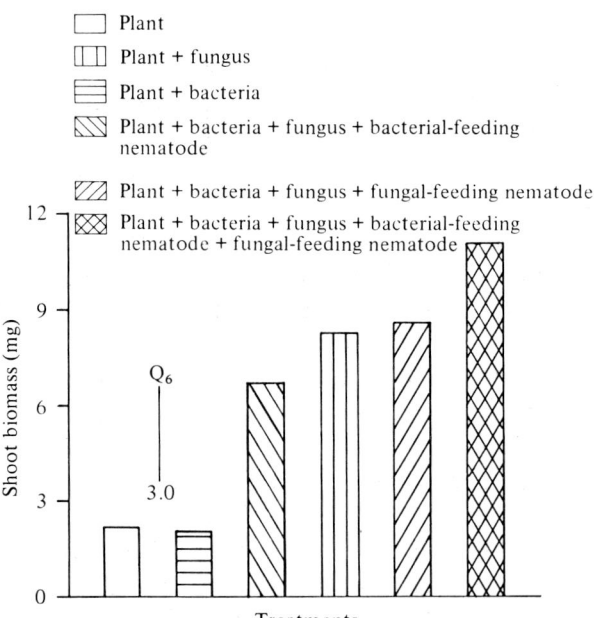

Fig. 2.15. Shoot biomass of *Bouteloua gracilis* grown in unamended soil with different biological treatments.

Discussion

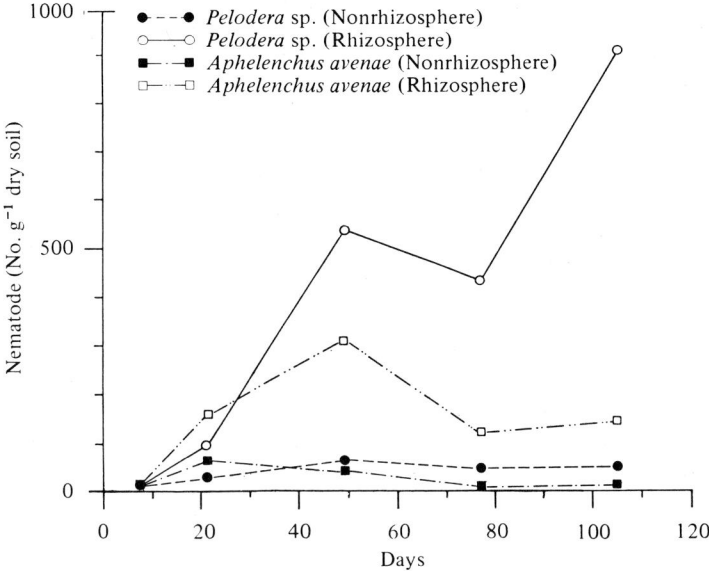

Fig. 2.16. Nematode populations in chitin amended soil from non-rhizosphere and rhizosphere areas of *Bouteloua gracilis*. (From Ingham *et al.*, unpublished.)

Faunal effects on microbial population dynamics

An important interaction between soil microflora and their respective grazers is the effect of the animals on the population sizes and turnover rates of the microflora. Most earlier studies have found that bacteriophagic nematodes significantly reduce bacterial numbers (Anderson *et al.*, 1981b; Coleman *et al.*, 1977, 1978; Whitford *et al.*, 1982), yet recently Abrams & Mitchell (1980) and Trofymow & Coleman (1982) have reported increases in bacterial numbers with bacterial-feeding nematodes compared to identical treatments without nematodes. Ingham and co-workers observed that these nematodes had no significant effect on bacterial population dynamics in non-rhizosphere soil. Yet in the rhizosphere, where bacterial and nematode population densities were much higher, bacterial-feeding nematodes significantly stimulated bacterial production (Fig. 2.17). Bacterial populations were also increased by the presence of a fungal-feeding nematode and this increase was, again, most marked in the rhizosphere.

Only a few studies have been carried out on effects of faunal feeding on root-associated, or mycorrhizal, fungi. Shafer, Rhodes & Riedel (1981) added the fungal-feeding nematode *Aphelenchus bicaudatus* to

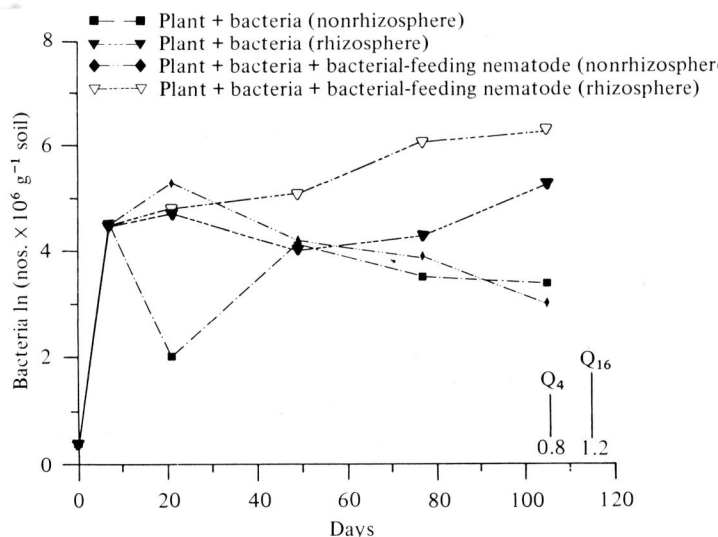

Fig. 2.17. Natural log of bacterial numbers in rhizosphere and non-rhizosphere soil of *Bouteloua gracilis* grown in chitin amended soil with and without *Pelodera* sp. (From Ingham *et al.*, unpublished.)

agar cultures of five species of ericoid mycorrhizal fungi. They observed destruction of aerial hyphae, and reduced mass of all species to half or two-thirds of control, ungrazed fungal cultures. Other workers (Hussey & Roncadori, 1981) have shown little or no effects of fungal-grazing nematodes on vesicular–arbuscular (VA) mycorrhizal biomass or subsequent plant growth. Further studies of nematode (including root-feeders)–mycorrhizal interactions are reviewed by Hussey & Roncadori (1982).

Information on microarthropod effects on mycorrhizal growth are even more scarce. Warnock, Fitter & Usher (1982) followed growth responses of *Allium porrum* in pots with and without the VA mycorrhizal fungus *Glomus fasciculatum*, and with and without the collembolan *Folsomia candida*. No direct feeding of the *Folsomia* was observed on external VA hyphae, but plant growth was significantly reduced in the plant–fungus–grazer pots than in those in which the Collembola were absent.

Recent work in our laboratory by John Moore and Robert Ames (unpublished) has shown consumption of hyphae and chlamydospores of several VA mycorrhizal fungi, in particular *Gigaspora rosea* and

Glomus fasciculatum, by the euedaphic collembolan *Onychiurus encarpatus*.

Microbe–faunal interaction effects on nutrient cycling

The more rapid cycling rate often observed in the systems which have feeding by predators upon microbial grazers may be very important to various primary producers. This is particularly true in systems having comparatively short (or episodic) pulses during the growing season, where nutrient availability may be critical over quite short time periods. Perhaps various lengths of food chains have differing effects on nutrient availability. The early work of Smith (1969) on even- and odd-numbered food chains should be considered in this context. He found that one-, three- and five-membered food chains had one sort of pattern (resources concentrated in the primary producer component), whereas those which had two, four, six or even-numbered members had a different primary control (greater activity of secondary predators, more free resources). These studies have been extended to terrestrial systems (Pimm & Lawton, 1977), including roles of omnivores (Pimm & Lawton, 1978).

Other studies in field systems have selectively removed active predatory forms (using biocides) that may play a role in enhancing or altering nutrient cycling. For example, the field studies of Santos & Whitford (1981), Santos, Phillips & Whitford (1981) and Elkins & Whitford (1982) may have some generality for soil systems. In plots with longer food chains, i.e. predatory mites feeding on bacterial-feeding nematodes, which in turn consumed bacteria in leaf litter bags in the Chihuahuan Desert of New Mexico, there was considerably greater litter decomposition, carbon dioxide evolution and nitrogen and phosphorus mineralization than in plots where the predatory mites were eliminated.

Implications and opportunities for future research

Trophic interaction processes, while important, are only a few of the many which occur in various terrestrial ecosystems. They should be viewed in the context of nutrient availability as well as short and long-term processes of organic matter formation and turnover. A small proportion, perhaps only 1 or 2% of the total soil organic matter is microbial (Jenkinson & Ladd, 1981), but a sizeable proportion of the total labile nitrogen and phosphorus is directly associated with that in the microbial pool. There are several experiments under way in various countries to examine the roles of soil fauna such as protozoa and nematodes in mineralization phenomena. This may be important in the standard ten-day incubation after treatment with ethanol-free chloroform (Jenkinson & Powlson, 1976).

There seem to be several similarities in the behaviour of roots, microbes, and fauna in using aggregations of organic matter as places in which to grow and proliferate. St John, Coleman & Reid (1983) have noted very similar behaviour of root growth and VA mycorrhizal growth in relation to soil OM. In each case, the roots and hyphae were distributed at random in open soil, then proliferated in small pockets (naturally occurring, or experimentally manipulated) of organic matter. The abundances of root and VAM closely approximated a negative binomial distribution.

A further aspect of interest is the apparent bi-directional flow of carbon in soil between roots on the one hand, and via gaseous loss of CO_2 on the other. Sparling, Cheshire & Mundie (1982) noted considerable microbial-derived CO_2 in barley root cell wall materials after seedlings grew in labelled organic matter for 42 days. These developments show that caution should be exercised in describing the extent of root–microbial–faunal interactions in soil systems.

We suggest that concentrating on localized areas of considerable activity within the mosaic of substrate types and pore space, such as the rhizosphere, will pay large dividends in further understanding the many processes which occur in soils, and will relate studies to a more holistic approach of nutrient cycling in entire ecosystems. Considering a whole spectrum of microflora, fauna, the plant root, and the soil abiotic physical substrate is likely to lead to important conceptual advances in this exciting area of ecology.

References

Abrams, B. I. & Mitchell, M. J. (1980). Role of nematode–bacterial interactions in heterotrophic systems with emphasis on sewage sludge decomposition. *Oikos*, **35**, 404–10.

Anderson, R. V., Coleman, D. C. & Cole, C. V. (1981a). Effects of saprotrophic grazing on net mineralization. In *Terrestrial Nitrogen Cycles*, ed. F. E. Clark & T. Rosswall, pp. 201–15. Ecological Bulletin 33. Stockholm: Swedish Natural Science Research Council.

Anderson, R. V., Elliott, E. T., Coleman, D. C. & Cole, C. V. (1981b). Effect of the nematodes *Acrobeloides* sp. and *Mesodiplogaster lheritieri* on substrate utilization and nitrogen and phosphorus mineralization in soil. *Ecology*, **62**, 549–55.

Bååth, E., Lohm, U., Lundgren, B., Rosswall, T., Söderström, B., Sohlenius, B. &

References

Wiren, A. (1978). The effect of nitrogen and carbon supply on the development of soil organism populations and pine seedlings: A microcosm experiment. *Oikos*, **31**, 153–63.

Bååth, E., Lohm, U., Lundgren, B., Rosswall, T., Söderström, B. & Sohlenius, B. (1981). Impact of microbial-feeding animals on total soil activity and nitrogen dynamics: A soil microcosm experiment. *Oikos*, **37**, 257–64.

Bornebusch, C. H. (1930). The fauna of the forest soil. *Forstlige Forsoksvaesen I Danmark*, **11**, 244.

Bowen, G. D. & Smith, S. E. (1981). The effects of mycorrhizas on nitrogen uptake by plants. In *Terrestrial Nitrogen Cycles*, ed. F. E. Clark & T. Rosswall, pp. 237–47. Ecological Bulletin 33. Stockholm: Swedish Natural Science Research Council.

Brown, M. E. (1975). Rhizosphere microorganisms: Opportunists, bandits or benefactors? In *Soil Microbiology*, ed. N. Walker, pp. 21–36. New York and London: Wiley.

Brown, M. E. (1976). Microbial manipulation and plant performance. In *Microbiology in Agriculture, Fisheries and Food*, ed. F. A. Skinner & J. G. Carr, pp. 37–53. Society of Applied Bacteriology Symposium Series No. 4. New York: Academic Press.

Chaboud, A. & Rougier, M. (1981). Sécrétions racinaires mucilagineuses et rôle dans la rhizosphère. *Année Biologique*, **20**, 313–26.

Clarholm, M. (1981). Protozoan grazing of bacteria in soil – impact and importance. *Microbial Ecology*, **7**, 343–50.

Cole, C. V., Elliott, E. T., Hunt, H. W. & Coleman, D. C. (1978). Trophic interactions in soils as they affect energy and nutrient dynamics. V. Phosphorus transformations. *Microbial Ecology*, **4**, 381–7.

Coleman, D. C. (1976). A review of root production processes and their influence on soil biota in terrestrial ecosystems. In *The Role of Terrestrial and Aquatic Organisms in Decomposition Processes*, ed. J. M. Anderson & A. Macfadyen, pp. 417–34. Oxford: Blackwells.

Coleman, D. C., Cole, C. V., Anderson, R. V., Blaha, M., Campion, M. K., Clarholm, M., Elliott, E. T., Hunt, H. W., Schaefer, B. & Sinclair, J. (1977). Analysis of rhizosphere–saprophage interactions in terrestrial ecosystems. In *Soil Organisms as Components of Ecosystems*, ed. U. Lohm & T. Persson, pp. 299–309. Ecological Bulletin 25. Stockholm: Swedish Natural Science Research Council.

Coleman, D. C., Anderson, R. V., Cole, C. V., Elliott, E. T., Woods, L. & Campion, M. K. (1978). Trophic interactions in soils as they affect energy and nutrient dynamics. IV. Flows of metabolic and biomass carbon. *Microbial Ecology*, **4**, 373–80.

Coleman, D. C., Reid, C. P. P. & Cole, C. V. (1983). Biological strategies of nutrient cycling in soil systems. *Advances in Ecological Research*, **13**, 1–55.

Cutler, D. W., Crump, L. M. & Sandon, H. (1922). A quantitative investigation of the bacterial and protozoan population of the soil, with an account of the protozoan fauna. *Philosophical Transactions of the Royal Society Series*, **211**, 317–50.

Darbyshire, J. F. & Greaves, M. P. (1973). Bacteria and protozoa in the rhizosphere. *Pesticide Science*, **4**, 349–60.

Elkins, N. Z. & Whitford, W. G. (1982). The role of microarthropods and nematodes in decomposition in a semi-arid ecosystem. *Oecologia*, **55**, 303–10.

Elliott, E. T., Coleman, D. C. & Cole, C. V. (1979). The influence of amoebae on the uptake of nitrogen by plants in gnotobiotic soil. In *The Soil-Root Interface*, ed. J. L. Harley & R. S. Russell, pp. 221–9. London: Academic Press.

Foster, R. C. (1981). The ultrastructure and histochemistry of the rhizosphere. *New Phytologist*, **89**, 263–73.

Foster, R. C. & Martin, J. K. (1981). *In situ* analysis of soil components of biological origin. In *Soil Biochemistry*, vol. 5, ed. E. A. Paul & J. N. Ladd, pp. 75–111. New York & Basel: Marcel Dekker, Inc.

Gould, W. D., Bryant, R. J., Trofymow, J. A., Elliott, E. T., Anderson, R. V. & Coleman, D. C. (1981). Chitin decomposition in a model soil system. *Soil Biology and Biochemistry*, **13**, 487–92.

Hanlon, R. D. G. (1981). Influence of grazing by Collembola on the activity of senescent fungal colonies grown on media of different nutrient concentration. *Oikos*, **36**, 362–7.

Hussey, R. S. & Roncadori, R. W. (1981). Influence of *Aphelenchus avenae* on vesicular–arbuscular endomycorrhizal growth response in cotton. *Journal of Nematology*, **13**, 48–52.

Hussey, R. S. & Roncadori, R. W. (1982). Vesicular–arbuscular mycorrhizae may limit nematode activity and improve plant growth. *Plant Disease*, **66**, 9–14.

Ingham, R. E. (1981). The Role of Soil Nematodes in Ecosystem Regulation. Unpublished Ph.D. dissertation, Colorado State University.

Ingham, R. E. & Klein, D. A. (1982). Relationship between fluorescein diacetate-stained hyphae and oxygen utilization, glucose utilization and biomass of submerged fungal batch cultures. *Applied and Environmental Microbiology*, **44**, 363–70.

Jacot, A. P. (1940). The fauna of the soil. *Quarterly Review of Biology*, **15**, 28–58.

Jenkinson, D. S. & Ladd, J. N. (1981). Microbial biomass in soil : Measurement and turnover. In *Soil Biochemistry*, vol. 5, ed. E. A. Paul & J. N. Ladd, pp. 415–71. New York: Marcel Dekker.

Jenkinson, D. S. & Powlson, D. S. (1976). The effects of biocidal treatments on metabolism in soil. V. A method for measuring soil biomass. *Soil Biology and Biochemistry*, **8**, 209–13.

Jenny, H. & Grossenbacher, K. (1963). Root–soil boundary zones as seen in the electron microscope. *Soil Science Society of America, Proceedings*, **27**, 273–7.

Kloepper, J. W., Leong, J., Teintze, M. & Schroth, M. N. (1980*a*). Enhanced plant growth by siderophores produced by plant growth-promoting rhizobacteria. *Nature, London*, **286**, 885–6.

Kloepper, J. W., Leong, J., Teintze, M. & Schroth, M. N. (1980*b*). *Pseudomonas* siderophores: A mechanism explaining disease-suppressive soils. *Current Topics in Microbiology and Immunology*, **4**, 317–20.

Luxton, M. (1982). Quantitative utilization of energy by the soil fauna. *Oikos*, **39**, 342–54.

Lynch, J. M. & Panting, L. M. (1980). Cultivation and the soil biomass *Soil Biology and Biochemistry*, **12**, 29–33.

Macfadyen, A. (1968). The animal habitat of soil bacteria. In *The Ecology of Soil Bacteria*, ed. T. R. G. Gray & D. Parkinson, pp. 66–76. Toronto, Ontario: University of Toronto Press.

Macfadyen, A. (1969). The systematic study of soil ecosystems. In *The Soil Ecosystem*, ed. J. G. Sheals, pp. 191–7. Systematics Association Publication No. 8. London: The Systematics Association.

McClellan, J. F., Frey, J., Campion, M. K. & Coleman, D. C. (1981). Protozoan mineralization roles in soil ecosystems. In *Proceedings VI International Congress of Protozoology*, p. 241 Warsaw, Poland.

Marshall, K. C. (1976). *Interfaces in Microbial Ecology*. Cambridge, Massachusetts: Harvard University Press.

Melin, E. & Nilsson, H. (1952). Transport of labelled nitrogen from an ammonium source

to pine seedlings through mycorrhizal mycelium. *Svensk botanisk tidskrift*, **46**, 281–5.
Nikolyuk, V. F. & Geltzer, J. G. (1972). *Soil Protozoa of the USSR*. Tashkent, Uzbek. SSR: FAN Press.
Petersen, H. (1982). The total soil fauna biomass and its composition. *Oikos*, **39**, 330–9.
Pimm, S. L. & Lawton, J. H. (1977). Number of trophic levels in ecological communities. *Nature, London*, **268**, 329–31.
Pimm, S. L. & Lawton, J. H. (1978). On feeding on more than one trophic level. *Nature, London*, **275**, 542–4.
Rovira, A. D., Foster, R. C. & Martin, J. K. (1979). Note on terminology : Origin, nature and nomenclature of the organic materials in the rhizosphere. In *The Soil–Root Interface*, ed. J. L. Harley & R. S. Russell, pp. 1–4. London: Academic Press.
Santos, P. F., Phillips, J. & Whitford, W. G. (1981). The role of mites and nematodes in early stages of buried litter decomposition in a desert. *Ecology*, **62**, 664–9.
Santos, P. F. & Whitford, W. G. (1981). The effects of microarthropods on litter decomposition in a Chihuahuan desert ecosystem. *Ecology*, **62**, 654–63.
Schroth, M. N. & Hancock, J. G. (1982). Disease-suppressive soil and root-colonizing bacteria. *Science*, **216**, 1376–81
Shafer, S. R., Rhodes, L. H. & Riedel, R. N. (1981). *In vitro* parasitism of endomycorrhizal fungi of ericaceous plants by the mycophagous nematode *Aphelenchoides bicaudatus*. *Mycologia*, **73**, 141–9.
Shamoot, S., McDonald, L. & Bartholomew, W. V. (1968). Rhizo-deposition of organic debris in soil. *Soil Science Society of America, Proceedings*, **32**, 817–20.
Smerda, S. M., Jensen, H. J. & Anderson, A. W. (1971). Escape of Salmonellae from chlorination during ingestion by *Pristionchus lheritieri* (Nematoda : Diplogasterinae). *Journal of Nematology*, **3**, 201–4.
Smith, F. E. (1969). Effects of enrichment in mathematical models. In *Eutrophication: Causes, Consequences, and Correctives*, pp. 631–45. Washington, DC: National Academy of Sciences.
Söderström, B. E. (1977). Vital staining of fungi in pure cultures and in soil with fluorescein diacetate. *Soil Biology and Biochemistry*, **9**, 59–63.
Sohlenius, B. (1980). Abundance, biomass, and contribution to energy flow by soil nematodes in terrestrial ecosystems. *Oikos*, **34**, 186–94.
Sparling, G. P., Cheshire, M. V. & Mundie, C. M. (1982). Effect of barley plants on the decomposition of ^{14}C-labeled organic matter. *Journal of Soil Science*, **33**, 89–100.
Stewart, J. W. B. & McKercher, R. B. (1982). Phosphorus cycle. In *Experimental Microbial Ecology*, ed. R. G. Burns & J. H. Slater, pp. 221–38. Oxford: Blackwells.
St John, T. V., Coleman, D. C. & Reid, C. P. P. (1983). Growth and spatial distribution of nutrient-absorbing organs: Selective exploitation of spatial heterogeneity. *Plant and Soil*, **71**, 487–93.
Stout, J. D. (1980). The role of protozoa in nutrient cycling and energy flow. *Advances in Microbial Ecology*, **4**, 1–50.
Tisdall, J. M. & Oades, J. M. (1982). Organic matter and water-stable aggregates in soils. *Journal of Soil Science*, **33**, 141–63.
Trofymow, J. A. & Coleman, D. C. (1982). The role of bacterivorous and fungivorous nematodes in cellulose and chitin decomposition. In *Nematodes in Soil Ecosystems*, ed. D. W. Freckman, pp. 117–38. Austin: University of Texas Press.

Trofymow, J. A., Gurnsey, J. & Coleman, D. C. (1980). A gnotobiotic plant microcosm. *Plant and Soil*, **55**, 167–70.

Warnock, A. J., Fitter, A. H. & Usher, M. B. (1982). The influence of a springtail *Folsomia candida* (Insecta, Collembola) on the mycorrhizal association of leek *Allium porrum* and the vesicular–arbuscular mycorrhizal endophyte *Glomus fasciculatus*. *New Phytologist*, **90**, 285–92.

Whitford, W. G., Freckman, D. W., Santos, P. F., Elkins, N. Z. & Parker L. W. (1982). The role of nematodes in decomposition in desert ecosystems. In *Nematodes in Soil Ecosystems*, ed. D. W. Freckman, pp. 98–116. Austin: University of Texas Press.

Witkamp, M. (1960). Seasonal fluctuations of the fungus flora in mull and mor of an oak forest. *Mededelingen Instituut Toegepast Biolisch Onderzat Natuur (ITBON)*, **46**, 1–52.

Woods, L. E., Cole, C. V., Elliott, E. T., Anderson, R. V. & Coleman, D. C. (1982). Nitrogen transformations in soil as affected by bacterial–microfaunal interactions. *Soil Biology and Biochemistry*, **14**, 93–8.

Yeates, G. W. & Coleman, D. C. (1982). Nematodes and decomposition. In *Nematodes in Soil Ecosystems*, ed. D. W. Freckman, pp. 55–80. Austin: University of Texas Press.

3
Interactions between microorganisms and soil invertebrates in nutrient flux pathways of forest ecosystems

J. M. ANDERSON and P. INESON*
Wolfson Ecology Laboratory, Department of Biological Sciences, University of Exeter, Exeter EX4 4PS, UK

Key words: animal–microbial interactions; nutrient cycles; nitrogen mineralization; temperate forests; grazing; fungi; bacteria

Introduction

A fairly comprehensive body of information is available about the structure and functioning of forest ecosystems (Reichle, 1981), which, together with the widespread application of watershed techniques, has emphasized the role of soil biological processes in regulating the dynamics and conservation of nutrients in these systems. In particular, the key processes of nutrient immobilization, mineralization and release not only determine the availability of elements for root uptake but, at a gross level, regulate the response of the ecosystem to perturbations such as insect attacks (Zlotin & Khodashova, 1980) or the catastrophic effects of fire, wind-throw or clear-felling (Woodmansee & Wallach, 1983; Tamm, Holmen, Popović & Wiklander, 1974; Vitousek & Melillo, 1979; Khanna, 1981). The balance between nutrient immobilization and mineralization in relation to rates of accession may also be critical for the accumulation of a nitrogen capital sufficient to support a forest ecosystem (Reiners, 1981; Bradshaw, Marrs, Roberts & Skeffington, 1982); alternatively accumulation may represent effective removal of available nutrients from short-term cycles with the possibility that primary productivity becomes limited as the stand matures (Miller, 1979).

The functioning of decomposition and nutrient cycling processes at this gross level is reasonably well understood in natural ecosystems and this knowledge has been applied to watershed management and forestry

* Institute of Terrestrial Ecology, Merlewood Research Station, Grange-over-Sands, Cumbria, UK

practices (Swank & Waid, 1980; Aber, Botkin & Melillo, 1978, 1979; Miller, 1981). There has, however, been little explicit recognition of the complexity of the processes which regulate nutrient fluxes between litter, soil and plant roots in forest ecosystems (Frissel & Van Veen, 1982), and no general recognition, let alone quantification, of the involvement of soil invertebrates in these processes. The behaviour of nitrogen, for example, is frequently interpreted in terms of simple hypotheses based on one or two processes developed for agricultural systems (Heal, Swift & Anderson, 1982). There is evidence to support the concept of critical carbon:nitrogen ratios in agricultural soils (Scarsbrook, 1965; Alexander, 1977; Paul & Juma, 1981) where organic matter inputs tend to be high in nitrogen, low in lignin, and are associated with relatively homogeneous mineral soils with high bacterial activity. However, there are good reasons for supposing that the general concept may be inadequate to explain nitrogen dynamics in forest soils where litter has a low quality and the time course of decay is protracted; in these conditions there is an increased probability of nitrogen immobilization in intractable organic compounds (Heal, Swift & Anderson, 1982). A number of new models have been proposed relating the timing of nitrogen release to the initial nitrogen content and decomposition rate of litter (Aber & Melillo, 1982; Bosatta & Staaf, 1982), but the rationale is based entirely on microbial processes. Bosatta & Staaf (1982) specifically assume, as a requirement of their model, that fauna have insignificant roles in nutrient mobilization in coniferous and many other forest ecosystems.

It is axiomatic from the trophic organization of decomposer communities that soil invertebrates will interact with fungi and bacteria in nutrient flux pathways between decomposing organic matter and plant root systems. The functional level at which it is necessary to take account of animal–microbial effects in understanding the mechanisms of nutrient cycles is unclear, though it will certainly vary between systems and individual processes.

The central problem is quantifying both the direct and indirect effects of invertebrates on soil processes in a way that may be integrated into ecosystem models or nutrient budgets. The direct effects involve transfers between pools where the fluxes are mediated directly by animals. Thus saprophagy, bacteriophagy, mycophagy, and necrophagy (carnivory) will involve the release of nutrients through feeding activities, excretion and turnover of secondary production. The indirect effects involve feedbacks to lower trophic levels; for example, the effects of comminution and grazing on microbial activity and functional organization of bacterial and fungal communities, and the effects of predators on the population dynamics of their prey.

We will therefore firstly consider some of the evidence for the role of soil invertebrates in mineral element fluxes, identifying direct and indirect effects on microorganisms where possible; then attempt to quantify the process variables and, finally, consider the mechanisms involved in animal–microbial interactions.

This review is largely confined to considering the role of mesofauna and macrofauna in mineral element release from forest leaf litter and soil organic matter in view of the contributions by Swift & Boddy (Chapter 4) and Coleman, Ingham, McClellan & Trofymow (Chapter 2).

Direct effects of animals on nutrient fluxes

There is a very extensive literature on the role of soil animals in nutrient flux pathways of grassland and forest ecosystems which is covered by numerous symposia and review volumes; for example, Doeksen & van der Drift (1963), Vaněk (1975), Anderson & Macfadyen (1976), Dickinson & Pugh (1974), Lohm & Persson (1977).

The general approach to quantifying the direct contribution of animals to nutrient cycles is to budget mineral element transfers involved in the trophic and population dynamics of the fauna; see, for example, Edwards, Reichle & Crossley (1970), Reichle (1977), Anderson, Coleman & Cole (1981), Krivolutsky & Pokarzhevsky (1977), Persson et al. (1980), Persson (1983). Results of a number of studies suggest that in temperate soils with a faunal biomass of at least 5–10 g dry wt m^{-2}, the annual turnover of elements by the fauna can equal or exceed element inputs in leaf litter to the decomposer system (Satchell, 1963; Graff, 1971; Zajonc, 1971; Zlotin, 1971; Zlotin & Khodashova, 1980; Krivolutsky & Pokarzhevsky, 1977; Stachurski & Zimka, 1977).

The majority of studies calculate nutrient losses from a year class of litter and do not place the results within the context of an ecosystem budget. Syers, Sharpley & Keeney (1979) calculated that the contribution of earthworms to nitrogen mineralization in a New Zealand pasture was small in relation to *in situ* microbial mineralization of nitrogen in litter, and that the activities of the worms were insignificant in relation to the overall turnover of nitrogen in the ecosystem. The contribution of earthworms to this budget, however, is based on the mineral nitrogen content of *Lumbricus rubellus* casts. Since this is not predominantly a surface casting species (Edwards & Lofty, 1977) the direct effects of the worms on nitrogen mineralization, let alone the indirect effects on microbial activity, are probably underestimated.

Few of these studies differentiate between litter and microbial pools and it is therefore impossible to assess the comparative importance of the animal–microbial interface in these nutrient fluxes. There is circumstantial evidence from litter-bag experiments that the feeding activities of the microfauna (Protozoa and nematodes) and mesofauna (Collembola and mites) are more important in mobilizing nutrients than contributing to weight loss from leaves (Wood, 1974; Seastedt & Crossley, 1980; Anderson, Proctor & Vallack, 1983) but few field studies exist where this effect has been quantified.

One of the most detailed analyses of the trophic structure and biomass dynamics of soil organisms is a study on a 120-year-old Scots pine (*Pinus sylvestris*) forest in Sweden carried out by Persson *et al.* (1980). Northern coniferous forests normally have a much lower soil fauna biomass than 5–$10\,\mathrm{g\,m^{-2}}$ and it could be argued that the effect of the animals might be insignificant in relation to the biomass and activity of the microorganisms. Persson *et al.* (1980) calculated that although the soil fauna represented a biomass of only $1.7\,\mathrm{g\,m^{-2}}$ (in comparison with $120\,\mathrm{g\,m^{-2}}$ fungi and $39\,\mathrm{g\,m^{-2}}$ bacteria) and contributed only 4% of heterotroph respiration, they consumed 30–60% of microbial production in the litter and humus layers of this site. As a consequence of this the soil invertebrates directly contributed between 10 and 49% of total nitrogen mineralization ($28\,\mathrm{kg\,ha^{-1}\,yr^{-1}}$) (according to the value of nitrogen assimilation efficiency assumed in the model), of which 70% was excretion by bacterivores and fungivores.

Ausmus, Edwards & Witkamp (1976) also used a budgetry approach in investigating fluxes of carbon, nitrogen, phosphorus and potassium through the soil microbiota in a warm temperate deciduous forest in Tennessee (USA). Element concentrations were measured over one year in litter fall, roots, atmospheric inputs, fauna, microorganisms, leachates and soil organic matter. Highest levels of microbial immobilization of nutrients occurred in summer, and lowest levels during the spring period of maximum root growth. Ausmus *et al.* (1976) concluded that the fauna were critical for mediating the nutrient transfers and that grazing of up to 86% of the fungal production by the fauna was a key process. Laboratory microcosm studies were cited as evidence of these interactions in which it was demonstrated that snails and millipedes were integral to the transfers of ^{135}Cs from litter to plant seedlings (Patten & Witkamp, 1967) and that a reduction of microbial biomass was involved in the mobilization of potassium and magnesium (Witkamp & Frank, 1969, 1970). However, no analyses were made for nitrogen or phosphorus in these experiments, although these elements are characteristically immobilized by microorganisms involved in litter decomposition (Swift, Heal & Anderson, 1979).

Anderson, Ineson & Huish (1983a) found that the addition of millipedes to decomposing leaves increased ammonium-N losses by up to sixteen times control levels. This effect did not occur step-wise with the addition of animals and, when the animals were removed, nitrogen mineralization only returned to control levels after a period of several weeks (Fig. 3.1). The implication of these results is that indirect effects of the animals are more important in nitrogen mineralization than their direct contribution through feeding or excretion.

Indirect effects of animals through interactions with soil microorganisms

The direct contribution of soil invertebrates to total heterotrophic metabolism varies from about 1% to 15% (Edwards *et al.*, 1970;

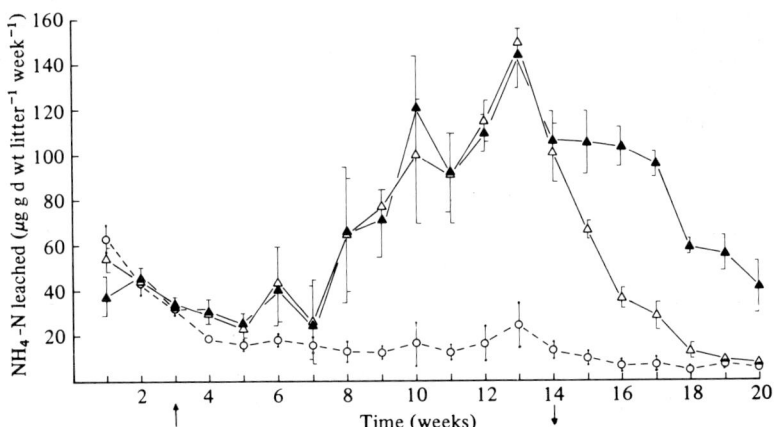

Fig. 3.1. The effects of adding and removing *Glomeris marginata* on ammonium-N release from oak leaf litter. Three sets of microcosms were set up containing 2 g dry wt oak leaf litter and incubated for three weeks at 15 °C before the animals (4 per chamber) were added to two treatments (▲, △). Controls (○) were maintained without animals. At week 14 the millipedes were removed from one of the animal treatments (△). Results are expressed as μg N g dry wt litter^{-1} week^{-1} ±1 SE; n = 4. (Anderson, Ineson & Huish, unpublished data.)

Reichle, 1977; Persson *et al.*, 1980), but their influence on microbial activity may be proportionally greater through the indirect effects of comminution, grazing and the dispersal of inocula (Macfadyen, 1963).

Standen (1978) found that experimental material held in mesh litter-bags enclosing Diptera larvae or enchytraeids had higher weight loss, attributable to greater microbial respiration, than control bags lacking animals. Laboratory experiments have shown, however, that the quantitative effects of animal feeding activities are complex and not predictable *a priori*. Woodlice and millipedes feeding on leaf litter were found to induce a non-linear response in microbial respiration with enhancement of up to 1.6 times control levels with optimal numbers of animals, yet at higher feeding intensities microbial respiration declined (Hanlon & Anderson, 1979). In addition, comminution may enhance microbial catabolism by increasing the surface area of the resource exposed to enzyme attack so that respiration is inversely related to litter particle size. Thus respiration increases in relation to the surface area to volume ratio of the resource, but the compaction of small particles into aggregates may inhibit fungal attack (Hanlon, 1981*a*, *b*). Scanning electron micrographs of artificial and faecal aggregates show that while bacteria are distributed throughout the comminuted litter, fungal hyphae are confined to the outer surface of the aggregate (Hanlon and Anderson, unpublished data). The results of these experiments suggest that the feeding activities of particular size groups of fauna may determine subsequent pathways and processes of decomposition, not only in the litter layers, but as the comminuted material is sorted by physical processes in the soil profile. The accumulation of finely particulate material in the humus layers might therefore result in inhibited decomposition and the immobilization of nutrients in a way which is not predictable from the original composition of the litter resource. Highly organic, moder humus-forms, where the soil organic matter is largely composed of microarthropod and enchytraeid faeces (Kubiena, 1953; Zachariae, 1965; Babel, 1972), can develop under tree species which produce litter readily attacked by soil animals and microorganisms in the litter layers (Anderson, 1973*a*). The causal relationship between litter comminution by mesofauna and the accumulation of soil organic matter is speculative but might be a gross manifestation of the phenomenon observed by Grossbard (1969) and others, that the faeces of cryptostigmatid mites appear to have a longer residence time in soils than the cellular structure of the parent litter from which they were derived.

Grazing involves the consumption of fungi and bacteria by microfauna or mesofauna which are of the size to select microbial tissues from the organic matter matrix. Their effects may therefore be very specific and can indirectly influence the composition and activities of the microbial community. A number of workers have demonstrated in the laboratory that protozoa and nematodes are able to regulate the growth dynamics of bacteria and hence the turnover and mineralization of nitrogen and phosphorus (Cole, Elliot, Hunt & Coleman, 1978; Anderson *et al.*, 1981; Bååth *et al.*, 1978, 1981; Clarholm *et al.*, 1981), and Coleman *et al.* (Chapter 2) suggest that microsite effects are of particular consequence in the rhizosphere.

Parkinson, Visser & Whittaker (1979) showed that grazing by Collembola may be species selective and that the differential effects of grazing intensities can alter the balance of fungal species in decomposing leaf litter. Newell (1980) also invoked grazing by Collembola to account for the microdistribution of *Mycena galopus* and *Marasmius androsaceus* mycelia in the soil profile; however, the functional consequences of these changes are unknown. Ineson, Leonard & Anderson (1982) showed that Collembola enhanced nitrogen mineralization in the early stages of litter decomposition but Visser, Whittaker & Parkinson (1981) found no evidence of grazing effects on nutrient release. On the other hand Seastedt & Crossley (1980) suggest that the stimulation of fungal growth by microarthropod grazing could lead to greater immobilization of nutrients. Similar conflicting effects of grazing on fungal biomass and respiration have been recorded (Hanlon & Anderson, 1979; Hanlon, 1981*c*; Addison & Parkinson, 1978; van der Drift & Jansen, 1977, and Bengtsson & Rundgren, 1983). The positive and negative responses of fungi to grazing may be reconciled by considering the growth dynamics of fungi and Collembola in leaf litter (Fig. 3.2*a*) in relation to nitrogen mineralization (Fig. 3.2*b*). In the ungrazed litter systems fungal development followed the classic growth curve exhibiting lag, exponential and stationary phases. In contrast, changes in fungal standing crop (living and dead material) were more complicated in treatments with animals, showing a stimulation of fungal growth (and reduced nitrogen losses) at low grazing intensities, similar to the results of Hanlon & Anderson (1979). At high grazing intensities, resulting from a population increase of Collembola after six weeks, a reduction in fungal standing crop occurred which was associated with increased nitrogen mineralization. Parkinson *et al.* (1979) and Hanlon & Anderson (1979) also observed that grazing rates may exceed the production of fungal hyphae, but the balance of these processes may be modified by the growth response of the fungus to increasing nitrogen concentrations (Park, 1976), the available nutrient supply (Hanlon, 1981c), and the food quality of the fungus for the Collembola (Booth & Anderson, 1979). The interaction of these variables in relation to the physical structure of the substrate has been demonstrated by Leonard (unpublished) using two- and three-dimensional bead matrices in which the fungus *Mucor plumbeus* was grown in media with defined nitrogen concentrations as asparagine, and grazed by the Collembola *Folsomia candida*. At a low nitrogen concentration (2 μg l^{-1}) the interaction was stabilized (Fig. 3.3a); the growth and reproduction of the Collembola

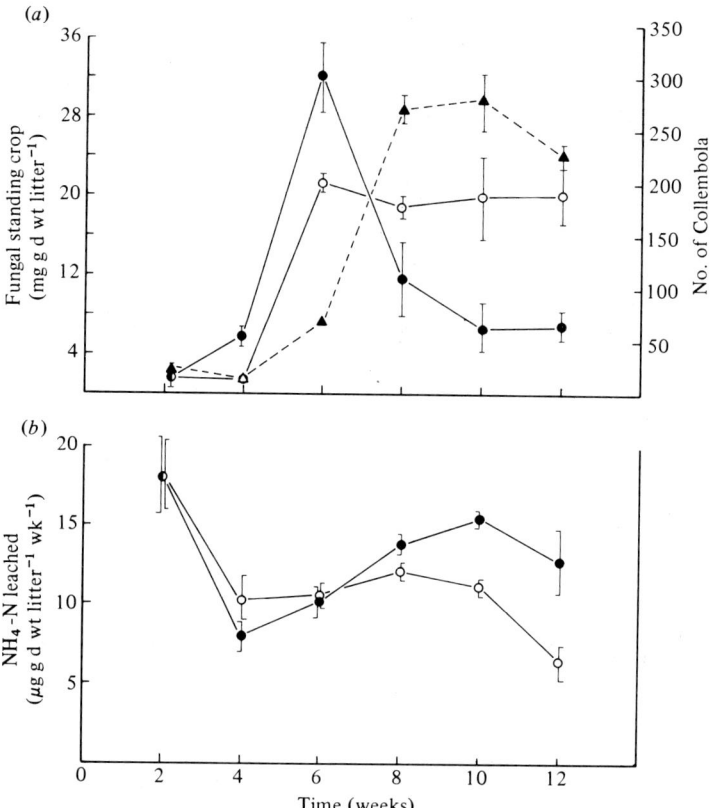

Fig. 3.2. (a) Fungal standing crop on oak leaf litter with (●) and without (○) the effects of grazing and changes in the Collembola populations (▲). (b) The mobilization of nitrogen, as ammonium, from grazed (●) and ungrazed (○) litter. Values shown are means ±1 SE; n = 4. (After Ineson, Leonard & Anderson, 1982.)

were more limited by the nutritional quality of the fungus than was the growth of the fungus by the quality of the medium. In the three-dimensional bead matrix, using the same amount and concentration of growth medium, the dynamics of collembolan and fungal growth were completely different: the Collembola population increased, where it had previously been static, and fungal growth was enhanced relative to ungrazed controls (Fig. 3.3b). The fungus used in these experiments showed an approximately linear growth response to increasing nitrogen, but further complexities of interaction would arise with fungi showing non-linear or negative responses to nitrogen concentration (Park, 1976). In addition to these specific effects on fungal growth, both grazing and comminution cause a gross shift from fungal to bacterial dominance in decomposing litter and soil organic matter (Hanlon & Anderson, 1979, 1980). An increase in bacteria on substrates associated with the feeding activities of mesofauna and macrofauna has been noted in many studies and is generally attributed to bacterial growth, both of ingested and endogenous gut microbiota, during the passage of material through the animal gut. This phenomenon is well documented for earthworms

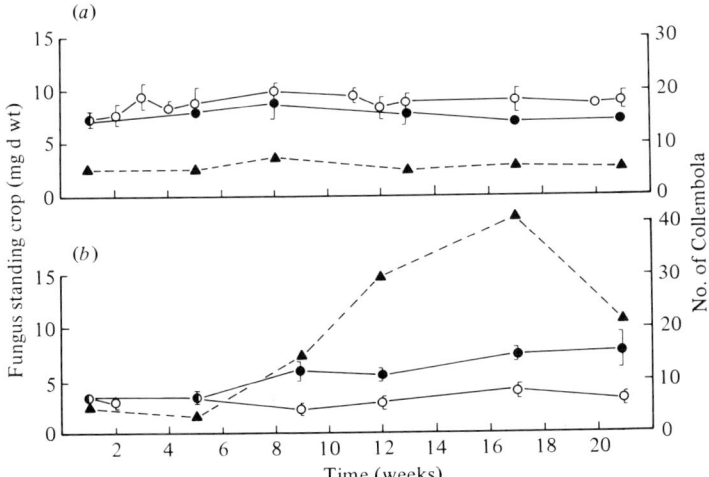

Fig. 3.3. The influence of spatial heterogeneity on the growth of *Mucor plumbeus* in (a) two-dimensional and (b) three-dimensional bead matrices with (●) and without (○) the effects of grazing by *Folsomia candida* populations (▲). The fungus was grown in a defined medium containing 2 mg l^{-1} N (as asparagine) and Collembola were added seven days after the fungal inoculum. Fungal mycelium was recovered by washing and sonic separation from the matrix, and the Collembola by floatation. Results are expressed as mean weight or counts per culture ±1 SE; n = 5. (Leonard, unpublished.)

(Ghilarov, 1963; Went, 1963; Parle, 1963; Atlavinyité & Lugauskas, 1971; Kozloskaja, 1971; Márialigeti, 1979) as well as for a wide range of arthropod groups including Diptera larvae (Szabo, Marton & Buti, 1969; Swift & Boddy, Chapter 4), isopods (Reyes & Tiedje, 1976; Anderson & Ineson, 1983), millipedes (MacBreyer, 1973; Anderson & Bignell, 1980; Anderson & Ineson, 1983; Szabo *et al.* 1983) and mites (Stefaniak & Seniczak, 1976). Thus, even in soils with low pH conditions where biological processes are considered to be dominated by fungi, the feeding activities of the soil fauna form a dynamic mozaic of microsites of intense bacterial activity.

Assessing the overall significance of soil invertebrates in nutrient cycles resolves on quantifying the indirect effects on animal-microbial interactions. The complexity of the processes involved precludes the formulation of a working model to evaluate their importance, but unless this can be done it is only possible to quantify the direct contribution of invertebrates to nutrient pools and fluxes. It is unlikely that this paradox will be resolved in the field at the organism-microsite level in the foreseeable future, and it is necessary to adopt an approach where processes and parameters are compartmentalized at a level where direct and indirect effects on nutrient transfers can be related to simple first order (empirical) variables.

This is essentially the value of the microcosm approach, defined by Patten & Witkamp (1967), where the components of a system can be studied at a realistic level of complexity and processes resolved into simple inputs and outputs from experimental materials in relation to specific treatments. There has been a proliferation of such microcosm studies during the 1970s, culminating in some extremely complex experiments carried out by Bååth *et al.* (1978, 1981), Clarholm *et al.* (1981) and Coleman *et al.* (Chapter 2). These studies have provided considerable insight into the dynamics of processes and mechanisms of animal-microbial interactions but cannot be quantitatively extrapolated into the field. The same limitations apply to our demonstrations of the effects of millipede and collembolan grazing on nutrient mobilization (Ineson *et al.*, 1982; Anderson *et al.*, 1983*a*).

Our current approach is to develop specific hypotheses for animal-microbial interactions in nutrient fluxes, which are testable in the context of processes operating in deciduous woodlands with acid, organic humus forms; particularly moders (*sensu* Kubiena, 1953) developed under sessile oak (*Quercus robur*), which is the commonest form of woodland system in Devon, UK. The effects of animals are measured in terms of minerals released into solution from litter and organic matter, where they are potentially available for root uptake. Such releases are related to simple process variables.

Process variables of animal–microbial interactions

Swift *et al.* (1979) identified the main variables of decomposition processes as the organisms (animals and microorganisms), resource quality (the physico-chemical attributes of the material), and the physical environmental conditions (particularly temperature and moisture) in which the organisms are acting. We use this framework for considering the process variables involved in animal–microbial interactions.

All the following experiments were carried out in chambers described by Anderson & Ineson (1982) which facilitate measurements of leaching losses from soil and litter materials in relation to animal treatments. Details of the preparation of materials and analysis are given in Anderson *et al.* (1983*a*). The experiments were designed to investigate the contribution of mesofauna and macrofauna to nutrient fluxes with nematodes and protozoa as uncontrolled variables.

Organisms: animal groups

The effects of a range of animal groups on nutrient mobilization from oak leaf litter were investigated by Anderson *et al.* (1983*a*) and the results are summarized in Table 3.1. Numbers of animals were chosen to demonstrate the maximum likely effects which might occur within aggregated field populations, and the results show that the animal treatments generally resulted in enhanced losses of all elements over controls, although individual effects for animals and groups were not significant in many cases. Animals had the most marked effect on ammonium fluxes, with increased leachate concentrations 10 to 20 times control levels for the millipedes and 60 times control levels for earthworms. The effects of Collembola and enchytraeids were comparatively small since treatments contained much lower biomass of these animals but, if the results are expressed as ammonium release per gram of animal biomass, the Collembola have nearly the largest effects of the groups examined (2.5 mg N g litter^{-1} g animal^{-1}). The overall effect of animal groups on ammonium-N release (N) in mg g litter^{-1} over a 6-week period relates linearly to animal biomass (B) according to the function:

$$N = 1.55B - 0.03.$$

Table 3.1. Effect of animal groups upon the release of elements from F layer materials[a]

	Control (without animals)	Enchytraeids	Tomocerus minor	Orchesella villosa	Polydesmus angustus	Iulus scandinavius	Glomeris marginata	Lumbricus rubellus
NH_4^+-N	0.03	0.02	0.10*	0.10*	0.38*	0.68***	0.54**	1.78***
NO_3^--N	0.002	0.002	0.004	0.004	0.006	0.008*	0.003	0.21***
Na^+	0.14	0.14	0.17*	0.19***	0.22*	0.18*	0.15	0.28***
K^+	0.13	0.13	0.20	0.17	0.29	0.44*	0.38***	0.46***
Ca^{++}	0.93	0.75**	0.84	0.81*	0.93	0.97	1.01	1.88***
Mean leachate pH	5.3	5.4	5.3	5.3	5.4	5.5	5.2	5.6
Wet weight of animals (g)	–	0.03	0.04	0.04	0.14	0.49	0.59	1.00
Number of animals	–	25	15	15	3	4	4	3

[a] Results expressed in mg g^{-1} dry wt of litter over a 6-week period. * $p < 0.005$, ** $p < 0.01$ and *** $p < 0.001$ for differences between animal treatments and controls; n = 3.

From Anderson, Ineson & Huish (1983a).

Such a function might form a basis for a process model through interactions with substrate quality (see below). This relationship also suggests that the taxonomic identity of the animals is not significant as a variable in relation to their biomass and activity. In the light of this we have chosen the millipede *Glomeris marginata* as a tool for modelling general animal effects, since this animal is particularly suitable for laboratory use.

The experiment described above was designed to simulate short-term, localized effects of feeding activities and the time scale was chosen to minimize nutrient immobilization through secondary production. The reproductive recruitment to these populations is longer term, and the effects of these changes will not normally affect local aggregations of individuals. The macroarthropod grazing effects, in particular, may be seen as a temporal perturbation, of variable intensity or duration, before the animals move on or the aggregation disperses, after which time the mineralization of nitrogen slowly returns to lower levels as discussed earlier. Collembola grazing had comparatively small effects on nutrient leaching from litter, but there was a five times larger transfer of nitrogen to animal biomass. This is a continuation of an effect seen for Protozoa and nematode grazing, where short generation times have the consequence that animal population levels closely follow local food supplies. Since weight specific metabolic rates and biomass turnover increase with decreasing body sizes, it may be that the faunal body size groupings of micro-, meso- and macrofauna reflect the increasing relative importance of indirect contributions to nutrient fluxes through animal–microbial interactions. Such relationships may be stabilized by the high spatial heterogeneity of the soil for invertebrates with a small body size (Fig. 3.3; Elliot, Anderson, Coleman & Cole, 1980; Couteaux & Pussard, 1983), while the macrofauna cause major changes to the microbial environment through litter comminution and the formation of soil aggregates (e.g. Jeanson-Luusinang, 1963).

Organisms: fungi

The selection of fungi by fungivores is a well documented phenomenon under laboratory conditions but less is known about the influence of fungi on the feeding activities of macrofauna. The fungal conditioning of litter appears to be a prerequisite for the onset of feeding by saprophagous animals, but the extent to which different species of fungi influence the scale and timing of effects requires investigation. If the dominant fungal species decomposing litter was a variable influencing the feeding activities of millipedes the broad, compartmental approach to investigating the effects of animals on mineralization processes suggested here would be invalidated.

Experiments were carried out in which oak leaf litter was autoclaved and then inoculated with one of six species of fungi: the basidiomycetes *Coriolus versicolor, Phallus impudicus, Marasmius androsaceus,* and *Flammulina velutipes*; *Mucor ramannianus* (Zygomycotina), and *Cladosporium herbarum* (Deuteromycotina). These were incubated for eight weeks at 20 °C before being transferred to the experimental chambers, which were allocated to treatments of 0 or 4 *G. marginata* per chamber (the optimum number of animals for demonstrating feeding effects in these systems). Ammonium-N leaching was then determined at intervals of one week over a period of nine weeks. Cumulative results for this period are summarized in Fig. 3.4. The data reflect the wide variation in

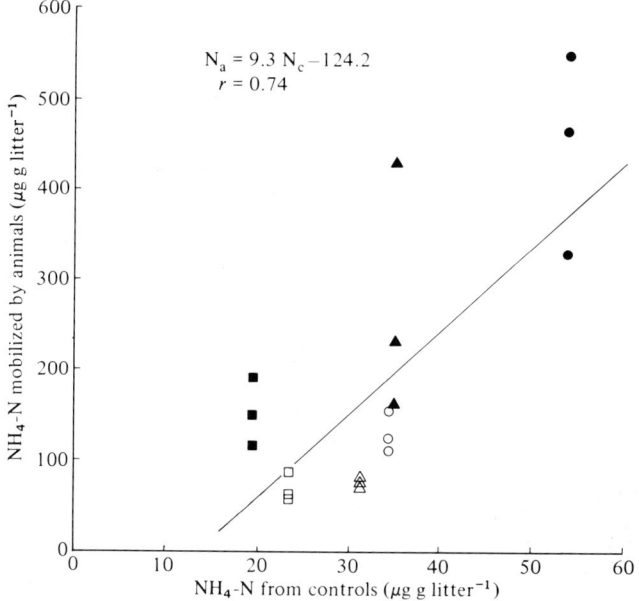

Fig. 3.4. The relationship between nitrogen mineralization by different species of litter-decaying fungi and the mobilization of nitrogen by millipede feeding activities. Oak leaf litter was inoculated with *Coriolus versicolor* (●), *Phallus impudicus* (▲), *Marasmius androsaceus* (■), *Mucor ramannianus* (○), *Cladosporium herbarum* (△) and *Flammulina velutipes* (□), and incubated for eight weeks at 20 °C before being transferred to microcosms and allocated to animal (4 *Glomeris marginata*) or control (no animals) treatments. Ammonium-N release was then determined weekly for nine weeks. Results are cumulative for this period.

ammonium-N release by the fungi, with *C. versicolor* achieving the highest levels of ammonium-N release (reflected by extensive structural breakdown of the litter), while *F. velutipes* showed extensive mycelial development on the litter but with little visual sign of catabolic activity. The response of these monospecific fungal cultures to animal grazing, as nitrogen mineralization (N_a), was significantly correlated (P = 0.001, f = 16) with nitrogen mineralized by controls (N_c) according to the function:

$$N_a = 9.3\ N_c - 124.2.$$

Thus, the effect of millipede feeding can be related to the activities of the fungus, and again transcends the specific identity of the organism involved. If this assumption is valid for mixed cultures then animal–microbial interactions in nitrogen mineralization can be predicted at the process level for a particular resource. We next consider resource as a variable.

Resource type

Litter (L), fermentation layer (F) and humus layer (H) materials were collected from three oak woodlands in Devon, and mineral element losses were compared for treatments with 0 and 4 *G. marginata* over a period of six weeks. Losses of sodium and potassium were increased in all cases but the effects were small in relation to the abiotic leaching processes which dominate the fluxes of these elements (Anderson *et al.*, 1983*b*). The effects of the animals on nitrogen mineralization were more marked (Table 3.2).

Leachates from the L-layer materials contained small concentrations of nitrate-N and the ammonium-N dominated nitrogen fluxes showed the expected enhancement by animals from about 10 to 15 times control levels. The F-layer material from Perridge and Brook Woods showed higher control levels of ammonium-N than the L-layer materials but the Stoke Woods material had similar leachate concentrations of ammonium-N and nitrate-N. The millipedes enhanced ammonium-N losses to about eight times normal levels in these experiments, while the enhancement of nitrogen as nitrate was insignificant. Another series of experiments with Stoke Woods F-layer material showed an enhancement of both forms of mineral-N, so that the ratio of ammonium-N to nitrate-N in leachate was similar in the control and animal systems. The H-layer material from Stoke Woods (Table 3.2) showed a dominance of nitrate-N in leachate from controls. This was maintained in the animal

Table 3.2. *Effects of Glomeris feeding on nitrogen leaching from different organic soil layers of three oak woodlands in Devon*

Site:		Stoke Woods		Perridge Wood		Brook Wood	
	Treatment	Control	Animals	Control	Animals	Control	Animals
Litter layer (L)	NH_4-N	18.8 ± 5.7***	183.3 ± 10.3	23.7 ± 3.2*	251.1 ± 81.3	38.0 ± 19.5***	591.3 ± 69.6
	NO_3-N	1.6 ± 0.4*	3.5 ± 0.7	1.7 ± 0.4	3.7 ± 1.0	1.1 ± 0.5	5.5 ± 1.5
Fermentation layer (F)	NH_4-N	37.2 ± 8.4***	259.3 ± 24.3	217.9 ± 50.7**	513.9 ± 48.2	716.4 ± 194.4*	1333.5 ± 90.9
	NO_3-N	46.8 ± 28.2	16.4 ± 4.4	12.7 ± 9.5	4.4 ± 0.6	17.6 ± 5.2	61.2 ± 31.6
Humus layer (H)	NH_4-N	15.3 ± 2.1*	188.7 ± 50.7	131.0 ± 7.2***	214.7 ± 4.1	107.1 ± 6.5**	188.8 ± 17.5
	NO_3-N	116.1 ± 0.5	384.5 ± 50.3	2.3 ± 0.3	2.1 ± 0.2	13.9 ± 3.2	9.5 ± 2.0

Results are expressed as mean cumulative concentrations of mineral nitrogen (μg g dry wt litter^{-1} ± SE; n = 3) mobilized as ammonium-N or nitrate-N over a period of five weeks after the addition of four animals to the animal treatments. Values for controls (without animals) are shown for the same period of time. The experiments were incubated at 15°C. Significant differences between animal treatments and controls are shown as *** (P = 0.001), ** (P = 0.01) and * (P = 0.05).

treatments, although the effects of the animals were larger on the ammonium-N fluxes (enhanced 12 times) than on the nitrate-N fluxes (enhanced three times). Ammonium-N was the main form of inorganic nitrogen in H-layer leachates from the other two sites and animals produced a similarly small enhancement of the F-layer materials.

It is often assumed that acid soils have a predominantly ammonium-N economy because nitrifying bacteria have higher pH optima than ambient conditions. However, there is increasing evidence, as illustrated by these experiments, that this is an overgeneralization (Robertson, 1982); e.g. it takes no account of heterotrophic nitrification which is carried out by a wide range of bacteria (including actinomycetes) as well as many fungi (Fochte & Verstraete, 1977). In order to investigate the enhancement of nitrification by animals, H-layer materials were collected from nine oak woodlands and treated as in the previous experiment. Nitrate-N concentrations in leachates were expressed relative to ammonium-N concentrations as the Relative Nitrification Index (RNI), calculated by dividing the nitrate-N concentration by the ammonium-N concentration (Robertson, 1982). The nine humus layer samples showed a wide range of RNI values, from less than 1 (ammonium-N dominance) to nearly 20 (nitrate-N dominance).

In all cases the animals reduced the RNI, by enhancing nitrogen losses as ammonium-N. However, it was also noted that the RNI before and after grazing was related to cation leaching. The correlations between RNI and animal-induced calcium leaching are highly significant ($r = 0.90$, $p < 0.001$) and it is possible to predict the release of calcium (C) in μg g litter^{-1} by animals by measuring the RNI (N) of controls using the function:

$$C = 11.5N + 3.6$$

We attribute this effect to the release of hydrogen ions as nitrification proceeds, releasing calcium from cation exchange sites. Organic matter is the main pool of exchange sites in acid soils and the effects of animals in enhancing nitrogen mineralization processes through their feeding activities can therefore have wider implications for ionic balances under these conditions.

Physical variables: temperature and moisture

Temperature is an important determinant of metabolic activity, and soil processes mediated by animals and microorganisms might be expected to show a predictable response to temperature changes within their physiological optima. A convenient expression of this response is the Q_{10} coefficient which indicates the fold-increase of a response over a 10 °C change in temperature. In temperate regions soil animal respiration (Reichle, 1971; Persson *et al.*, 1980) and the respiratory activity of soil and litter microorganisms (Anderson, 1973*b*; Schlesinger, 1977) show a Q_{10} relationship of 2 to 3 over the normal annual soil temperature range of about 5 °C to 20 °C. Many biological and physical parameters limit this response. Witkamp & Frank (1970) concluded that mineral element release effected by animal feeding activities would be expected to follow a $Q_{10} = 2$ relationship, but this would be an exception at higher temperatures under field conditions because of interactions between temperature and moisture. They observed that high temperatures and low moisture tensions inhibited microbial and millipede activities. Upon re-wetting, millipedes continued feeding at the same rates as before drying, but there was a characteristic flush of microbial activity and element release followed by immobilization during the subsequent growth phase of the microorganisms. Nutrient immobilization during the wetting and drying processes has been shown in the laboratory to be more significant than biotic processes for nutrient mobilization at high temperatures (Witkamp & Barzansky, 1968; Witkamp, 1969), but their significance for nutrient losses from leaf litter in the field remains to be quantified.

In the microcosm studies of Anderson *et al.* (1983*b*) the moisture content of the experimental material was never limiting but, even so, the temperature effect on potassium and calcium losses from animal treatments was much less than a twofold increase within a temperature range of 5 to 20 °C. This is consistent with the hypothesis that abiotic processes are mainly involved in the leaching of simple cations.

Both Clarholm *et al.* (1981) and Anderson *et al.* (1983*b*) found larger responses for nitrogen mineralization than can be predicted from simple temperature responses. Clarholm *et al.* (1981) did not differentiate between animal or microbial processes involved in these fluxes but recorded a fiftyfold increase in mineral nitrogen release between 5 °C and 20 °C from pine-forest soil material incubated at high moisture levels. Millipedes feeding on deciduous forest litter, however, produced a sevenfold increase in leaching of ammonium-N between 5 °C and 15 °C but only a threefold increase from 10 to 20 °C (Anderson *et al.*, 1983*b*). Whatever the causes of these effects, the consequences are that temperature responses are unpredictable with our current understanding of these systems. Increases in nitrate-N were small but significant and were attributed to enhanced nitrification under the warm and moist experimental conditions. This inflated effect of temperature on nitrogen mineralization is strongly indicative of indirect, synergistic interactions between animals and microorganisms, rather than direct metabolic transfers involved in ammonification. We conclude that far more information is needed on the temperature response of animal–microbial interactions for the formulation of any descriptive or predictive model of these processes.

In our current approach to quantifying the role of animal–microbial interactions under field conditions we recognize the problems of quantifying indirect and direct effects, particularly the turnover of animal biomass, at similar scales where the results can be combined into estimates of total nutrient fluxes. We are attempting to resolve these problems by using different scales of experimental systems, ranging from 0.5 m by 0.5 m field chambers for ecosystem studies down to laboratory microcosms for investigating microsite processes, towards formulating a multivariate model integrating faunal biomass, resource quality and temperature effects on animal–microbial interactions in nutrient fluxes.

Mechanisms of nitrogen mineralization and release

The role of soil animals in effecting the release of nutrients from decomposing plant materials is generally attributed to comminution increasing the susceptibility of simple cations to leaching or, particularly for limiting nutrients such as nitrogen and phosphorus, the reduction of microbial biomass through grazing. The nutrients in microbial tissues are then released over a longer time scale by the turnover of heterotroph tissues or by excretion (egesta and metabolic products) over a short time scale. The importance of metabolic nitrogen excretion as a process of nitrogen mineralization is emphasized in the reviews by Anderson *et al.* (1981) and Persson (1983).

Our interest in understanding the mechanisms of animal–microbial interactions in nutrient fluxes centres on the necessity for identifying the nutrient pools which are mobilized directly and indirectly by animal feeding activities, in order to develop quantitative models of the processes. Results of our experiments up to 1981 led us to advance the following hypothesis to account for the short term effects of millipedes on the release of nitrogen from leaf litter as ammonium (Anderson & Ineson, 1983).

It was assumed that during the early stages of litter decomposition net mineralization of nitrogen is low as a consequence of immobilization in fungal tissues. This is consistent with the low release of nitrogen from our ungrazed systems as well as with the general theory of the time course of litter decomposition (Swift *et al.*, 1979). In laboratory systems, without animals, the release of nitrogen from the fungal pool may not occur until carbon becomes limiting, as predicted by the C:N theory, but in natural systems animal feeding activities will accompany microbial colonization and the proliferation of facultative saprophytic fungi after leaf fall.

The mycelial growth form is very susceptible to disruption by animal feeding activities, more so than the unicellular bacterial thallus. Even so, 60 to 80% of a ^{14}C-label was assimilated from both bacteria and fungi by *G. marginata* (Anderson & Bignell, 1982; Bignell, personal communication). There is, however, extensive bacterial growth of both the residual litter flora and endogenous gut bacteria in the hind-gut of this millipede, so that the faeces contain several hundred times higher bacterial counts than the food litter (Anderson & Bignell, 1980; Anderson & Ineson, 1983) with increased abundance of ammonifiers as well as ureolytic and uricolytic forms. The importance of these bacteria in the nutrition of the animal has not been determined but the specific activities of these groups can be related to the degradation of amino acids in food materials and gut secretions, urea stored in fungal hyphae, and to uric acid excretory products of the animal which are voided into the gut (Anderson & Ineson, 1983). The end product of these processes is manifested by high concentrations of ammonium in the faeces of *G. marginata* (Bocock, 1963) and nitrogen immobilized in bacteria rather than fungi. We hypothesize that the low levels of fungal activity in these microsites together with carbon limitation of bacterial growth reduces nitrogen immobilization so that the ammonium is available for root uptake. Clarholm *et al.* (1981) found that addition of glucose to microcosms containing pine seedlings limited nitrogen availability to the plants because of increased microbial immobilization. The addition of glucose to our millipede-grazed systems had similar effects on ammonium leaching rates (Table 3.3).

The validity of these mechanisms, and the relative importance of direct and indirect effects of animals, resolves on the contribution of animal excretion to ammonium in the faeces and the mobilization of nitrogen from fungal tissues.

The excretory component of faecal nitrogen was investigated in experiments where *G. marginata* labelled with ^{15}N were allowed to

Table 3.3. *Nitrogen immobilization by glucose enrichment*

Treatments (n = 4)	NH_4-N mobilized (μg g dry wt litter^{-1} week^{-1} ± SE)	
No animals, no glucose	2.03 ± 1.18	} $P < 0.01$
Animals, no glucose	143.7 ± 30.3	
Glucose, no animals	5.90 ± 2.0	} $P = 0.05$
Glucose plus animals	79.7 ± 25.5	

Oak leaf litter was incubated in chambers at 15 °C until week 3 when four *Glomeris* were added to animal treatments. Glucose additions (1% w. v.) were not made until week 11 when the animal treatments showed enhanced ammonium-N mobilization. Results are expressed as mean concentrations for weeks 12 and 13.

graze on non-labelled litter. Animals were labelled by feeding them on *Cladosporium herbarum* grown with double-labelled ^{15}N ammonium nitrate (Prochem, London) as the sole nitrogen source in a defined medium. The labelled animals were added to unlabelled leaf litter and the leachates analysed for mineral nitrogen and the ratio of ^{14}N:^{15}N. It was calculated, based on the assumption that the animals were uniformly labelled, that 92.5% of the nitrogen released after adding the animals was derived from the litter and only 7.5% from excretion. Parallel experiments using ^{15}N-labelled litter and unlabelled animals showed the same general pattern. Thus, the short term significance of the animal effects lies predominantly in their indirect mineralization from the litter microbial pool, rather than the direct contribution through excretion.

Experiments were carried out in which litter was treated with nystatin to suppress fungal growth (Fig. 3.5). The experiment was monitored over 19 weeks at 15 °C and dilution series were prepared from litter controls, with and without nystatin, on five occasions during the course of the experiment and plated out on nutrient agar. Bacterial counts ranging from 2 to 7×10^8 colonies g litter^{-1} were not significantly different in the two treatments. The nystatin was effective in suppressing fungal growth since no fungi were isolated from the nystatin treated litter, whereas an average of 15×10^6 colony-forming units g litter^{-1} were found in the untreated controls. Unexpectedly, ammonium-N concentrations in leachates were similar from nystatin treated and untreated leaves even after animal addition (Fig. 3.5), suggesting that either the animals have similar effects in reducing bacterial or fungal immobilization of nitrogen (a conclusion supported by the ^{14}C-tracer experiments), or that the fungal nitrogen pool is insignificant in relation to nitrogen released from litter. But the net effect of feeding activities is that the animals assimilate only a small proportion of the nitrogen mobilized from the litter–microbial complex as a consequence of gut passage.

Conclusions

It would appear from the studies reviewed here that soil invertebrates have important roles in nutrient cycles of temperate grasslands and deciduous forests on base rich soils where their biomass equals or exceeds that of microorganisms and the decomposition rates of high quality litters are fairly rapid. In intensively farmed arable systems the applications of fertilizer usually override any necessity to account for animal or animal–microbially mediated effects on nutrient fluxes, though there is evidence for the importance of earthworms in

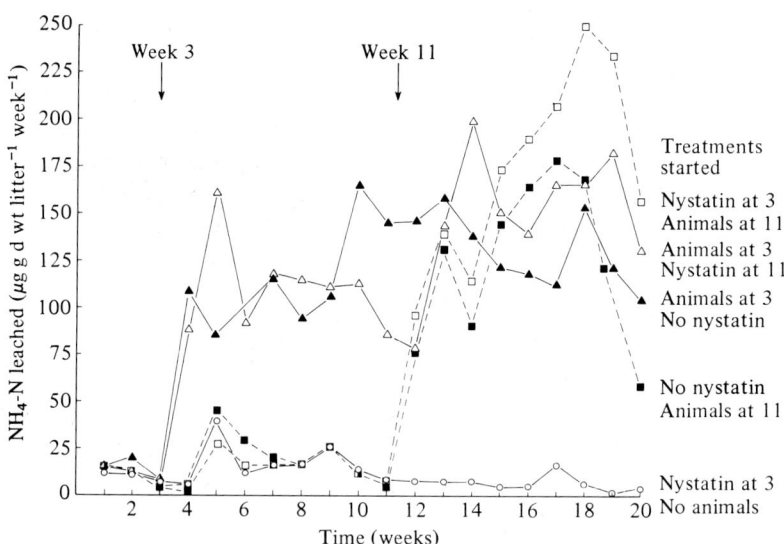

Fig. 3.5. The effects of nystatin treatments on ammonium-N release from litter by the feeding activities of *Glomeris marginata*. Chambers were set up containing oak leaf litter and treatments of 0 or 4 millipedes, with or without nystatin (50 µg ml^{-1}), were initiated at week 3 or week 11 as shown.

promoting the root growth of crops through their burrowing activities in minimum tillage systems (Edwards & Lofty, 1978).

In acid forest soils, particularly those developed under conifers, the immobilization of nitrogen and phosphorus in organic matter has economic implications for forestry practices and watershed management. Miller (1981) showed that in a pine plantation at Culbin Sands the application of 200 kg N ha^{-1} and a continued rate of mineralization of 4% only increased the available nitrogen from 60 to 63 kg ha^{-1} yr^{-1}. A shift in the mineralization rate for the soil organic pool (approximately 900 kg N ha^{-1}) from 4% to 4.5% would increase availability to 71 kg N ha^{-1} yr^{-1}, an amount roughly equivalent to the rate of immobilization resulting in nitrogen deficiency for the stand.

Net mineralization of nitrogen in a black spruce stand at Fairbanks (Alaska) is about 3% per annum of a 3070 kg ha^{-1} soil nitrogen pool (Van Cleve & Alexander, 1981). The data presented by Flanagan & Van Cleve (1977) for this site show that the bacterial and fungal standing crop (5.7 g m^{-2}) contains an insignificant fraction of the soil nitrogen pool, but the amount of nitrogen turned over through microbial productivity (117 g m^{-2}) amounts to about 6 g m^2 yr^{-1} or approximately 60% of the net annual mineralization reported by Van Cleve & Alexander (1981).

Both of these estimates of nitrogen mineralization, and the increase required at Culbin to support production, appear to be within the scale of the direct nitrogen transfers calculated by Persson (1983) for the Swedish Scots pine stand and the indirect effects demonstrated by us. It would therefore appear to merit serious consideration that animal–microbial interactions are functionally important in these sites and that manipulation of the soil fauna through management practices (Hill, Metz & Farrier, 1975; Lundkvist, 1977; Huhta, 1979) could provide a means of altering the balance of immobilization/mobilization processes.

We have little quantitative understanding of the role of soil fauna in mobilizing nitrogen from soil organic matter pools in acid deciduous forest soils. Most of the meso- and macrofaunal populations and feeding are associated with the litter and fermentation layers rather than with the organic humus horizons which contain the bulk of the immobilized nutrient pool. We see instead the soil fauna as agents of key processes determining the transfers of organic matter and mineral nutrient fluxes to the slower nutrient cycles associated with the humus pool. The feeding activities of animals not only enhance the existing patterns of nitrogen mineralization by the microorganisms, but also disrupt the time

Table 3.4. *Effects of* Glomeris *feeding activities on nitrogen and cation release from oak leaf litter and uptake by beech mycorrhizae–root systems in the field*

Treatments	Minerals in leachate (μeq g dry wt litter^{-1} week^{-1})				
	NH_4^{+1}	NO_3^-	Na^+	K^+	Ca^{2+}
No animals, no roots	3.84	0.13	0.36	0.33	1.61
Animals, no roots	7.50	0.09	0.61	0.49	1.78
No animals, roots	3.20	0.06	0.30	0.50	2.05
Animals, roots	3.61	0.04	0.63	0.33	2.11

scale of nitrogen mineralization in the surface litter layers so that nitrogen is released into solution in the sub-horizons where ectotrophic mycorrhizae are active.

The potential significance of these transfers has been demonstrated in experiments where mycorrhizal–root systems of oak trees have been introduced through ports in experimental chambers under field conditions. Ammonium-N released from the leaf litter by the feeding activities of millipedes was taken up by the root complex (Table 3.4) under conditions where the C:N ratio of the litter was approximately 100:1. Hence, the fauna are effecting transfers of nitrogen under conditions where net mineralization of nitrogen by microorganisms is thought to be insignificant.

The relative importance of the microbial and animal–microbial mediated processes is indicated by the results of the experiment shown in Fig. 3.1. Cumulative nitrogen mineralization over 15 weeks from the controls amounted to 3% of the total nitrogen capital (25.5 mg) in the litter compared with 9.7% mobilized through animal effects. Thus models of nitrogen mineralization from leaf litter which exclude the disruptive effects of animals on the time course of microbial decomposition processes must be considered unrealistic for temperate forest soils.

Acknowledgments. We would like to express our thanks to Sally Huish for her assistance with many aspects of the project, and for her patience and good humour in working with us! Peter Splatt also provided excellent technical support. We are grateful to Mike Leonard for the use of unpublished data. The research programme is funded by the Natural Environment Research Council.

References

Aber, J. D., Botkin, D. B. & Melillo, J. M. (1978). Predicting the effects of different harvesting regimes on forest floor dynamics in northern hardwoods. *Canadian Journal of Forest Research*, **8**, 308–16.

Aber, J. D., Botkin, D. B. & Melillo, J. M. (1979). Predicting the effects of different harvesting regimes on productivity and yield in northern hardwoods. *Canadian Journal of Forest Research*, **9**, 10–14.

Aber, J. D. & Melillo, J. M. (1982). Nitrogen immobilization in decaying hardwood litter as a function of initial nitrogen and lignin content. *Canadian Journal of Botany*, **60**, 2263–69.

Addison, J. A. & Parkinson, D. (1978). Influence of collembolan feeding activities on soil metabolism at a high arctic site. *Oikos*, **30**, 529–38.

Alexander, M. (1977). *Introduction to Soil Microbiology*, 2nd Edn. New York & London: John Wiley.

Anderson, J. M. (1973a). The breakdown and decomposition of sweet chestnut (*Castanea sativa* Mill.) and beech (*Fagus sylvatica* L.) leaf litter in two deciduous woodland soils: I. Breakdown, leaching and decomposition. *Oecologia (Berl.)*, **12**, 251–74.

Anderson, J. M. (1973b). Carbon dioxide evolution from two deciduous woodland soils. *Journal of Applied Ecology*, **10**, 361–78.

Anderson, J. M. & Bignell, D. E. (1980). Bacteria in the food, gut contents and faeces of the litter-feeding millipede *Glomeris marginata* (Villers). *Soil Biology and Biochemistry*, **12**, 251–4.

Anderson, J. M. & Ineson, P. (1982). A soil microcosm system and its application to measurements of respiration and nutrient leaching. *Soil Biology and Biochemistry*, **14**, 415–16.

Anderson, J. M. & Ineson, P. (1983). Interactions between soil arthropods and microbial populations in carbon, nitrogen and mineral nutrient fluxes from decomposing leaf litter. In *Nitrogen as an Ecological Factor*, ed. J. A. Lee, S. McNeill & I. Rorison, pp. 413–42. Oxford: Blackwell Scientific Publications.

Anderson, J. M., Ineson, P. & Huish, S. A. (1983a). Nitrogen and cation release by macrofauna feeding on leaf litter and soil organic matter from deciduous woodlands. *Soil Biology and Biochemistry*, **15**, 463–7.

Anderson, J. M., Ineson, P. & Huish, S. A. (1983b). The effects of animal feeding activities on element release from deciduous forest litter and soil organic matter. In *New Trends in Soil Biology*, ed. Ph. Lebrun, H. M. Andre, A. de Medts, C. Gregoire-Wibo & G. Wauthy, pp. 87–100. Louvain-la-Neuve: Dieu-Brichart.

Anderson, J. M. & Macfadyen, A. (eds.) (1976). *The Role of Terrestrial and Aquatic Organisms in Decomposition Processes*. Oxford: Blackwell Scientific Publications.

Anderson, J. M., Proctor, J. & Vallack, H. W. (1983). Ecological studies in four contrasting lowland rain forests in Gunung Mulu National Park, Sarawak. III. Decomposition processes and nutrient losses from leaf litter. *Journal of Ecology*, **71** (in press).

Anderson, R. V., Coleman, D. C. & Cole, C. V. (1981). Effects of saprotrophic grazing on net mineralization. In *Terrestrial Nitrogen Cycles*, ed. F. E. Clark & T. Rosswall, pp. 201–15. Ecological Bulletin 33. Stockholm: Swedish Natural Science Research Council.

Atlavinyité, O. & Lugauskas, A. (1971). The effect of lumbricidae on soil microorganisms. In *Organisms du Sol et production primaire*, IV Colloquium

Pedobologiae, Dijon, 14–19 September 1970, pp. 73–80. Paris: Institut National de la Recherche Agronomique.

Ausmus, B. S., Edwards, N. T. & Witkamp, M. (1976). Microbial immobilisation of carbon, nitrogen, phosphorus and potassium: implications for forest ecosystem processes. In *The Role of Terrestrial and Aquatic Organisms in Decomposition Processes*, ed. J. M. Anderson & A. Macfadyen, pp. 397–416. Oxford: Blackwell Scientific Publications.

Bååth, E., Lohm, U., Lundgren, B., Rosswall, T., Söderström, B., Sohlenius, A. & Wiren, A. (1978). The effect of nitrogen and carbon supply on the development of soil organism populations and pine seedlings: A microcosm experiment. *Oikos*, **31**, 153–63.

Bååth, E., Lundgren, B., Rosswall, T., Söderström, B. & Sohlenius, A. (1981). Impact of microbial-feeding animals on total soil activity and nitrogen dynamics: a soil microcosm experiment. *Oikos*, **37**, 257–64.

Babel, U. (1972). *Moderprofile in Wäldern: Morphologie und Umsetzungsprozesse*. Hohenheimer Arbeiten 60, Stuttgart: Verlag Eugen Ulmer.

Bengtsson, G. & Rundgren, S. (1983). Respiration and growth of a fungus *Mortierella isabellina* in response to grazing by *Onychiurus armatus* (Collembola). *Soil Biology and Biochemistry*, **15**, 469–73.

Bocock, K. L. (1963). The digestion and assimilation of food by *Glomeris*. In *Soil Organisms*, ed. J. Doeksen & J. van der Drift, pp. 85–91. Amsterdam: North Holland Publishing Company.

Booth, R. G. & Anderson, J. M. (1979). The influence of fungal food quality on the growth and fecundity of *Folsomia candida* (Collembola: Isotomidae). *Oecologia (Berlin)*, **38**, 317–23.

Bosatta, E. & Staaf, H. (1982). The control of nitrogen turnover in forest litter. *Oikos*, **39**, 143–51.

Bradshaw, A. D., Marrs, R. H., Roberts, R. D. & Skeffington, R. A. (1982). The creation of nitrogen cycles on derelict land. *Philosophical Transactions of the Royal Society of London. Series B*, **296**, 557–61.

Clarholm, M., Popovic, B., Rosswall, T., Soderstrom, B., Sohlenius, B., Staaf, H. & Wiren, A. (1981). Biological aspects of nitrogen mineralization from a pine forest podsol incubated under different moisture and temperature conditions. *Oikos*, **37**, 137–45.

Cole, C. V., Elliot, E. T., Hunt, H. W. & Coleman, D. C. (1978). Trophic interactions in soils as they affect energy and nutrient dynamics. V. Phosphorus transformations. *Microbial Ecology*, **4**, 381–7.

Couteaux, M. M. & Pussard, M. (1983). Nature du régime alimentaire des Protozoaires du sol. In *New Trends in Soil Biology*, ed. Ph. Lebrun, H. M. André, A. de Medts, C. Gregoire-Wibo & G. Wanthy, pp. 179–95. Louvain-la-Neuve: Dieu-Brichart.

Dickinson, C. H. & Pugh, G. J. F. (eds.) (1974). *Biology of Plant Litter Decomposition*. London and New York: Academic Press.

Doeksen, J. & van der Drift, J. (eds.) (1963). *Soil Organisms*. Amsterdam: North Holland Publishing Company.

Edwards, C. A. & Lofty, J. R. (1977). *Biology of Earthworms* (2nd edn). London: Chapman & Hall.

Edwards, C. A. & Lofty, J. R. (1978). The influence of arthropods and earthworms upon root growth in direct drilled cereals. *Journal of Applied Ecology*, **15**, 789–95.

Edwards, C. A., Reichle, D. E. & Crossley, D. A. (1970). The role of soil invertebrates in turnover of organic matter and nutrients. In *Ecological Studies. Analysis and Synthesis* vol. 1, ed. D. E. Reichle, pp. 147–72. Berlin: Springer Verlag.

Elliot, E. T., Anderson, R. V., Coleman, D. C. & Cole, C. V. (1980). Available pore space and microbial trophic interactions. *Oikos*, **35**, 327–35.
Flanagan, P. W. & Van Cleve, K. (1977). Microbial biomass, respiration and nutrient cycling in a black spruce taiga ecosystem. In *Soil Organisms as Components of Ecosystems*, ed. U. Lohm & T. Persson, pp. 261–73. Ecological Bulletin 25. Stockholm: Swedish Natural Science Research Council.
Fochte, D. D. & Verstraete, W. (1977). Biochemical ecology of nitrification and denitrification. *Advances in Microbial Ecology*, **1**, 135–214.
Frissel, M. J. & Van Veen, J. A. (1982). A review of models for investigating the behaviour of nitrogen in soil. *Philosophical Transactions of the Royal Society of London. Series B*, **296**, 341–9.
Ghilarov, M. (1963). On the interrelations between soil dwelling invertebrates and soil micro-organisms. In *Soil Organisms*, ed. J. Doeksen & J. van der Drift, pp. 255–9. Amsterdam: North Holland Publishing Company.
Graff, O. (1971). Stickstoff, Phosphor und Kalium in der Regenwurm losung auf der Wiesen versuchsfläche des Sollingsprojektes. In *Organisms du Sol et production primaire*, IV Colloquium pedobiologiae, pp. 503–12. Paris: Institut National de la Recherche Agronomique.
Grossbard, E. (1969). A visual record of the decomposition of ^{14}C-labelled fragments of grasses and rye added to soil. *Journal of Soil Science*, **20**, 38–51.
Hanlon, R. D. G. (1981a). Some factors influencing microbial growth on soil animal faeces. I. Bacterial and fungal growth on particulate oak leaf litter. *Pedobiologia*, **21**, 257–63.
Hanlon, R. D. G. (1981b). Some factors influencing microbial growth on soil animal faeces. II. Bacterial and fungal growth on soil animal faeces. *Pedobiologia*, **21**, 264–70.
Hanlon, R. D. G. (1981c). Influence of grazing by Collembola on the activity of senescent fungal colonies grown on media of different nutrient concentration. *Oikos*, **36**, 362–7.
Hanlon, R. D. G. & Anderson, J. M. (1979). The effects of collembola grazing on microbial activity in decomposing leaf litter. *Oecologia (Berlin)*, **38**, 93–9.
Hanlon, R. D. G. & Anderson, J. M. (1980). Influence of macroarthropod feeding activities on microflora in decomposing oak leaves. *Soil Biology and Biochemistry*, **12**, 255–61.
Heal, O. W., Swift, M. J. & Anderson, J. M. (1982). Nitrogen cycling in United Kingdom forests: the relevance of basic ecological research. *Philosophical Transactions of the Royal Society of London. Series B*, **296**, 427–44.
Hill, S. B., Metz, L. J. & Farrier, M. H. (1975). Soil mesofauna and silvicultural practices. In *Forest Soils and Forest Land Management*, ed. B. Bernier & C. H. Winget, pp. 119–35. Quebec: Les Presses de l'Université Laval.
Huhta, V. (1979). Effects of liming and deciduous litter on earthworm (Lumbricidae) populations of a spruce forest, with an inoculation experiment on *Allolobophora caliginosa*. *Pedobiologia*, **19**, 340–5.
Ineson, P., Leonard, M. A. & Anderson, J. M. (1982). Effect of collembolan grazing on nitrogen and cation leaching from decomposing leaf litter. *Soil Biology and Biochemistry*, **14**, 601–5.
Jeanson-Luusinang, C. (1963). Etude experimentale de l'action de *Lumbricus herculeus* Savigny (Oligochaeta, Lumbricidae) sur la microflore totale d'un milieu artificiel. In *Soil Organisms*, ed. J. Doeksen & J. van der Drift, pp. 266–70. Amsterdam: North Holland Publishing Company.
Khanna, P. K. (1981). Leaching of nitrogen from terrestrial ecosystems – patterns, mechanisms and ecosystem responses. In *Terrestrial Nitrogen Cycles*, ed. F. E.

Clark & T. Rosswall, pp. 343–52. Ecological Bulletin 33. Stockholm: Swedish Natural Science Research Council.

Kozloskaja, L. S. (1971). Der Einfluss der Wirbellosen auf die Tägigkeit der Mikroorganismen in Torfböden. In *Organisms du Sol et production primaire* IV. Colloquium Pedobiologiae, pp. 81–8. Paris: Institut National de la Recherche Agronomique.

Krivolutzsky, D. A. & Pokarzhevsky, A. D. (1977). The role of soil animals in nutrient cycling in forest and steppe. In *Soil Organisms as Components of Ecosystems*, ed. U. Lohm & T. Persson, pp. 253–60. Ecological Bulletin 25. Stockholm: Swedish Natural Science Research Council.

Kubiena, W. C. (1953). *The Soils of Europe*. London & Madrid: Thomas Murby.

Lohm, U. & Persson, T. (eds.) (1977). *Soil Organisms as Components of Ecosystems*. Ecological Bulletin 25. Stockholm: Swedish Natural Science Research Council.

Lundkvist, H. (1977). Effects of artificial acidification on the abundance of Enchytraeidae in a Scots pine forest in northern Sweden. In *Soil Organisms as Components of Ecosystems*, ed. U. Lohm & T. Persson, pp. 570–3. Ecological Bulletin 25. Stockholm: Swedish Natural Science Research Council.

Macfadyen, A. (1963). The contribution of the microfauna to total soil metabolism. In *Soil Organisms*, ed. J. Doeksen & J. van der Drift, pp. 3–17. Amsterdam: North Holland Publishing Company.

McBreyer, J. F. (1973). Exploitation of deciduous leaf litter by *Apheloria montana* (Diplopoda: Eurydesmidae). *Pedobiologia*, **13**, 90–3.

Márialigeti, K. (1979). On the community structure of the gut-microbiota of *Eisenia luscens* (Annelida, Oligochaeta). *Pedobiologia*, **19**, 213–20.

Miller, H. G. (1979). The nutrient budgets of even-aged forests. In *The Ecology of Even-aged Forest Plantations*, ed. E. D. Ford, D. C. Malcolm & J. Atterson, pp. 221–56. Cambridge: Institute of Terrestrial Ecology.

Miller, H. G. (1981). Forest fertilization: some guiding concepts. *Forestry*, **54**, 157–67.

Newell, K. (1980). 'The effect of grazing by litter arthropods on the fungal colonization of leaf litter.' Unpublished Ph.D. Thesis, University of Lancaster.

Park, D. (1976). Nitrogen level and cellulose decomposition by fungi. *International Biodeterioration Bulletin*, **12**, 95–9.

Parkinson, D., Visser, S. & Whittaker, J. B. (1979). Effects of collembolan grazing on fungal colonization of leaf litter. *Soil Biology and Biochemistry*, **11**, 529–35.

Parle, J. N. (1963). A microbiological study of earthworm casts. *Journal of General Microbiology*, **31**, 13–22.

Patten, B. C. & Witkamp, M. (1967). Systems analysis of ^{134}Cesium kinetics in terrestrial microcosms. *Ecology*, **48**, 813–24.

Paul, E. A. & Juma, N. G. (1981). Mineralization and immobilization of soil nitrogen by microorganisms. In *Terrestrial Nitrogen Cycles*, ed. F. E. Clark & T. Rosswall, pp. 179–94. Ecological Bulletin 33. Stockholm: Swedish Natural Science Research Council.

Persson, T. (1983). Influence of soil animals on nitrogen mineralization in a northern Scots pine forest. In *New Trends in Soil Biology*, ed. Ph. Lebrun, H. M. André, A. de Medts, C. Gregoire-Wibo & G. Wauthy, pp. 117–26. Louvain-la-Neuve: Dieu-Brichart.

Persson, T., Bååth, E., Clarholm, M., Lundkvist, H., Söderström, B. E. & Sohlenius, B. (1980). Trophic structure, biomass dynamics and carbon metabolism of soil organisms in Scots pine forest. In *Structure and Function of Northern Coniferous Forests – An Ecosystem Study*, ed. T. Persson, pp. 419–59. Ecological Bulletins 32. Stockholm: Swedish Natural Science Research Council.

Reichle, D. E. (1977). The role of invertebrates in nutrient cycling. In *Soil Organisms as*

References

Components of Ecosystems, ed. U. Lohm & T. Persson, pp. 145–54. Ecological Bulletin 25. Stockholm: Swedish Natural Science Research Council.
Reichle, D. E. (ed.) (1981). *Dynamic Properties of Forest Ecosystems.* Cambridge University Press.
Reiners, W. A. (1981). Nitrogen cycling in relation to ecosystem succession. In *Terrestrial Nitrogen Cycles*, ed. F. E. Clark & T. Rosswall, pp. 507–28. Ecological Bulletins 33. Stockholm: Swedish Natural Science Research Council.
Reyes, V. G. & Tiedje, J. M. (1976). Ecology of the gut microbiota of *Tracheoniscus rathkei. Pedobiologia*, **16**, 67–74.
Robertson, G. P. (1982). Nitrification in forested ecosystems. *Philosophical Transactions of the Royal Society of London. Series B*, **296**, 445–57.
Satchell, J. E. (1963). Nitrogen turnover by a woodland population of *Lumbricus terrestris*. In *Progress in Soil Biology*, ed. O. Graff & J. E. Satchell, pp. 102–19. Amsterdam: North Holland Publishing Company.
Scarsbrook, C. E. (1965). Nitrogen availability. In *Soil Nitrogen*, ed. W. V. Bartholomew & F. E. Clark, pp. 481–502. New York: Academic Press.
Schlesinger, W. H. (1977). Carbon balance in terrestrial detritus. *Annual Reviews of Ecology and Systematics*, **8**, 51–81.
Seastedt, T. R. & Crossley, D. A. (1980). Effects of microarthropods on the seasonal dynamics of nutrients in forest litter. *Soil Biology and Biochemistry*, **12**, 337–42.
Stachurski, A. & Zimka, J. (1977). Release of macronutrients from decomposing litter in *Pino–Quercetum* and *Carici elongatae–Alnetum* associations. The role of litter microorganisms and saprophages in releasing processes. *Bulletin de l'Académie polonaise des sciences. Classe II. Série des Sciences biologiques*, **24**, 655–62.
Standen, V. (1978). The influence of soil fauna on decomposition by microorganisms in blanket bog litter. *Journal of Animal Ecology*, **47**, 25–38.
Stefaniak, O. & Seniczak, S. (1976). The microflora of the alimentary canal of *Achipteria coleoptrata* (Acarina, Oribitei). *Pedobiologia*, **16**, 185–94.
Swank, W. T. & Waide, J. B. (1980). Interpretation of nutrient cycling research in a management context: evaluating potential effects of alternative management strategies on site productivity. In *Forests: Fresh Perspectives from Ecosystem Analysis*, ed. R. Waring, pp. 137–58. Corvallis: Oregon State University Press.
Swift, M. J., Heal, O. W. & Anderson, J. M. (1979). *Decomposition in Terrestrial Ecosystems.* Oxford: Blackwell Scientific Publications.
Syers, J. K., Sharpley, A. M. & Keeney, D. R. (1979). Cycling of nitrogen by surface-casting earthworms in a pasture ecosystem. *Soil Biology and Biochemistry*, **11**, 181–5.
Szabo, I. M., Jáger, K., Contreras, E., Marialigeti, K., Dzingov, A., Barabás, G. & Pobozsny, M. (1983). Composition and properties of the external and internal microflora of millipedes – Diplopoda. In *New Trends in Soil Biology*, ed. Ph. Lebrun, H. M. André, A. de Medts, C. Gregoire-Wibo & G. Wauthy, pp. 197–206. Louvain-la-Neuve: Dieu-Brichart.
Szabo, F., Marton, M. & Buti, F. (1969). Intestinal microflora of the larvae of St Mark's fly. IV. Studies on the intestinal bacterial flora of the larval population. *Acta microbiologica Academiae scientiarum hungaricae*, **16**, 381–97.
Tamm, C. O., Holmen, H., Popović, B. & Wiklander, G. (1974). Leaching of plant nutrients from soils as a consequence of forestry operations. *Ambio*, **3**, 211–21.
Van Cleve, K. & Alexander, V. (1981). Nitrogen cycling in tundra and boreal ecosystems. In *Terrestrial Nitrogen Cycles*, ed. F. E. Clark & T. Rosswall, pp. 375–404. Ecological Bulletin 33. Stockholm: Swedish Natural Science Research Council.
van der Drift, J. & Jansen, E. (1977). The grazing of springtails on hyphal mats and its influence on fungal growth and respiration. In *Soil Organisms as Components of*

Ecosystems, ed. U. Lohm & T. Persson, pp. 302–9. Ecological Bulletin 25. Stockholm: Swedish Natural Science Research Council.

Vaněk, J. (1975). (ed.) *Progress in Soil Biology*. Prague: Academia.

Visser, S., Whittaker, J. B. & Parkinson, D. (1981). Effects of collembolan grazing on nutrient release and respiration of a leaf litter inhabiting fungus. *Soil Biology and Biochemistry*, **13**, 215–18.

Vitousek, P. M. & Melillo, J. M. (1979). Nitrate losses from disturbed forests: patterns and mechanisms. *Forest Science*, **25**, 605–19.

Went, J. C. (1963). Influence of earthworms on the number of bacteria in the soil. In *Soil Organisms*, ed. J. Doeksen & J. van der Drift, pp. 260–5. Amsterdam: North Holland Publishing Company.

Witkamp, M. (1969). Environmental effects on microbial turnover of some mineral elements. 1. Abiotic factors. *Soil Biology and Biochemistry*, **1**, 167–76.

Witkamp, M. & Barzansky, B. (1968). Microbial immobilization of ^{137}Cs in forest litter. *Oikos*, **19**, 392–5.

Witkamp, M. & Frank, M. L. (1969). Cesium-137 kinetics in terrestrial microcosms. In *Proceedings of the Second National Symposium on Radioecology*, ed. D. J. Nelson & F. C. Evans, pp. 635–43. Virginia: United States Department of Commerce.

Witkamp, M. & Frank, M. L. (1970). Effects of temperature, rainfall, and fauna on transfer of cesium-137, K, Mg and mass in consumer–decomposer microcosms. *Ecology*, **51**, 465–74.

Wood, T. G. (1974). Field investigations on the decomposition of leaves of *Eucalyptus delegatensis* in relation to environmental factors. *Pedobiologia*, **14**, 343–71.

Woodmansee, R. G. & Wallach, L. S. (1983). Effects of fire regimes on bio-geochemical cycles. In *Fire Regimes and Ecosystem Properties*, ed. H. Mooney, T. M. Bonnicksen, N. L. Christensen, J. E. Lotan & W. A. Reiners. Washington, DC: USDA Forest Series General Technical Reports (in press).

Zachariae, G. (1965). Spurentierischer Tätigkeit im Boden des Buchenwaldes. *Forestwissenschaftliche Forschungen*, **20**, 1–68.

Zajonc, I. (1971). Participation des lombrics (Lumbricidae) dans la libération des éléments minéraux des feuilles mortes d'une forêt de hêtres et de chênes. In *Organisms du Sol et production primaire*, IV Colloquium pedobiologiae, pp. 387–98. Paris: Institut National de la Recherche Agronomique.

Zlotin, R. I. (1971). Invertebrate animals as a factor of the biological turnover. In *Organisms du Sol et production primaire*, IV Colloquium pedobiologiae, pp. 455–62. Paris: Institute National de la Recherche Agronomique.

Zlotin, R. I. & Khodashova, K. S. (1980). *The Role of Animals in Biological Cycling of Forest–Steppe Ecosystems*. Stroudsburg, Pennsylvania: Dowden, Hutchinson & Ross.

4
Animal–microbial interactions in wood decomposition

M. J. SWIFT and LYNNE BODDY*

Department of Botany, University of Zimbabwe, PO Box M.P.167, Mount Pleasant, Harare, Zimbabwe

*Department of Microbiology, University College, Newport Road, Cardiff CF2 1TA, Wales, UK

Key words: microbial attractants; animal grazing; colonization; decomposition; mineral nutrient dynamics; humification; nutritional quality; enzymatic degradation

Introduction

Rodin & Basilevic (1967) point out that between 92 and 99% of above-ground biomass in temperate deciduous and tropical rain forests is in the form of wood. Furthermore a high proportion of the root system of trees is also woody. As such, wood constitutes the food resource of highest potential availability to heterotrophs in forest ecosystems. For two main reasons, however, the expectation that wood should form the basis of most detritus food chains may not be realized. Firstly, woody tissues tend to be conserved by plants; litter fall studies show that, in contrast to photosynthetic tissues, only a small proportion of the woody biomass is shed each year (Bray & Gorham, 1964; Boddy & Swift, 1983a). Nonetheless, the majority of litter fall studies underestimate the input of wood to decomposition since they seldom measure any but the smallest of wood components and only rarely incorporate root or whole tree input and decomposition prior to fall.

Secondly, wood is a low quality resource (*sensu* Swift, Heal & Anderson, 1979) due to extensive lignification, low content of soluble sugars and mineral nutrients, often high content of allelopathic compounds, hardness, protection by bark which is suberized and itself contains high concentrations of allelopaths, and, in some cases, the massive size of woody 'units'. The overall consequence is that the range of organisms capable of exploiting wood (other than cell contents) in an *unaltered* form as food is narrow and specific. Such organisms are largely confined to the Basidiomycotina and Insecta (mainly Isoptera and Coleoptera).

Many basidiomycetes are capable, by virtue of their enzymatic capacity, of totally reducing the substance of wood to carbon dioxide, water and mycelium. Under natural conditions wood units are usually decomposed by the joint action of a community of organisms. Within these communities interactions between microbes and animals regularly occur which are crucial in the decomposition process. Among these interactions are some which have attracted considerable study; critical consideration of some of these are given by Dowding (Chapter 5), Webber & Brasier (Chapter 10) and Martin (Chapter 6). Our intention here is not to describe any one association in detail but rather to consider the significance of the various interactions between animals and microbes in relation to broad patterns of decomposition.

Our approach will be to describe a particular case, that of small branch wood decomposing within the environment of the temperate deciduous forest, at the British International Biological Programme site at Meathop Wood, Cumbria and also at an abandoned beech–oak coppice at Blean Woods, Kent, UK. Here, decomposition follows a general pattern in which the primary stages of decay are brought about by basidiomycete fungi. Most of the wood is then invaded by larvae of specialized wood-boring animals, members either of the Coleoptera or Diptera. It is our contention that the events of the terminal 40% of decomposition are wholly determined by this interaction between animals and fungi. The significance of this for the forest ecosystem lies in the fact that it is during these terminal stages of decomposition that the events of nutrient release and soil organic matter formation occur.

From this base we will consider the decomposition of other woody materials (e.g. twigs, larger branches, main stems, stumps, roots etc.); similarly the decomposition of branch wood in different ecosystems (e.g. savanna, conifer forests, tropical rain forests) which may have widely differing rates and characteristics. Our contention is, however, that in all cases the interaction between animal and fungus will be a crucial event of fundamental significance.

The process of wood decomposition

The major events of the decomposition process in temperate deciduous woodlands are illustrated in Fig. 4.1. Decomposition above ground is initiated in the canopy where present indications are that basidiomycetes and xylariaceous ascomycetes rapidly become established in branches near the end of their life (Boddy and Rayner, 1981, 1983, unpublished data). At early stages animals do not appear generally to have a significant role. However, considerable numbers of microfauna, such as nematodes, can sometimes be found (Boddy and Rayner, unpublished), and there are some striking examples of initiation of decay and of branch and whole tree death resulting from animal activity; for example, over 1 million oaks were killed in an area of 7300 ha as a direct result of gypsy moth defoliation in New Jersey in 1968–70 (Kegg, 1970, cited in Rexrode, 1971). Many other microorganisms may be present at this stage, particularly in wounds, but the basidiomycetes and xylariaceous ascomycetes usually remain the most active agents of decomposition until branch fall.

At Meathop Wood the mean relative density (RD; g cm^{-3}, see Healey & Swift, 1971) of branches at fall was 0.318 g cm^{-3}, which represents a loss of 43% of weight compared with living wood (Fig. 4.2a, c). However, there was large variation in the RD values and in some cases complete or almost complete decomposition had occurred prior to fall. This situation particularly occurs in trees such as oak

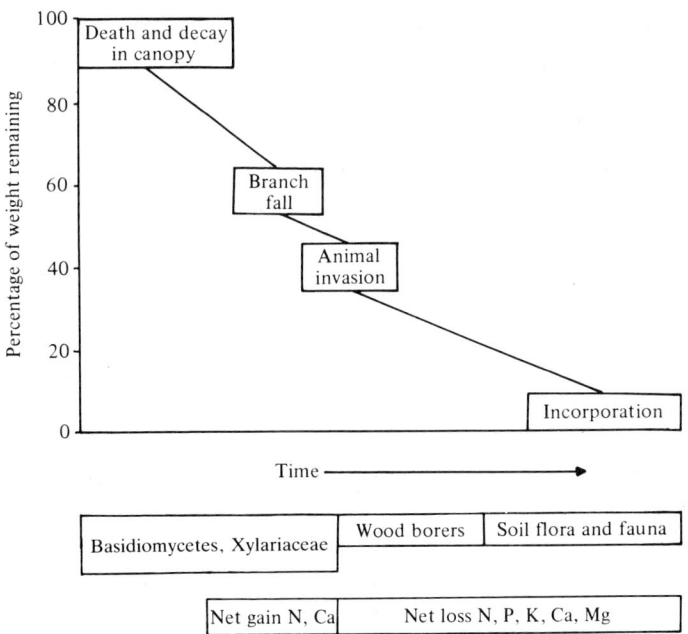

Fig. 4.1. The major events of the wood decomposition process in temperate deciduous woodlands showing the dominant organisms and the mineral nutrient characteristics of each phase.

(*Quercus* spp.) where large branches remain attached to the tree, by central heartwood, for longer periods. Under such circumstances the decayed wood may well become invaded by insects in an analogous fashion to that more often observed on the forest floor (see below). Nonetheless, the pattern of weight loss at branch fall observed at Meathop Wood was found to be similar at two other sites; at Blean Woods (Boddy & Swift, 1983a) and in tropical rain forest at Barro Colorado Island, Panama (Healey and Swift, unpublished). In the latter case a small proportion (14%) of the branches showed animal invasion, mainly by termites, before fall (Fig. 4.2d).

Previously established fungi probably continue to dominate decomposition, at least for a short time, in the litter layer. However, the branches soon become susceptible to invasion by fungi from the litter layer and by specialized wood-boring insects. At Meathop and Blean Woods this attack was predominantly by larvae of the crane-fly *Tipula flavolineata* (Diptera) which were found respectively in 42% and 39% of branches in the litter layer (Fig. 4.2e; Swift, Boddy & Healey, 1983). The rate of wood decay in the litter at Meathop was shown to be about double that in the canopy; conditions in the experimental branches, however, may have been considerably different from those in attached

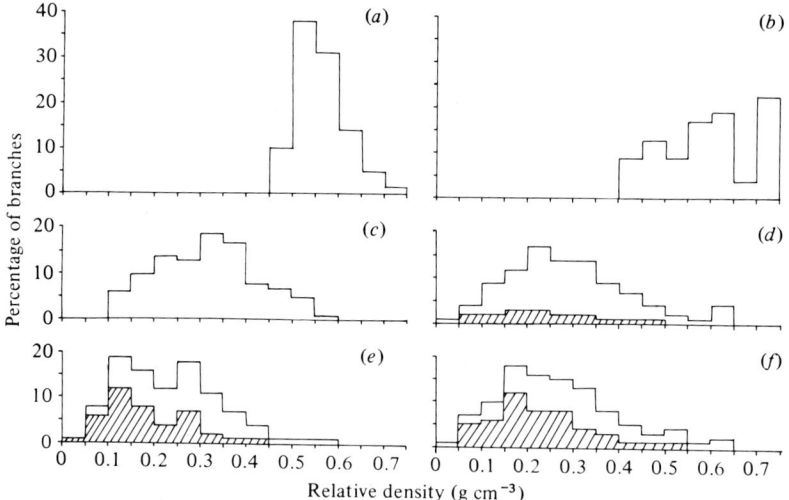

Fig. 4.2. Distribution of RD (g cm^{-3}) in small branch wood at two sites; (a), (c), (e) Meathop Wood, Cumbria, UK. (b), (d), (f), Barro Colorado Island, Panama. (a) and (b) are of living wood; (c) and (d) are of wood at branch fall; (e) and (f) are of wood on the forest floor. Hatched areas are wood invaded by animals; clear areas are uninvaded.

branches. It was not possible to distinguish between the pre- and post-invasion phases in the canopy (see below).

The event of invasion, by *T. flavolineata*, of branches inhabited by decay fungi is of considerable significance in terms of the subsequent progress of decomposition. Before discussing this, specific features of the influence of the microbes on the animals and *vice versa* will be considered.

The influence of the resident microflora on the development of the animal community

In the study at Meathop Wood, illustrated in Fig. 4.2, 42% of branches on the forest floor were invaded by animals. At Blean the proportion was 60% and at Barro Colorado 50%. Below an RD of 0.2 g cm^{-3}, i.e. at about 65% weight loss, the proportions were 57, 73 and 64% respectively on the three sites. All these branches of lower RD are inhabited by a diverse community of animals. A census of the branch-inhabiting mesofauna was carried out at Blean by Swift *et al.* (1983) and a list of the dominant fauna is given in Table 4.1. This community has a broadly similar composition to that found by Fager (1968) in his very detailed study of the invertebrates of oak branches. General observation showed a similar pattern at Meathop. A characteristic of the Blean fauna was that conclusive evidence of past and/or present activity of *T. flavolineata* was found in 65% of the invaded branches, larvae being present in 27%. The former is a minimum figure as many of the branches recorded as 'non-tipulid' were among the most heavily decayed, and contained the faeces of a wide range of animals which may have obscured traces of earlier colonizers such as *Tipula*. More importantly, the branches of higher RD (i.e. above 0.2 g cm^{-3}) which showed evidence of invasion invariably had evidence of *Tipula* and often had larvae present. The mean RD of branches with *Tipula* larvae present was 0.22 g cm^{-3} with a standard error ± 0.01 g cm^{-3}. This provides circumstantial evidence for *T. flavolineata* as a pioneer invader of branches, the diversification of the animal community only being a secondary development. *Tipula* seems to play a similar role at Meathop, but Fager's (1968) conclusions, using 'synthetic' oak logs, suggest no definite pioneers at Wytham Wood. However, Elton (1966) suggested that *T. flavolineata* might be one of a number of larvae occupying this niche at the same site. In branches from tropical rain forest, termites and beetle larvae occupied an equivalent niche to *Tipula* (Healey and Swift, unpublished).

Table 4.1. *Dominant mesofauna in oak branches decomposing on the forest floor at Blean Woods*

Orders	Species
Diptera	*Limoniid* spp.
	Mycetophilid spp.
	Tipula flavolineata
Coleoptera	*Cylindronotus* sp.
Chilopoda	*Lithobius variegatus*
Diplopoda	*Cylindroiulus punctatus*
	Polydesmus angustus
Isopoda	*Oniscus asellus*
Oligochaeta	*Lumbricus rubellus*
	Lumbricus terrestris

From Swift *et al.*, 1983.

The pattern described so far suggests two main features; firstly that attack by *Tipula* and equivalent Diptera and Coleoptera larvae occurs only relatively late in decomposition, and moreover secondly that it may occur within only a relatively narrow range of RD. This suggests that animal invasion is favoured by some prior microbial 'conditioning', which could take one of four possible forms: production of attractant substance(s) by the microorganisms; softening of the wood due to enzymatic action; destruction of substances allelopathic to wood-inhabiting animals; improvement of the nutritional quality of the resource.

Microbial attractants

The possibility that *Tipula* attack is related to some specific attractant produced by the wood or wood plus fungus was tested experimentally at Meathop Wood (Swift and Healey, unpublished). Thirty-two branches decayed by basidiomycetes were taken from the canopy and each cut transversely into eight units which were laid out, in a random pattern, in the litter layer in March 1968. After 0, 0.25, 0.5, 0.75, 1.0, 1.5, 2.0 and 3.0 years 32 units (i.e. one from each branch) were examined for the presence of *T. flavolineata* and other animals, and the RD was estimated (Fig. 4.3). The experiment was thus intended to simulate the events following branch fall. The branches had a spread of decay not dissimilar to that at branch fall (cf. RD distributions in Figs. 4.3a and 4.2c) but the mean extent of decomposition was somewhat less (21% vs. 43%) so that the distribution is located more to the right in Fig. 4.3 than in Fig. 4.2. At the end of the experiment the mean RD was 0.279 g cm^{-3} compared with that of 0.236 g cm^{-3} found in the survey described in Fig. 4.2e.

There was no evidence of preferential selection, by animals, of the three tree species used in the experiment: oak (*Quercus robur* + *petraea*), ash (*Fraxinus excelsior*) (8 branches each) and hazel (*Corylus avellana*) (16 branches). In the final count 23% of the invaded units were oak, 28% ash, and 49% hazel. The 32 branches were invaded by seven different basidiomycetes which occurred with varying frequency. By the end of the experiment at least one of the eight units, into which

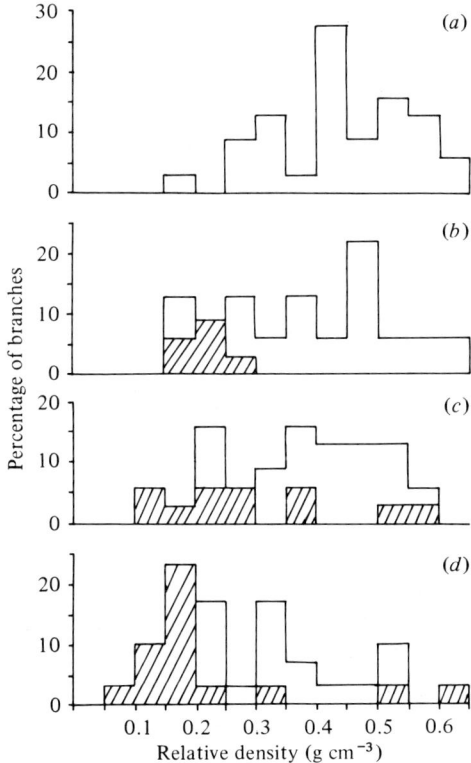

Fig. 4.3. Progressive change in RD (g cm^{-3}) of small wood during decomposition. (*a*) Branches at start of the experiment; branches after (*b*) 1 year, (*c*) 2 years, (*d*) 3 years in the litter layer. Hatched areas are branches invaded by *T. flavolineata*. See text for details of experiment.

Table 4.2. *The distribution of fungi and animal invasion between 32 branches decaying on the forest floor at Meathop Wood* (see text for details)

Fungus	No. of branches	No. of branches invaded	% 'units' invaded
Stereum hirsutum	11	8	34
Stereum sp.	4	4	25
Coriolus versicolor	7	5	24
Tyromyces sp.	3	3	21
Resupinate sp. 1	3	2	44
Basidiomycete sp. 1	3	2	27
Basidiomycete sp. 2	1	1	13

each branch was divided, had been invaded without any apparent bias in the distribution of invasion (Table 4.2).

The main oviposition period for *T. flavolineata* at Blean is in June and July (Swift *et al.*, 1983). No evidence has been found to suggest that there is any specific chemical attraction to rotting branches. Indeed isolation data for the fungi suggest that at Blean, as at Meathop, wood colonized by all of the common basidiomycetes is equally likely to become invaded by the larvae. One possible exception to this was *Armillaria* sp. which was commonly found in oak branches but rarely in association with *Tipula*. It is probably reasonable to hypothesize that successful oviposition and colonization by this insect is largely non-selective.

Many examples are, however, known where highly specific positive and/or negative chemical stimuli are exerted by a resident fungus on invading invertebrates. Perhaps the best documented examples are those related to termites. For instance several authors have reported that termites are attracted to certain brown-rot fungi, whereas wood decayed by white-rot fungi such as *Ganoderma applanatum* was sometimes unattractive or even toxic to them. Both the wood decayed by *Gloeophyllum trabeum (Lenzites trabea)* and the mycelium of this fungus in culture contain termite trail-following, arrestant and feeding stimulants which attract species of *Reticulitermes*, *Heterotermes*, *Coptotermes and Kalotermes* (Kovoor, 1964; Ritter & Coenen-Saraber, 1969; Amburgey & Smythe, 1977*a*, 1977*b*). Trail-following effects on *Coptotermes formosanus* were also produced by extracts of wood colonized by *Daedalea dickinsii*, *G. trabeum*, *Serpula lacrimans* and *Tyromyces palustris* (Matsuo & Nishimoto, 1974). Becker & Lenz

(1972) demonstrated that the odour of such fungi may stimulate termites to eat more sound wood and build more galleries. Esenther, Allen, Casida & Shenefelt (1961) noted that runways of *Reticulitermes* spp. on trees tend to go straight to decaying wood, and they suggested that a gradient of attractive compounds may aid them in finding their food supply.

With regard to white-rot fungi, wood decayed by *Ganoderma applanatum* was toxic to *Microcerotermes edentatum* although *Pleurotus ostreatus* was attractive and non-toxic (Becker, 1965). *Bjerkandera adusta*, *Coriolus versicolor* and *Stereum sanguinolentum* were all toxic to *Reticulitermes santonensis* (Schultze-Dewitz & Unger, 1972). Under natural conditions pine stumps decayed by white-rot fungi are mostly free from termites (Becker, 1969).

Boddy & Rayner (1981, 1983) described a characteristic assemblage of basidiomycetes which decomposed attached oak branches in the canopy. They found occasional evidence of animal activity in decay columns occupied by most basidiomycete species, but insect activity was consistently associated with *Phlebia radiata*, *Schizopora paradoxa*, and *Hyphoderma setigerum*. It is unclear as to whether animal activity was associated with these fungi as a result of some specific feature relating to the fungi or decay they produce, or solely as a result of the fact that these fungi are usually correlated with well decayed wood, or even whether the fungi colonized wood invaded by animals. The last may be true for *H. setigerum* which is often found fruiting in beetle tunnels (Fig. 4.8e), but it is also very common in conditions prone to drying out such as often occur after insect invasion.

The intricate relationship of attractant chemistry to the physiological state of the fungus is demonstrated by an intriguing observation made recently by Boddy, Coates & Rayner (1983). They noted the attraction of fungus gnats of the genus *Bradysia* (Mycetophilidae) to mycelium of wood-rotting basidiomycetes and xylariaceous fungi in culture. The flies were attracted specifically to zones of intraspecific antagonism, consisting largely of dead or dying hyphae, between somatically incompatible colonies. The observations suggest the production of a specific attractant which is a product of dying mycelium. The adaptive significance of this can only be speculated on.

Enzymatic softening by microorganisms

Few animals have the enzymatic capacity to break down lignin. Similarly, few produce the enzymes required for complete cellulolysis although some contain gut symbionts which are able to do this, or employ an exosymbiotic co-digestion strategy to cope with the problem (see Martin, Chapter 6). However, even if the necessary enzymes are available to the animals, unless they are able to gain entry to the wood or are able to remove it to their nests they will not be able to make use of the wood as food. Thus in many instances microbial enzymatic conditioning of wood is probably necessary prior to animal invasion.

Enzymatic degradation of wood has two important effects: release of readily assimilable substrates, and softening which makes it easier to chew. As the two factors are directly correlated it is often difficult to differentiate which of the changes is of most importance to the wood fauna.

It may be hypothesized that invasion occurs below a specific threshold RD, i.e. only after a given amount of 'softening' by decomposition has occurred. This is illustrated in Fig. 4.4 where the cumulative percentages of branch 'units' invaded by animals in the previously described experiment at Meathop are plotted as a function of RD. Over 75% of

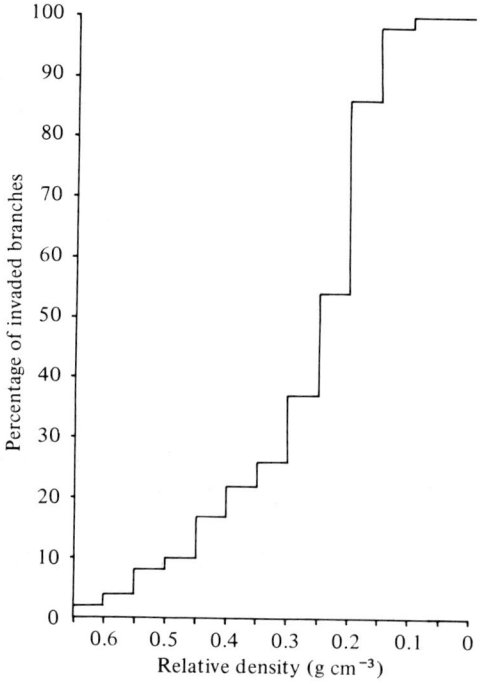

Fig. 4.4. Cumulative frequency distribution of RD (g cm^{-3}) for branches invaded by animals at Meathop Wood.

the invaded units have an RD below 0.3 g cm^{-3}. The experiment cannot provide unequivocal evidence, for clearly one of the effects of invasion is to reduce RD. Nevertheless, it would seem that invasion is much less frequent above this value.

The 'softening' effect has seldom been demonstrated unequivocally but one example where it may be a primary factor is that of woodwasps (*Siricidae*: Hymenoptera). As described by Martin (Chapter 6), these insects inoculate wood with basidiomycete fungi at the time of oviposition. The inoculated oidia germinate and produce a mycelium which rapidly initiates enzymatic breakdown of both cellulose and lignin. The larvae tunnel into and feed in the wood which has been partly decayed by the fungus.

Another instance where prior softening by fungi seems to be prerequisite for animal invasion, but far removed from those discussed so far, is that of nest building by woodpeckers. Conner, Miller & Adkisson (1976) examined over 90 nest trees (of various tree and woodpecker species) in the USA and found that the heartwood in all had been softened by fungal decay, mainly by a white-rot polypore *Spongipellis pachyodon*. The woodpeckers were apparently able to identify suitably decayed trees, presumably by sounding with their bills. On one or two occasions there was evidence that abortive attempts had been made on trees with undecayed heartwood. In this instance enzymatic preconditioning is purely a softening prerequisite. This example is not completely irrelevant to decomposition, for woodpeckers may deposit considerable amounts of wood on the forest floor, and the fragmentation of the wood may well accelerate its rate of decay.

Microbial destruction of allelopathic compounds

Wood may be initially repellent or unpalatable to animals because of the presence of compounds which have an allelopathic effect. Fungal enzymatic activity may however result in the destruction of these compounds or their reduction to a non-active level. The effectiveness of fungi in doing this has been demonstrated on a number of occasions; for instance Tattar, Shortle & Rich (1971) found a decrease of about 70% in the total phenolic content of sugar maple (*Acer saccharum*) wood following attack by a variety of fungi, and wood decayed by *Oxyporus populinus* (= *Fomes connatus*) contained only 5% of its original phenolic content. Most of the well documented examples again concern termites; Williams (1965) found that in central America *Coptotermes niger* used, as its preferred feeding and nest site, the heartwood of standing trees of *Pinus caribaea* attacked by the white-rot fungus *Lentinus pallidus*. The sound heartwood has a natural resistance to termite attack, being both toxic and repellent to them due to the high turpentine and resin content; the fungus degrades these compounds and thus, together with other aspects of enzymatic conditioning, permits termite attack. A similar example is given by Becker (1975) who found that *Coptotermes sjöstedti* attacked the very resistant heartwood of Mukulunga (*Autranella congolensis*) in Africa after fungal decay. With respect to invasion by *T. flavolineata* we have no evidence of the involvement of allelopathic compounds.

Improvement of nutritional quality

There is considerable evidence that wood partially decayed by fungi provides a higher quality food resource for animals than undecayed wood. The nature of the nutritional effects brought about by the fungi may be complex, embracing the release of a wide range of assimilable organic molecules by catabolism of polymers, the transformation of specific molecules to palatable form, and the process of enrichment of the mineral nutrient elements. It is the latter feature with which we shall deal here. The basis of this hypothesis is that wood is initially a resource with a low availability of the major mineral nutrient elements. This is not simply a matter of content, as Park (1976) has pointed out. One litre of wood may contain as much as 2 g of nitrogen, enough to satisfy the growth requirements of considerable numbers of animals. More specifically, the low availability of nitrogen and other elements lies in their low concentrations relative to the overall organic mass of the resource. A convenient way of representing this is as the C : nutrient (e.g. C : N) ratio. In the case of nitrogen this may be in the range of from 100 : 1 to 500 : 1. In Table 4.3 data are given in this form for a sample of 20 small (diameter 2–4 cm) branches from Meathop Wood (Swift, 1977). The sample was equally divided between the four tree species oak (*Quercus petraea*), ash (*Fraxinus excelsior*), birch (*Betula pendula* + *pubescens*) and hazel (*Corylus avellana*), but the data given in the first row of Table 4.3 are the bulked means for the whole sample.

Also given in Table 4.3 are values for the C : nutrient ratios of decayed wood on the assumption that whilst carbon is lost there is no net gain or loss of the other elements. Whether this is a reasonable assumption will be discussed later, but this theoretical progression enables us to see the changing position with regard to mineral nutrition

Table 4.3. C : nutrient ratios for undecayed wood together with calculated values for decayed wood on the assumption that there has been no net gain or loss of the mineral nutrient element. (Values for C : nutrient ratios of decomposer fungi and animals are also given)

	RD (g cm^{-3})	Weight loss %	C : N	C : P	C : K	C : Ca	C : Mg
Undecayed wood	0.560	0	247 : 1	3643 : 1	343 : 1	135 : 1	1787 : 1
Decayed wood	0.400	29	175 : 1	2587 : 1	243 : 1	96 : 1	1288 : 1
Decayed wood	0.300	46	133 : 1	1967 : 1	185 : 1	73 : 1	965 : 1
Decayed wood	0.200	64	89 : 1	1311 : 1	123 : 1	49 : 1	643 : 1
Decayed wood	0.100	82	44 : 1	655 : 1	62 : 1	24 : 1	322 : 1
Fungal mycelium	—	—	35 : 1	505 : 1	115 : 1	60 : 1	485 : 1
Insecta	—	—	6 : 1	52 : 1	67 : 1	157 : 1	235 : 1

From Swift, 1977, except Insecta – from Allen, Grimshaw, Parkinson & Quarmby, 1974.

for a wood-boring animal like *T. flavolineata*. It is clear from Table 4.3 that an animal eating wood that has been decayed by fungi to the extent of losing about two-thirds of its carbon will have a diet much richer in nitrogen and other elements than one eating undecayed wood. To put it another way; in order to obtain 10 μg of nitrogen an animal would have to consume 5.3 mg of undecayed wood but only 1.9 mg of wood decayed to an RD of 0.2 g cm^{-3}. The demand this puts on the digestive efficiency of the animal can be assessed roughly from the information in the bottom row of Table 4.3 where it is shown that the required ratio of C : N in invertebrates such as *Tipula* is approximately 6 : 1.

A considerable fraction of the nitrogen and other elements consumed by a wood-boring animal when feeding on wood decayed by basidiomycetes will be within the fungal mycelium. Swift (1978) described the quantitative pattern of growth of a basidiomycete in decaying oak wood (Fig. 4.5). By the time the wood had been two-thirds decomposed by the fungus, mycelium had accumulated to a value of about 185 mg g^{-1} of wood. Fungus mycelium has a high nitrogen content compared with wood (Table 4.3) and an easy calculation shows that at this level of growth the fungal biomass should theoretically contain about 46% of the total N in the wood. For the other elements, the values are P = 48, K = 20, Ca = 15, and Mg = 24%. In a separate experiment involving only seven branches Swift (1977) found that the proportions of nutrients in the mycelium were even higher, probably due to net loss of nitrogen from the branches prior to fungal immobilization. It is thus apparent

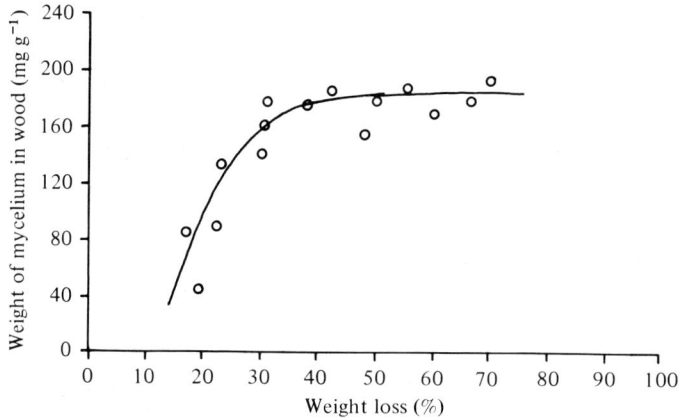

Fig. 4.5. The accumulation of mycelium during progressive decomposition of oak wood by the basidiomycete *Stereum hirsutum*. (After Swift, 1978.)

that not only may the animal benefit nutritionally in an indirect manner from the activity of the fungus, i.e. by the food as a whole being enriched, but it may also be advantaged directly if its diet contains a high proportion of mycelium.

It is in the fungus-growing termites (Macrotermitinae) that this benefit is gained to the most complete extent. In this case there is a mutualistic association between the termites and white-rot basidiomycetes of the genus *Termitomyces*. The termites harvest wood with great efficiency (Collins, 1981) but only partially digest it during initial passage through the gut of the foragers. The faeces are deposited within the nest to form a 'comb' on which *Termitomyces* grows (Wood, 1976). The basidiomycete is responsible for a major part of the cellulose and lignin breakdown of the wood, and in the process nutrient rich bodies (mycotêtes) and mycelium are formed (Table 4.4; Thomas, 1981). The termites in the nest feed directly on the mycotêtes as well as on decomposed comb, thus consuming a food which is much higher in nutrients than the original wood or the other parts of the comb.

A classic example of the nutritional benefit of feeding on fungally decayed wood is that of the Death Watch beetle, *Xestobium rufovillosum* (Anobiidae), studied by Fisher (1940, 1941). This animal causes serious damage to timber in buildings. The problem originates, however, in wood on the standing tree which is invaded by the beetle before harvesting. The insect is found naturally in the decaying branches or trunks of a number of trees, the two most common being oak (*Quercus* spp.) and willow (*Salix* spp.) (Hickin, 1963). It is rarely found in fallen wood. Attack by the beetle is associated with wood decayed by basidiomycetes such as *Laetiporus sulphureus*, *Fistulina hepatica* and *Phellinus cryptarum* in oak, and *Bjerkandera fumosa*, *Coriolus versicolor*, and *Coniophora puteana* in willow. Fisher (1941) investigated the length of time taken by the larvae to complete their life cycle in oak and willow timber decayed to different extents by a variety of these fungi. He found that the length of the life cycle and rate of disintegration of wood due to larval boring was related to the extent of fungal decay (Fig. 4.6). In undecayed willow the larvae either died soon after they had penetrated the surface or, in a few instances, developed very slowly – reaching only 3 mm in length after nine and a half years. In welldecayed wood the life cycle was completed in 10 to 17 months. In slightly decayed oak sapwood the life cycle was similarly prolonged whereas in wood extensively decayed by *Phellinus cryptarum* the life span was 14 to 24 months. In the field in oak attacked by the same

Table 4.4. *Mineral nutrient content (% of oven dry weight) of food, foodstore, fungus comb and mycotêtes of Termitomyces in nests of* Macrotermes bellicosus

	n	%N	C:N	%P	C:P	%K	C:K	%Ca	C:Ca	%Mg	C:Mg
Food	14	0.28 ± 0.03	168	0.15 ± 0.05	313	0.20 ± 0.08	235	2.47 ± 0.14	19	0.22 ± 0.01	214
Food store	5	0.58 ± 0.04	81	0.12 ± 0.02	392	0.16 ± 0.05	294	2.39 ± 0.20	20	0.29 ± 0.01	162
Fresh fungus comb	5	0.85 ± 0.08	55	0.19 ± 0.04	247	0.06 ± 0.01	788	2.89 ± 0.14	16	0.34 ± 0.01	138
Old fungus comb	5	0.82 ± 0.06	57	0.10 ± 0.02	470	0.06 ± 0.00	783	3.10 ± 0.08	15	0.39 ± 0.02	121
Mycotêtes	1	6.68	7	0.46	102	1.37	34	0.22	214	0.02	2350

Values are mean ± SE of mean.
C : nutrient ratios calculated assuming 47% C content

From Thomas, 1981.

fungus the life span varied from 3 to 7 years. Similarly, the larvae failed to establish themselves in undecayed oak heartwood but when decayed by *P. cryptarum* the life cycle was as short as 16 months.

That physical factors are important is shown by the fact that although the beetles have a faster generation time in willow than in oak in relation to per cent weight loss, if RD is considered (willow is less dense than oak) the results are almost identical for the two species (Fig. 4.6). Nonetheless, nutritional factors are probably also important. Fisher (1941) suggests that the larvae are able to process a larger volume of decayed wood than undecayed wood and are consequently able to obtain nitrogen more rapidly.

Becker (1965) found that, in wood blocks rotted to varying extents by brown-rot fungi, dry wood termites such as *Kalotermes flavicollis* and *Heterotermes indicola* fed approximately twice as much on decayed

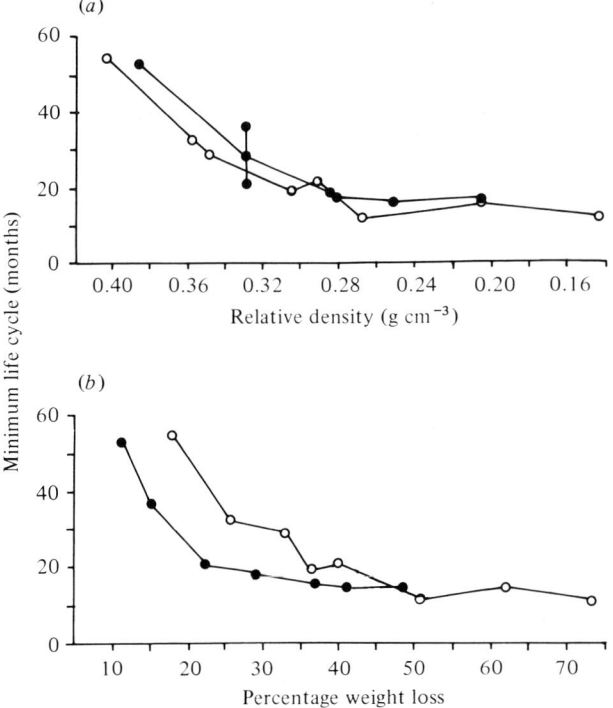

Fig. 4.6. Relationship between duration of life cycle of *Xestobium rufovillosum* and (*a*) RD (g cm^{-3}); (*b*) weight loss (% of original dry weight.) ○──○ oak sapwood decayed by *Phellinus cryptarum* (white rot); ●──● willow decayed by *Coriolus versicolor* (white rot). (After Fisher, 1941; Crown Copyright.)

blocks as on those without decay. The size of the termite colonies was more than five times larger in decayed than in undecayed wood at the end of an 18 month experiment. Decay of between about 5 and 20% weight loss was most favourable for the development of the termites. In this instance however, the highest extent of decay was the least favourable. The distinction here from earlier examples is that the termites have a very active cellulolytic fauna in their guts and extensive fungal breakdown of wood polysaccharides may thus be of no benefit.

The influence of animals on the development of the microbial community

The microflora of small branches after invasion by animals is quite different from that preceding invasion (Fig. 4.7). Prior to invasion basidiomycetes were readily isolated together with a number of characteristic sterile mycelia. These fungi were also occasionally isolated from branches which had been invaded by *T. flavolineata* and other animals,

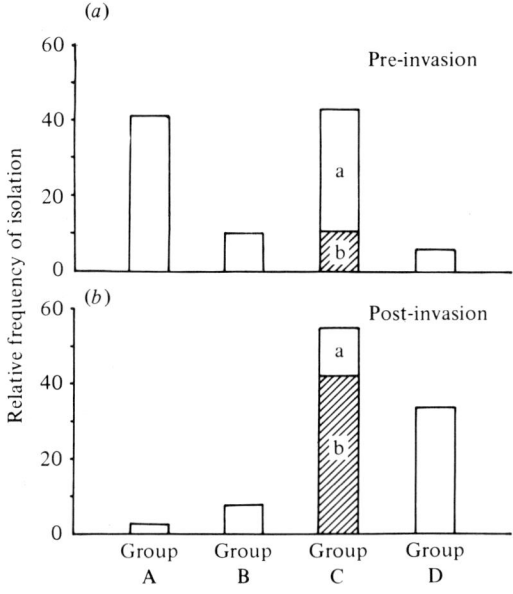

Fig. 4.7. The relative frequency of isolation of different groups of fungi from (*a*) uninvaded and (*b*) invaded beech branches taken from the litter layer at Blean Wood. Relative frequency of isolation expresses the percentage of plates on which fungi of each group were isolated (15 plates per branch; 32 branches). Group A is identified basidiomycetes and sterile hyaline hyphae; Group B is sterile dematiaceous hyphae; Group C is Moniliales (see text for distinction between *a* and *b*); Group D is Mucorales.

but with a very low frequency (Swift, 1982, Swift and Nesbitt, unpublished). In contrast members of the Mucorales were only rarely found in uninvaded wood but several species, such as *Mucor hiemalis*, *Mortierella rammaniana* and *M. isabellina*, were isolated with high frequency from invaded branches. Deuteromycetes were common in both types of branch but with a somewhat higher frequency after invasion. The species composition also changed markedly; species of *Humicola*, *Acremonium*, *Torula*, *Cladosporium* and a number of unidentified dematiaceous types were common in the uninvaded branches (Fig. 4.7a), whilst invaded branches were dominated by *Trichoderma* species (*alba*, *koningii* and *viride*), and *Penicillium* spp., and also yielded fungi such as *Sporotrichum* sp. and *Cordana pauciseptata* (Fig. 4.7b). Another feature of the invaded branches was that several yeasts (*Rhodotorula* spp.) and 'yeast-like' (*Trichosporon* spp.) organisms were isolated with high frequency. Bacterial numbers were also higher in invaded than in uninvaded wood (Swift, 1982; T. R. G. Gray, personal communication).

It is evident from these observations that there is present in invaded branches a group of fungi that is not present in uninvaded branches. These organisms, Mucorales, *Penicillium* and *Trichoderma* spp., are among those that grow most readily on simple media on agar plates and it is unlikely that if present they would not be isolated. However, the latter is not so certain for basidiomycetes. The evidence is nonetheless strong enough to support the hypothesis that invasion by *T. flavolineata* results in a significant shift in the inhabitant microflora from one dominated by a specialized 'wood-rotting' flora to one more characteristic of a generalized 'soil' microflora.

A number of interactive phenomena involving insects and microbes may help to explain this shift: destruction of the resident microflora by the feeding action of the animal; carriage and inoculation of spores and other propagules by the animal; creation of microenvironmental or nutritional conditions favouring growth of one group of organisms rather than another.

The effect of animal grazing on the resident microflora

The impact of animal grazing on microbial communities is discussed elsewhere in this volume (Anderson & Ineson, Chapter 3; Coleman *et al.*, Chapter 2) and no further elaboration of the mechanisms involved is required here. We have already advanced the suggestion that a wood-consuming animal such as *Tipula* may well be deriving

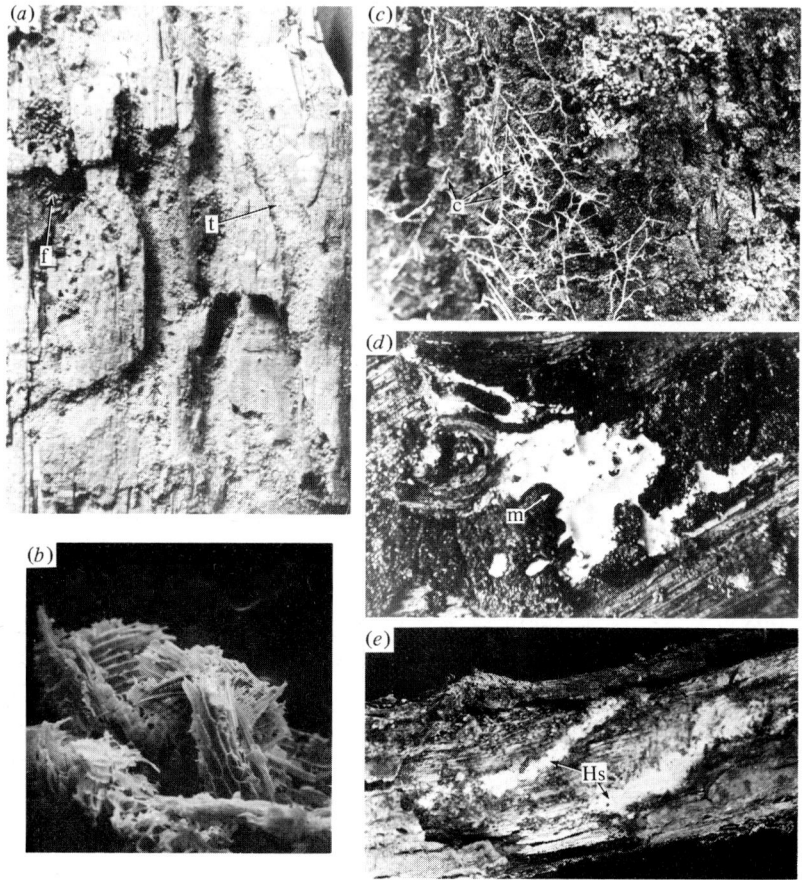

Fig. 4.8. (*a*) Section through a branch showing Tipulid tunnels (t) and frass (f). (*b*) SEM of freshly deposited faeces of *T. flavolineata* removed from a tunnel in beech wood. (*c*) and (*d*) Bark removed from suppressed oak trees revealing basidiomycete cords (c) and mycelium (m) growing over insect frass. (*e*) *Hyphoderma setigerum* (Hs) fruiting in longhorn beetle galleries.

a considerable proportion of its nutrition from fungal mycelium. Inevitably this will lower the biomass of viable basidiomycete mycelium in the branch. Feeding results in the creation of particulate 'frass' (Fig. 4.8*a*, *b*) within which fragments of mycelium are often clearly visible. Attempts to isolate basidiomycetes from *Tipula* frass even onto media selective for basidiomycetes were unsuccessful.

Animal carriage and inoculation of microbial propagules

All animals carry large numbers of microbial propagules. We are here concerned specifically with situations where the carriage of these cells or spores results in the inoculation of a resource with a microorganism which then colonizes and exploits the resource – in our case wood. Specific instances of the role of animals as vectors are considered by Dowding (Chapter 5) and Webber & Brasier (Chapter 10). Here a general outline of salient features will be given.

Carriage and dispersal of inoculum. Carriage can be divided into two types according to whether the propagules are passively (or accidentally) or actively ('deliberately') carried. In the former case the propagules may be found on any part of the animal's body. In the latter case carriage may be confined to specialized morphological structures.

The former occurs, to a greater or lesser extent, with all animals that come into contact with fungal sporophores or bacterial colonies. The types of animals involved are diverse (e.g. nematodes, slugs, earthworms, springtails, flies, beetles, wasps, mammals) as are the fungi involved (e.g. Zygomycotina, Ascomycotina and Basidiomycotina including Agaricales, Aphyllophorales and Phallales). The basidiocarps of white-rot fungi are used for food and breeding grounds by a large number of mycetophagous insects, one of the most common of these being *Ciid* beetles of which 74 species have been found inhabiting white-rot fungi (mainly Polyporaceae) in N. America (Lawrence 1967). Fly larvae are also common fruit-body feeders (Russell-Smith, 1979) and slugs are found feeding on larger agarics. In the latter case the spores appear to pass uninjured through the gut and germinate freely (Ingold, 1971). Interestingly, one type of spore produced by *Ganoderma* must pass through a fly larva gut, before it will readily germinate (Nuss, 1982). Nematodes carry spores of the fungus *Dilophospora* to the susceptible growing point of cereal shoots (Anatasoff, 1925) and it may be that carriage also occurs in trees. Boddy & Rayner (unpublished) have observed nematodes in the vessels of recently dead, decaying and living attached branches and twigs of oak (*Q. robur*). Exposed fungal stromata of *Cryptostroma corticale* (which causes sooty bark disease of Sycamore, *Acer pseudoplatanus*) are often extensively scored by teeth marks of squirrels which carry numerous conidia on claws and in the stomach and buccal cavity and may well act as vectors of the disease (Abbott, Bevercombe & Rayner, 1977). Other stromatal fungi, such as *Eutypa acharii*, are also extensively grazed

(Bevercombe, 1980; Bevan & Greenhalgh, 1983), and what appear to be squirrel teeth marks occur on the resupinate fruit-bodies of the basidiomycete *Vuilleminia comedens* on attached oak branches (Boddy, unpublished).

Active translocation of fungal propagules is a characteristic of a number of insects inhabiting or closely associated with wood. In all cases the animal derives some benefit from association with the fungus. The classic examples are Ambrosia beetles (Scolytidae; principally of the tribe *Xyleborini*) and other bark-feeding beetles, woodwasps (Siricidae, Xiphyriidae), and in tropical climes the higher termites (Macrotermitinae). Ambrosia beetles have been the subject of study for over 150 years, because of their economic importance. There are many species each of which is specifically associated with one (or sometimes more) species of Ascomycotina or Deuteromycotina. There is now increasing evidence that *Fusarium* is the major genus involved (Kok, 1979; Norris, 1979). Both adults and larvae inhabit tunnel systems formed mainly in sapwood although some also burrow into heartwood. The beetles carry the fungal propagules in small pockets in their external skeleton (termed mycangia or mycetangia) where a unicellular, or yeast-like, form of cellular multiplication occurs which differs from the mycelial growth form exhibited on culture media or on the walls of the brood galleries within wood. The mycetangia are generally present only in the females, and in different beetle species they differ in structure and location (Francke-Grosmann, 1967; Norris, 1979). The fungal propagules are deposited in the beetle tunnels when they invade the wood but how they reach the mycetangia prior to flight is not known.

Bark beetles (Scolytidae) feed on the phloem, a habitat rich in easily digestible food, but many of them are found in constant association with a specific fungal flora. Propagules of these fungi are transported by the adult beetles when they fly to breed in new trees. There are many records of association of bark beetles and stain fungi (reviewed by Mathiesen-Käarik, 1953; Mathre, 1964) most of which belong to the Ascomycete genus *Ceratocystis*. The *Ceratocystis* species are well adapted for endozoic and epizoic dissemination in that they have slimy spores which may adhere to the outside of the beetle or pass out of the gut undigested. Some are simply transmitted in the punctures of the integument and others in the oral pouch, whilst the beetle *Dendroctonus frontalis* possesses a capacious fungal tube at the anterior margin of the prothorax which resembles the complex mycetangia of certain Ambrosia beetles (Francke-Grosmann, 1967).

Woodwasps (Siricidae and Xiphydriidae) attack weakened trees or freshly cut logs, as do many Ambrosia beetles, but they differ from these in that the imago woodwasp is on the wing during the whole adult stage and the only wood-inhabiting stage is the larva. The adult female woodwasp deposits her eggs in moist wood through a long slender ovipositor at the base of which are a pair of pouches containing oidia. These are thus deposited in the wood during egg laying. Fungi associated with the woodwasp include the Basidiomycotina *Sterum sanguinolentum*, *S. chailletii*, *Amylostereum areolatum*, and *Daedalea unicolor* (Francke-Grosmann, 1967; Madden & Coutts, 1979).

Within the Macrotermitinae certain genera and species act as carriers of their symbiont between one colony generation and the next (Johnson, Thomas, Wood & Swift, 1981). Reproductive alates of *Macrotermes bellicosus* and five species of *Microtermes* carry conidia of *Termitomyces* in their guts at the time of flight from the nest at the onset of rains. These conidia act as inoculum for the establishment of the fungal comb in any new nest founded by successful alates. Interestingly, alates of *Macrotermes subhyalinus* and species of *Ancistrotermes* and *Odontotermes* do not carry conidia and the combs appear to be established from basidiospores collected by foraging workers.

From this brief review it is clear that the co-evolution of animals and microorganisms, utilizing wood as a habitat and a resource, has included the development of mechanisms which ensure the synchronization of colonization by both partners. In the case of mutualistic symbioses this is clearly an advantage in the maintenance of their relationship. In less developed associations the dispersal function may be more opportunistic but nonetheless important. For instance carriage of spores may be an important, or even essential, dispersal mechanism for many resupinate wood-rotting basidiomycetes forming sporophores on the underside of branches and logs on the forest floor where air currents are unlikely to be sufficient for aerial dispersal (Talbot, 1952).

Facilitation of colonization. In addition to the dispersive function animals may assist microbial colonization, during feeding and other activities, by breaching barriers to invasion of the resource.

In contrast to *Tipula* many other animals feed on wood at all stages of the decomposition process. For instance bark and Ambrosia beetles feed on healthy twigs and trunks, and on felled or fallen branches and logs and stumps in all stages of decay. Any activities which penetrate the tree's protective coating of bark result in a potential colonization court for decomposer microorganisms. The size and location of the court can range between the holes in the bark resulting from woodpecker activity, nibbling on twigs by beetles, stripping of bark from small branches and trunks by squirrels and deer, to the excavation of large tunnels which penetrate deep into the heartwood. Of the tunnelling animals those which create external access to internal parts are likely to be more important initially in producing colonization courts than those with closed burrows (e.g. *Tipula flavolineata* and woodwasps). The latter may however inoculate microorganisms at the time of oviposition, and at emergence large tunnels with external access may be provided.

Two well-known examples of microbial colonization through animal infection courts are provided by *Ceratocystis* (= *Ophiostoma*) *ulmi* (see Webber & Brasier, Chapter 10) and *Ceratocystis fagacearum* which are transmitted by beetles. Transmission of *C. fagacearum*, which causes oak wilt, is thought to be principally via sap-feeding beetles (Gibbs & French, 1980). *Pseudopityophthorus* spp. of beetle feed on new twig growth making wounds which are ideal infection courts for *C. fagacearum*, and infection occurs when contaminated beetles feed on healthy oaks (Rexrode & Jones, 1970).

The tunnels of Ambrosia beetles provide infection courts initially for the inoculated symbiotic microflora whose phase of development is brief. The time from tunnelling and egg laying to abandonment of brood galleries is sometimes as short as 30 days, after which Ambrosia fungi are rapidly replaced by other fungi.

A variety of beech tree diseases resulting in bark necrosis have been reported and of the many insects and fungi isolated, the most consistent association is between the felted beech scale insect (*Cryptococcus fagisuga*) and *Nectria coccinea* (Lonsdale, 1980*a, b*). The larvae of *C. fagisuga* establish themselves on the bark surface and the adults feed on sap by inserting their stylets into living parenchyma cells. When a number of insects colonize a small area, groups of parenchyma cells die, resulting in the development of minute cracks which allow entry of *N. coccinea* whose mycelium develops in the bark and cambium, killing infected tissues. Not only do the insects aid the entry of *N. coccinea* into bark but they also appear to predispose bark to infection by this fungus: the size of lesions was related to the severity of *C. fagisuga* infestation, an effect not connected with stylet-holes or fissures. It appears that *Nectria* infection depends upon a reduction in resistance of bark, which may be brought about by *C. fagisuga* infestation and probably also by abiotic factors such as drought (Lonsdale, 1980*b*).

Gibbs (1982) found that peck marks made by woodpeckers on healthy trees are quickly colonized by *Fusarium* and *Ceratocystis* spp. He also found gall midge larvae associated with the peck marks but these apparently had little to do with the fungi which were deleterious to the larvae. Boddy and Rayner (unpublished) have also occasionally found gall midge larvae and nematodes under the bark on oak branches in lesions extending for several metres. Associated with the area of animal occupation was a large area of stain in adjacent wood, presumably due to fungi which gained entry through the animal infection court.

In a contrasting environment Cragg and Swift (unpublished) have shown how the tunnelling activity of shipworms (Teredinidae) in mangrove roots greatly increase the potential for colonization of the wood by spores of wood-rotting Ascomycotina.

Although animal colonization courts may be implicated for a number of stain fungi (e.g. Dowding, 1973 and Chapter 5) there is less information concerning basidiomycetes. In attached oak branches, although animal activity may be considerable and certain basidiomycetes were consistently associated with animal activity (Boddy & Rayner, 1983), there was no indication that animal infection courts played a significant role in the establishment of these microorganisms (Boddy and Rayner, unpublished).

The examples so far have largely concerned standing trees and roots; animal courts may also be important sources of microbial colonization of wood decaying on the forest floor. Leach, Orr & Christensen (1937) studied the insects and fungi in logs of felled Norway spruce (*Picea abies*) and their data indicated that the presence of larvae of *Monochamus scutellatus* and *M. notatus* (Cerambicidae) markedly influenced the rate of decay of the heartwood. This was due both to their burrowing activities and to *Phlebia gigantea* which was often correlated with their presence. The beetles make large burrows through the sapwood and into the heartwood and these operate as colonization courts. The beetles do not appear directly to inoculate *P. gigantea* into the wood as no basidiomycetes were found, although over half of the beetle eggs tested showed surface contamination by fungi (predominantly *Sporotrichum* sp. and a variety of species of *Cladosporium*, *Fusarium*, *Macrosporium* and *Penicillium*). Similarly, female mouthparts, ovipositors and hatched larvae yielded a variety of microorganisms but no basidiomycetes. *P. gigantea* is therefore in all probability wind disseminated and gains access to the wood by chance deposition at the animal court. Once established the fungus advances rapidly in a longitudinal direction but only slowly radially and tangentially. In the presence of large numbers of insects and tunnels, however, invasion is rapid in both radial and tangential directions.

During later stages of decay abandoned animal tunnels may still provide sites of access for late colonizers such as soil and litter microorganisms. It is likely that their colonization is facilitated not by the specialized wood-boring insects but by secondarily invading arthropods such as Acari and Collembola which rapidly colonize bore holes in large numbers (Fager, 1968). This may well be the case for the *Tipula*-invaded branches for there is little evidence of any specific association of the larvae with the fungi characteristic of invaded branches.

Swift and Healey (unpublished) attempted to determine the relationship between the internal microflora of tipulid larvae and that developing in faeces. Larvae were dissected and the microflora of the different gut regions dispersed in 100 μl aliquots of 0.1% peptone water. Dilutions were prepared and plated out onto a general fungal

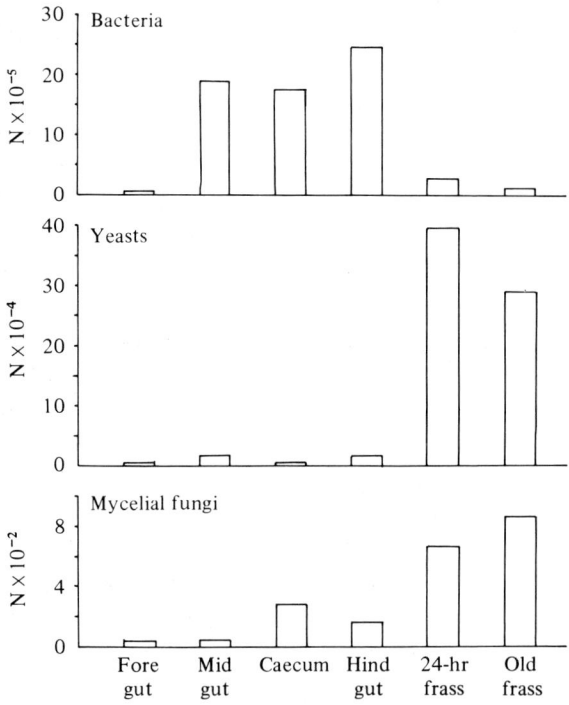

Fig. 4.9. Distribution of microbial propagules in the gut and faeces of *T. flavolineata* (see text for details).

medium (2% wt/vol. malt extract agar plus streptomycin) and a general bacterial medium (2% wt/vol. nutrient broth agar plus cycloheximide). Counts were made of bacteria, yeasts and mycelial fungi. Similar counts were made from 24 hour- and 7-day-old faeces (Fig. 4.9). The results indicated that conditions in the gut of the larva are inimical for fungi, although maintaining large numbers of bacteria. The latter do not persist in the faeces (the high numbers referred to earlier in this chapter appear to be characteristic of much older faeces), which are rapidly colonized by mycelial fungi, presumably by growth from wood and faeces in the same tunnels. The mycelial flora of the early faeces is rather different from that of the older. *Sporotrichum* sp. and *Cordana pauciseptata* formed 40% of the isolates from the former but were infrequent from the latter, where *Trichoderma* and *Penicillium* spp. were much more common. Species of Mucorales were common in both.

Influence of change in physical microenvironment

Besides the direct interactions discussed so far, animals can have considerable indirect effects on microorganisms (particularly fungi) inhabiting wood via their influence on the microclimate and location in which the wood decomposes.

Wood microclimate (temperature, moisture content and gaseous composition) influences decay rate by its effect upon the decomposer microorganisms (Boddy, 1983*a*, *b*, *c*). These may in turn be affected by the presence of animals mainly as a result of their tunnelling and chewing activities which result in loosening and loss of bark, and the formation of channels and cavities within the wood which may be full of frass and faeces. Loss of bark may affect the temperature and aeration of the underlying wood but also particularly its moisture content which is subject to more rapid and extreme fluctuations (Boddy, 1980, 1983*b*). This may be of relatively less significance on the forest floor where drying out is prevented, to some extent, by the buffering effect of leaf litter and vegetation. Approximately 10% of oak and 20% of beech branches at Blean were found to have lost most of their bark before fall (Boddy & Swift, 1983*a*) and Boddy & Thompson (1983) found that the bark on trunks of standing dead suppressed oak trees on three sites in the Forest of Dean was usually very loose, and often less than 80% remained. A variety of animals were found beneath the bark with bark beetle activity in most evidence. The large number of decorticated standing dead elms bears witness to their activity. Although invertebrates are one of the main causes of bark loss squirrels can also rapidly decorticate branches. Fungi may also loosen the bark through subcortical growth and animal activity in this region may often be as a result of prior fungal presence. Fig. 4.8c, d, e illustrates the subcortical activity of animals and fungi.

Many wood-decay basidiomycetes are relatively intolerant of low water potentials (Boddy, 1983a) and thus loss of bark may result in considerably reduced decay rates, and possibly a change in the fungal community structure. In oak two species of Basidiomycotina, *Schizopora paradoxa* and *Hyphoderma setigerum*, have been found predominantly in aerial wood without bark and where animal activity is plentiful (Boddy & Rayner, 1983). Neither of them are able to grow at low water potentials but *Schizopora paradoxa* is able to survive for a long time at low moisture contents and is presumably able to resume growth when conditions improve (Theden, 1961; Boddy, 1983a). Whether their predominance under these conditions is due to their ability to survive adverse moisture conditions or whether there is some other association is as yet unclear.

Tunnels can also affect the water relationship of decomposing wood by increasing the moisture holding capacity. Their main effect on microclimate is, however, probably in the alteration of the gaseous environment leading to better aeration owing to readier diffusion of gases through the tunnels connecting with the external environment. Boddy (1983c) has shown that reducing the length of 2 cm diameter branches by a factor of two often resulted in an increase in decay rate (as measured by CO_2 evolution) of over 40%, probably due to a decrease in the path length for the diffusion of gases. On the other hand tunnelling animals may lead to an increase in CO_2 (and decrease in O_2) as a result of metabolic activity. However, Paim & Beckel (1963) found no significant difference between the gaseous content of sealed drilled holes, containing *Orthosoma brunneum* (Cerambicidae) larvae in beech (*Fagus grandifolia*) branches 50–120 cm long, and that of uninvaded wood.

Animal activity often results in a change of location for the decomposition of wood which can in turn affect the decay rate. This may be on a small scale such as the expulsion of comminuted wood particles from tunnels. On a larger scale Boddy & Rayner (1983) found that quite considerable amounts of decayed wood were rapidly removed, by animal activity, from between heartwood wings of attached oak branches. On a larger scale still, animal tunnels of various kinds may be responsible for considerable weakening of branches and trunks, thus accelerating their fall to the forest floor and hence into the different microenvironmental conditions of that habitat.

Another physical feature that may be of considerable importance is that of the reduction of resource to a particulate form. The particles which fill the *Tipula* tunnels are typically about 44–70 μm by 40–120 μm. This implies the creation of a very large surface area for colonization. A number of authors have commented on how bacteria are much better adapted for the occupation of materials of this kind than are mycelial fungi (e.g. Parr, Parkinson & Norman, 1967), and Hanlon & Anderson (1980) have suggested that basidiomycetes in particular may be poorly suited to colonization of particulate resources. However, some microfungi with rapid mycelial extension and spore germination rates, a short inter-sporulation generation time and heavy spore production may be better adapted, and indeed constitute many of the fungi of the post-invasion microflora, particularly Mucorales, *Trichoderma* and *Penicillium* spp. Swift (1976) suggested that their pattern of life cycle could be designated as an r-type of strategy and contrasted with the k-type of the basidiomycetes. Experimental verification of this idea is still awaited.

Influence of change in the chemical environment
Passage of wood through the animal gut will result in a number of chemical changes which may influence the composition and activity of decomposers that subsequently utilize it. This does not appear to have been well documented for organisms in decaying wood, although Baker, Laidlaw & Smith (1970) have performed detailed analyses on chemical composition of wood and frass of *Anobium punctatum* feeding on undecayed Scots pine (*Pinus sylvestris*) sapwood. A number of features may be postulated which will exert a selective influence on the microflora. In particular, faecal material may be expected to be permeated by specific materials such as uric acid (this was not found in *A. punctatum* and it has been suggested that in some anobiid beetles yeast-like symbionts utilize uric acid) and may contain a relatively high content of chitin as a result of partial digestion of microbial cells.

So far much emphasis has been placed on the possible ways in which a shift from basidiomycetes to microfungi might be brought about by animal activities. It is now appropriate to mention a remarkable example in which this process is reversed. This again concerns the Macrotermitinae–*Termitomyces* symbiosis. When initially removed from the nest the fungal comb carries what is in effect a pure culture of

Termitomyces. Viable spores of other fungi, particularly species of *Aspergillus*, have been shown to be present (Thomas, 1981) but these fail to germinate within the nest. Once the comb is removed from the influence of the termites however, these fungi quickly overgrow the comb. The nature of the selective influence by the termites is uncertain but it may be due to secretion of an anti-fungal agent or to a physical 'grooming' effect (Thomas, 1981).

The influence of animal–microbial interactions on the process of decomposition

In the case-study described in this paper a specialized wood-boring animal, *Tipula flavolineata*, invades wood that is substantially decomposed by Basidiomycotina. The consequences of this in terms of changes in community structure have already been described. We now consider the effects of these changes on decomposition processes in terms of three main issues: rate of decomposition, particularly carbon flux; mineral nutrient dynamics; and formation of soil organic matter.

Effects on the rate of decomposition

Swift *et al.* (1976), using a negative-exponential model for decay, calculated that the mean rate of decomposition of small branches of all species in the litter layer at Meathop Wood was about 17% per year; Boddy & Swift (1983b) found a range between 15 and 36% per year for various categories of small beech branch at Blean. In neither case was any distinction made between invaded and uninvaded branches. However, in one small experiment at Meathop, 53 branches were taken from the forest floor, sub-sampled for RD, categorized as invaded or uninvaded and returned to their previous site (Healey and Swift, unpublished). After one year 40 of the branches were recovered and the RD remeasured. The weight losses are shown in Fig. 4.10 plotted against initial RD (IRD). The mean weight lost in one year for uninvaded branches (31.3 ± 3.8 (SE; n=22)) was significantly different ($P < 0.01$) from that of invaded branches (13.2 ± 1.8 (n=18)). Despite considerable variation about the regression line there is a highly significant linear relationship between annual weight loss and IRD. However, it should be noted that neither Swift *et al.* (1976) nor Boddy & Swift (1983b), with much larger samples, found any such significant relationship. Reference to Fig. 4.10 shows that the invaded branches are all in the lower RD category. The important issue is therefore whether

Table 4.5. *Regression parameters for the relationship between initial RD (X) and weight loss (Y) as a linear function* $Y = bX + a$

	a	b	F
Total branches	−2.20	0.082	53.97 ***
Uninvaded branches only	−6.49	0.093	15.93 ***
Invaded branches only	3.39	0.052	8.79 **

The F test for explained variance in regression is also shown; ***P = 0.001, **P = 0.01.

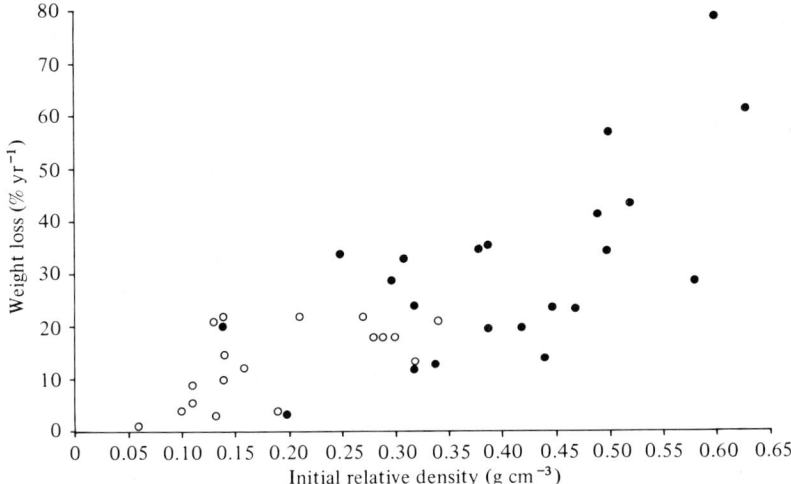

Fig. 4.10. The relationship between weight loss and initial state of decay (i.e. RD; g cm^{-3}) for wood that is uninvaded (●) or invaded (○) by animals. The regression equation for uninvaded, invaded and total wood is given in Table 4.5.

there is any difference in the relationship between weight loss and IRD for the two subsamples. As the regression equations in Table 4.5 indicate there is no evidence for this and we may conclude that any decrease in weight loss as decomposition proceeds is simply a justification of the negative exponential model (i.e. that weight loss is constant in proportion to the initial weight) rather than to any specific events, such as animal invasion, that occur in the later stages of decomposition.

This type of measurement of decomposition rate in field experiments gives at best an estimate of 'averaged' weight loss for a sizeable branch, within which may exist a variety of chemical and physical microenvironments, supporting decomposer communities active at widely differing levels. The influence of an individual animal may therefore not be apparent. Swift and Healey (unpublished) collected *Tipula* faeces from nine separate branches at Blean. In each case the faeces were taken from the 5 cm of tunnel immediately behind the animal and it was assumed that they had been recently deposited and were of even age. A subsample was taken from each and the moisture content measured. This ranged from 26 to 48% of the fresh weight (which was 31 to 57% of the moisture holding capacity (m.h.c.)). Samples of the faeces (2 g fresh weight) were weighed and adjusted to 60% m.h.c. with sterile distilled water. The samples were incubated at 15 °C in small pots for 4 months and then reweighed. The percentage mean loss in dry weight was 21.8 ± 0.46 (SE). Whilst no generalizations can be drawn from this single experiment under specific and artificial conditions, it can at least be concluded that in this instance the natural microflora is capable of decomposing *Tipula* faeces at a rate some four times higher than that typically encountered for intact branches in the field. Also, Ausmus (1977) reported an increase in the rate of decomposition of branches of *Liriodendron tulipifera* over a 5 year period, associated with narrowing of C : N ratio, and, by implication, animal invasion.

Effects on mineral nutrient dynamics

Earlier in this chapter the idea was advanced that by delaying attack on wood until it has been substantially decomposed by fungi, an animal may gain a nutritional advantage, in that mineral nutrient elements are effectively concentrated during decomposition due to release of carbon as CO_2. This possibility may be examined by considering data reported by Swift (1977) obtained from 51 branches collected from Meathop Wood, categorized as either invaded or uninvaded, and analysed for five major mineral nutrient element contents (Fig. 4.11). There is some evidence of consistently different patterns of nutrient retention and release between invaded and uninvaded branches. The pattern is most distinct in the case of nitrogen and calcium in which the phase of basidiomycete attack is usually characterized by a gain in concentration over that expected, or occasionally by close agreement with the predicted elemental concentration. The predominance of the former case is shown by the data in Table 4.6. Here the mean elemental concentrations have been expressed in terms of unit volume (to compensate for loss in weight). Thus a difference between means for

Animal–microbial interactions in decomposition 121

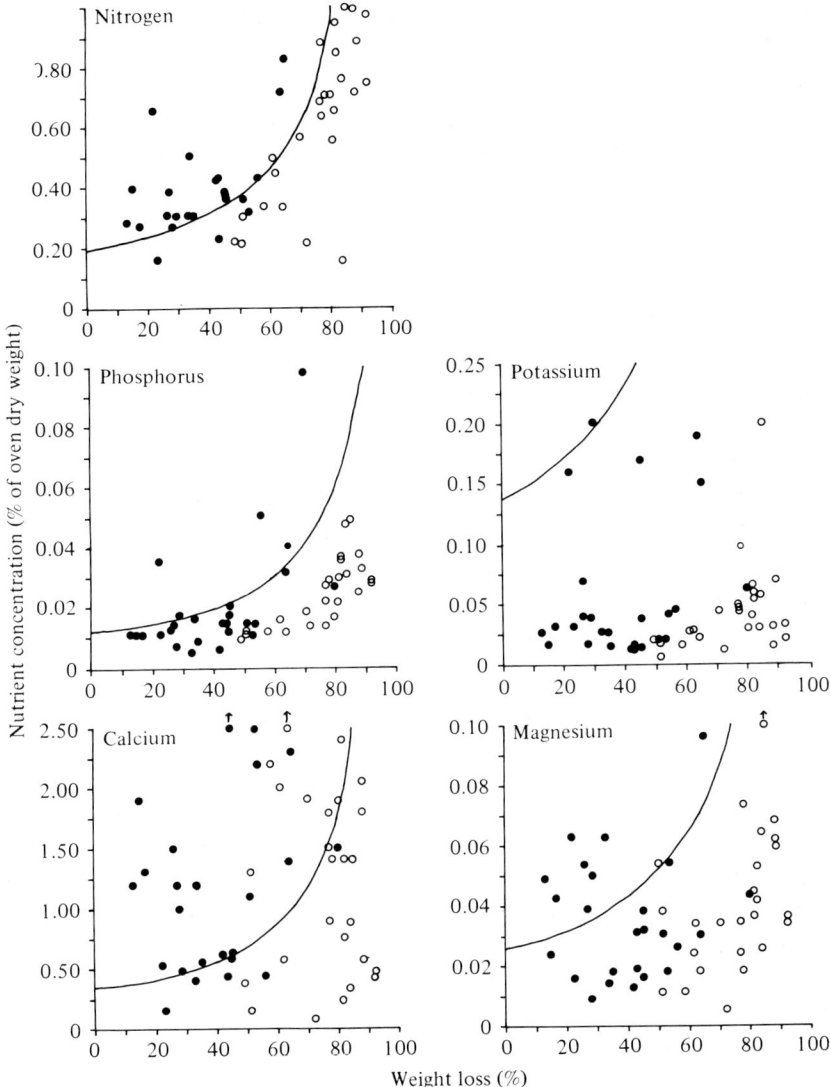

Fig. 4.11. The relationship between mineral nutrient concentration (as % of oven dry weight) and weight loss (% of original dry weight) of uninvaded (●) and invaded (○) branch wood collected from the litter layer. The line on each graph indicates the predicted change in elemental concentration as decomposition proceeds, assuming no net gain or loss. (From Swift, 1977.)

Table 4.6. *Nutrient content ($\mu g\ cm^{-3}$) of small branches from Meathop Wood*

	n	N	P	K	Ca	Mg
Living wood	20	1057 ± 28	70 ± 4	767 ± 63	1935 ± 132	146 ± 6
Uninvaded wood	26	1304 ± 87*	54 ± 6	141 ± 29*	3871 ± 499*	118 ± 14
Invaded wood	25	710 ± 47*	28 ± 2*	71 ± 17*	1777 ± 301*	58 ± 10*

Values are mean ± SE of mean.
*Indicates a significant difference ($P < 0.05$; ANOVA) between the mean value and the previous mean value in the same column.
From Swift, 1977.

living and fungally decayed wood indicates that there is a net gain, from external sources, of nitrogen and calcium (and a net loss of potassium). After animal invasion the situation is reversed with all five elemental concentrations being significantly lower than both undecomposed and fungally decayed wood. Reference to Fig. 4.11 indicates that the position is more complex than this; many of the fungally decayed branches show net loss, whilst others show net gain. Nonetheless, deviations from the expected are more marked in invaded branches. For phosphorus and potassium the majority of decayed branches show net loss but again this is greater from invaded branches than uninvaded with fungal decay.

The overall conclusion is that the intervention of animals into the process of wood decomposition results in the initiation or acceleration of the process of release of mineral nutrient elements from branches on the forest floor. The mechanisms involved are unclear but may include losses due to animal emigration, fungal sporulation and translocation, or increased leachability. With regard to the first of these, Healey (unpublished) found the mean weight of emergent craneflies, from beech branches at Blean, to be 45.1 ± 3.6 (SE, n = 5) mg for males and $122 = 8.4$ (SE, n = 6) mg for females. This would be equivalent to the complete removal of N from 4.2 or 11.6 cm^3 of wood respectively. Thus, emigration of *T. flavolineata* or other animals may contribute significantly to nutrient depletion.

Formation of fruit-bodies of Basidiomycotina can be important in the export of nutrients (Dowding, 1976; Swift, 1982). However, they are unusual on small branches and there is no evidence that they are more frequent on invaded than uninvaded branches. Asexual spores may be carried out on the bodies of emergent animals as a possible additional source of export. Translocation of nutrients by cords of the dry rot fungus *Serpula lacrimans* has been demonstrated (Watkinson, 1971). Cords of wood-rotting Basidiomycotina connect between decomposing branches, twigs, stumps, litter etc. on the forest floor (Thompson & Rayner, 1982, 1983; Thompson & Boddy, 1983) and are almost certainly a feature of the net gain characteristic of some nutrients during the fungal phase. There is correspondingly no reason why this should not be reversed as C : nutrient ratios narrow but no specific animal stimulus springs to mind. Increased leachability is a distinct possibility for comminuted wood but there is likely to be a complex interaction with the extent of bark cover; this feature awaits further investigation.

Effects on humification

The mechanisms of synthesis of humus molecules are uncertainly understood. Modern evidence suggests that there may be a variety of pathways leading from a high diversity of precursors of both plant and microbial origin. It seems well established, however, that one source of the aromatic core of humic acid, fulvic acid and humin is the modified products of lignin degradation, and as such wood is potentially a good source of humus formation. The visual character of wood at the terminal stages of decomposition supports this suggestion; at this stage intact bark often contains a dark mass of amorphous particulate matter inhabited by a wide range of soil organisms including lumbricids and millipedes as well as microarthropods and nematodes.

It is tempting to hypothesize that wood decomposition may be the major pathway of humus synthesis in woodlands and that the interaction of animals with microbes promotes this. There is, to our knowledge, no evidence to support this quantitative assertion (although Allison (1973) asserts that wood is a poor source for humus), but there are some indirect pointers to the potential importance of animal–microbe interactions in humus formation.

Fungi are now generally acknowledged to have a central role in humus synthesis but the involvement of Basidiomycotina seems quite different from that of other fungi (Haider, Martin & Filip, 1975). The latter have been identified as the main agents of synthesis for the pigmented aromatic molecules which appear to contribute to 'normal' humus, whilst the former tend to produce a bleached form of humus ('white-rot humus', Hintikka, 1970) which differs in a number of respects from 'normal' humus: it is more acid, has a higher nutrient content and is more readily degradable. This difference would be consistent with two distinct types of residue resulting from the uninvaded and invaded categories of wood. In the latter case the introduction of imperfect fungi and other components of the soil flora is the factor diverging the pathway from that of 'white-rot' humus to that of 'normal' humus.

That animals may themselves play a part in humification, both because of their comminutive action and because of chemical effects on microbial residues during passage through the gut, has been postulated (e.g. see Kononova, 1966; Babel, 1975). Striganova (1967) has shown, for instance, increased levels of humic and fulvic acids in wood fragments after passage through the gut of *Dendrobaena rubida*.

Conclusion

In the introduction to this paper we listed some of the main factors which determine the natural resistance of wood. Among these features, some, such as the presence of allelopathic compounds, are specific to certain types of wood, and often to particular tissues within distinct species of trees. These factors will thus have a selective effect on decomposers at a specific level and probably account for many of the observed 'resource-specific' associations between fungi and plant litter (Swift, 1976). Other features, notably extensive lignification and low mineral nutrient content, are general to all wood. There appear to have been two distinct adaptations in response to this selective pressure. The first strategy is that of exploiting only the living cells within wood. This is characteristic of the staining fungi and of certain animals such as bark beetles; in the Ambrosia beetles it comprises an associative strategy. The basis of this strategy is to grow in cells that are non-lignified, have high contents of nitrogen and other elements (Merrill & Cowling, 1966) and, in many cases, of storage sugars and starch. In a sense this is a strategy which avoids the problems posed by wood as a resource – and a strategy that results in only very limited exploitation.

Full exploitation of wood is confined to organisms which can either metabolize lignin or metabolize cell wall polysaccharides in the presence of lignin, and also possess mechanisms for the accumulation of nutrient elements from the surrounding environment. The capacity to metabolize lignin appears to be almost totally confined to the white-rot Basidiomycotina and Xylariacae; a limited range of other organisms can apparently modify lignin structure sufficiently to utilize a limited amount of its carbon, e.g. some soft-rot fungi and some bacteria from the intestine of invertebrates (Ander & Eriksson, 1978; Kirk & Fenn, 1982), but only with a very low comparative efficiency. The capacity to metabolize cell wall polysaccharides despite the masking effect of lignin is again most efficiently displayed by Basidiomycotina, in this case the 'brown-rots'. This potential is also displayed by a number of the protozoal and bacterial gut symbionts of wood-boring beetle and dipteran larvae and termites. In these cases the comminution of the wood by the animal probably also has an effect in diminishing the inhibitory effect of lignin on polysaccharase activity.

Thus it is among the Basidiomycotina particularly that we may see the initial evolutionary response to the presence of wood as a potential food resource. Corner (1964) has pointed out the close association between the appearance of these essentially terrestrial heterotrophs and the emergence of the tree-form within the land flora. Lignification was of course an essential part of the solution to the problem of 'tree-making' (Raven, 1977; Lewis, 1980). It has been suggested recently by a number of authors (Malloch, Pirozynski & Raven, 1980; Watling, 1982) that the evolution of a land flora involved an essential symbiotic association with biotrophic fungi. The establishment of basidiomycetes as the mycotrophic partner in a significant component of the tree flora may have preceded their saprotrophic exploitation of a newly-created niche, that of wood decay. The accumulation of undecayed plant debris during the carboniferous period may pay testament to the evolutionary delay in the occupation of this niche.

The establishment of basidiomycetes as exploiters of wood provides, in its turn, the potential for the utilization of this material by a wider food web of organisms displaying different adaptations. At this stage the selective premium is on those organisms which predate (graze) Basidiomycotina, or utilize their enzymatic capacities (see Martin, Chapter 6) or biomass, or enter into mutualistic association with them. Thus we see, in the dual pathway of exploitation of small branch wood illustrated in Figs. 4.2 and 4.3, possible examples of the historic solution of the problem of wood-exploitation by heterotrophs. In the small proportion of the community in which exploitation is confined to basidiomycetes lies the essential solution; in the large proportion of branches which support a much more diverse food web lies the secondary evolutionary development – the adaptation of arthropods to the exploitation of basidiomycete fungi.

References

Abbott, R. J., Bevercombe, G. P. & Rayner, A. D. M. (1977). Sooty bark disease of sycamore and the grey squirrel. *Transactions of the British Mycological Society*, **69**, 507–8.

Allen, S. E., Grimshaw, H. M., Parkinson, J. A. & Quarmby, C. (1974). *Chemical analysis of ecological materials*. Oxford: Blackwell Scientific Publications.

Allison, F. E. (1973). *Soil organic matter and its role in crop production*. Amsterdam, London & New York: Elsevier Scientific Publishing Co.

Amburgey, T. L. & Smythe, R. V. (1977a). Factors influencing termite feeding on brown-rotted wood. *Sociobiology*, **3**, 3–12.

Amburgey, T. L. & Smythe, R. V. (1977b). Factors influencing the production of termite trail-following and arrestant stimuli by isolates of *Gloeophyllum trabeum*. *Sociobiology*, **3**, 13–25.

Anatasoff, A. A. (1925). The *Dilophospora* disease of cereals. *Phytopathology*, **15**, 11–40.

Ander, P. & Eriksson, K. E. (1978). Lignin degradation and utilisation by microorganisms. *Progress in Industrial Microbiology*, **14**, 1–58.

Ausmus, B. A. (1977). Regulation of wood decomposition rates by arthropod and annelid

References

populations. In *Soil organisms as components of ecosystems*, ed. U. Lohm and T. Persson, pp. 180–92. Ecological Bulletin 25. Stockholm: Swedish Natural Science Research Council.

Babel, U. (1975). Micromorphology of soil organic matter. In: *Soil Components. Vol. 1, Organic Components*, ed. J. E. Gieseking, pp. 369–473. New York: Springer Verlag.

Baker, J. M., Laidlaw, R. A. & Smith, G. A. (1970). Wood breakdown and Nitrogen utilization by *Anobium punctatum*. Deg. feeding on Scots pine sapwood. *Holzforschung*, **24**, 45–53.

Becker, G. (1965). Versuche über den Einfluss von Braunfaulepilzen auf Wahl und Ausnutzung der Holznährung durch Termiten. *Material und Organismen*, **1**, 95–156.

Becker, G. (1969). Über holzzerstörender Insekten in Korea. *Zeitschrift für angewandte Entomologie*, **64**, 152–61.

Becker, G. (1975). *Coptotermes* in the heartwood of living trees in Central and West Africa. *Material und Organismen*, **10**, 149–54.

Becker, G. (1976). Termites and fungi. *Material und Organismen*, **3**, 465–78.

Becker, G. & Lenz, M. (1972). Stimulierung der Frab- und Galeriebau-Tätigkeit von Termiten durch den Geruch des Mycels holzzerstörender pilze. *Zeitschrift für angewandte Zoologie*, **59**, 269–83.

Bevan, R. J. & Greenhalgh, G. N. (1983). Pyrenomycetes and loculoascomycetes on sycamore wood and bark in the northwest of England. *Transactions of the British Mycological Society*, **80**, 83–9.

Bevercombe, G. P. (1980). *Diseases affecting sycamore bark*. Unpublished Ph.D. thesis, University of Exeter.

Boddy, L. (1980). *Decomposition ecology of fallen branch-wood*. Unpublished Ph. D. thesis, University of London.

Boddy, L. (1983*a*). The effect of temperature and water potential on the growth rate of wood-rotting basidiomycetes. *Transactions of the British Mycological Society*, **80**, 141–9.

Boddy, L. (1983*b*). Microclimate and moisture dynamics of wood decomposing in terrestrial ecosystems. *Soil Biology and Biochemistry*, **15**, 149–57.

Boddy, L. (1983*c*). Carbon dioxide release from decomposing wood: effect of water content and temperature. *Soil Biology and Biochemistry*, **15**, 501–10.

Boddy, L., Coates, D. & Rayner, A. D. M. (1983). Attraction of fungus gnats to zones of intraspecific antagonism on agar plates. *Transactions of the British Mycological Society*, **81**, 149–51.

Boddy, L. & Rayner, A. D. M. (1981). Fungal communities and formation of heartwood wings in attached oak branches undergoing decay. *Annals of Botany*, **47**, 271–4.

Boddy, L. & Rayner, A. D. M. (1983). Ecological roles of basidiomycetes forming decay communities in attached oak branches. *New Phytologist*, **93**, 77–88.

Boddy, L. & Swift, M. J. (1983*a*). Wood decomposition in an abandoned beech and oak coppiced woodland in south-east England. 1. Patterns of wood-litter fall. *Holarctic Ecology*, **6**, 320–32.

Boddy, L. & Swift, M. J. (1983*b*). Wood decomposition in an abandoned beech and oak coppiced woodland in south-east England. III. Decay rate and turnover time of twigs and branches. *Holarctic Ecology* (in press).

Boddy, L. and Thompson, W. (1983). Decomposition of suppressed oak trees in even-aged plantations. 1. Stand characteristics and decay of aerial parts. *New Phytologist*, **93**, 261–76.

Bray, J. R. & Gorham, G. (1964). Litter production in forests of the world. *Advances in Ecological Research*, **2**, 101–57.

Collins, N. M. (1981). The role of termites in the decomposition of wood and leaf litter in the Southern Guinea Savanna of Nigeria. *Oecologia*, **51**, 389–99.

Conner, R. N., Miller, O. K., & Adkisson, C. S. (1976). Woodpecker dependence on trees infected by fungal heart rots. *Wilson Bulletin*, **88**, 575–81.

Corner, E. J. H. (1964). *Life of Plants*. London: Weidenfeld & Nicolson.

Dowding, P. (1973). Effects of felling time and insecticide treatment on the interrelations of fungi and arthropods in pine logs. *Oikos*, **24**, 422–9.

Dowding, P. (1976). Allocation of resources, nutrient uptake and utilisation by decomposer organisms. In *The role of aquatic and terrestrial organisms in decomposition processes*, ed. J. M. Anderson & A. Macfadyen, pp. 169–83. Oxford: Blackwell Scientific Publications.

Elton, C. S. (1966). *The pattern of animal communities*. London: Methuen.

Esenther, G. R., Allen, T. C., Casida, J. E. & Schenefelt, R. D. (1961). Termite attractant from fungus-infected wood. *Science*, **134**, 50.

Fager, E. W. (1968). The community of invertebrates in decaying oak. *Journal of Animal Ecology*, **37**, 121–42.

Fisher, R. C. (1940). Studies of the biology of the death-watch beetle (*Xestobuim rufovillosum* de Geer). III. Fungal decay in timber in relation to the occurrence and rate of development of the insect. *Annals of Applied Biology*, **27**, 545–57.

Fisher, R. C. (1941). Studies of the biology of the death-watch beetle (*Xestobium rufovillosum* de Geer). IV. The effect of type and extent of fungal decay in timber upon the rate of development of the insect. *Annals of Applied Biology*, **28**, 244–60.

Francke-Grosmann, H. (1967). Ectosymbiosis in wood-inhabiting insects. In *Symbiosis vol. II. Association of Invertebrates, Birds, Ruminants and other Biota*, ed. S. M. Henry, pp. 141–205. New York and London: Academic Press.

Gibbs, J. N. (1982). An oak canker caused by a gall midge. *Forestry*, **55**, 69–78.

Gibbs, J. N. and French, D. W. (1980). The transmission of oak wilt. USDA Forest service general technical report NC–185.

Haider, K., Martin, J. P. & Filip, Z. (1975). Humus biochemistry. In *Soil biochemistry*, ed. E. A. Paul & A. D. McLaren, vol. 4, pp. 195–244. New York: Marcel Dekker.

Hanlon, R. D. G. & Anderson, J. M. (1980). Influence of macroarthropod feeding activities on microflora in decomposing oak leaves. *Soil Biology and Biochemistry*, **12**, 255–61.

Healey, I. N. & Swift, M. J. (1971). Aspects of the accumulation and decomposition of wood in the litter of a coppiced beech–oak woodland. In *Organisms du Sol et production primaire*, IV Colloquium Pedobiologiae, Dijon, 14–19 September 1970, pp. 417–30. Paris: Institut Nationale de la Recherche Agronomique.

Hickin, N. E. (1963). *The insect factor in wood decay*. London: Hutchinson.

Hintikka, V. (1970). Studies on white-rot humus formed by higher fungi in forest soils. *Communications Instituti Forestalis Fenniae*, **69**, 1–68.

Ingold, C. T. (1971). *Fungal spores: their liberation and dispersal*. Oxford: Clarendon Press.

Johnson, R. A., Thomas, R. J., Wood, T. G. & Swift, M. J. (1981). The inoculation of the fungus comb in newly founded colonies of some species of the Macrotermitinae (Isoptera) from Nigeria. *Journal of Natural History*, **15**, 751–6.

Kegg, J. D. (1970). New Jersey forest pest reporter. Tranton, New Jersey: New Jersey Department of Agriculture.

Kirk, T. K. & Fenn, P. (1982). Formation and action of the ligninolytic system in basidiomycetes. In *Decomposer basidiomycetes*, ed. J. C. Frankland, J. N. Hedger & M. J. Swift, pp. 67–90. Cambridge University Press.

References

Kok, L. T. (1979). Lipids of Ambrosia fungi and the life of mutualistic beetles. In *Insect–fungus symbiosis*, ed. L. R. Batra. Proceedings of Second International Mycological Congress. New York: John Wiley & Sons.

Kononova, M. M. (1966). *Soil Organic Matter.* 2nd Edn. Oxford: Pergamon Press.

Kovoor, J. (1964). Modifications chimiques provoquées par un termitide (*Microcerotermes edentatus* Was.) dans du bois de peuplier sain ou partiellement dégrade par des champignons. *Bulletin biologique de la France et de la Belgique*, **98**, 491–510.

Lawrence, J. F. (1967). Host preference in Ciid beetles (Coleoptera: Ciidae) inhabiting the fruiting bodies of Basidiomycetes in North America. *Bulletin of Museum of Comparative Zoology Harvard University*, **145**, 163–212.

Leach, J. G., Orr, L. W. & Christensen, C. (1937). The inter-relationships of bark beetles and blue-staining fungi in felled Norway Pine timber. *Journal of Agricultural Research*, **49**, 315–42.

Lewis, D. H. (1980). Boron, lignification and the origin of vascular plants – a unified hypothesis. *New Phytologist*, **84**, 209–99.

Lonsdale, D. (1980*a*). *Nectria* infection of beech bark in relation to infestation by *Cryptococcus fagisuga* Lind. *European Journal of Forest Pathology*, **10**, 161–8.

Lonsdale, D. (1980*b*). *Nectria coccinea* infection of beech bark: variations in disease in relation to predisposing factors. *Annales des Sciences Forestieres*, **37**, 307–17.

Madden, J. L. & Coutts, M. P. (1979). The role of fungi in the biology and ecology of woodwasps (Hymenoptera: Siricidae). In *Insect–Fungus Symbiosis*, ed. L. R. Batra. Proceedings of Second International Mycological Congress. New York: John Wiley and Sons.

Malloch, D. W., Pirozynski, K. A. & Raven, P. H. (1980). Ecological and evolutionary significance of mycorrhizal symbioses in Vascular plants. *Proceedings of the National Academy of Sciences, USA*, **77**, 2113–18.

Mathiesen-Käarik, A. (1953). Eine Übersicht über die gewöhnlichen mit Borkenkäfern associerten Blaupilze in Schweden. *Meddelanden från Statens Skogsförskningsinst*, **43**, 1–74.

Mathre, D. E. (1964). Survey of Ceratocystis spp. associated with bark beetles in California. *Contributions. Boyce Thompson Institute for Plant Research*, **22**, 363–88.

Matsuo, H. & Nishimoto, K. (1974). Response of termite *Coptotermes formosanus* (Shiraki) to extract fractions from fungus-infected wood and fungus mycelium. *Material und Organismen*, **9**, 225–38.

Merrill, W. & Cowling, E. B. (1966). Role of nitrogen in wood deterioration: amounts and distribution of nitrogen in tree stems. *Canadian Journal of Botany*, **44**, 1555–80.

Norris, D. M. (1979). The mutualistic fungi of *Xyleborini* beetles. In *Insect–Fungus Symbiosis*, ed. L. R. Batra. Proceedings of Second International Mycological Congress, New York: John Wiley and Sons.

Nuss, I. (1982). Die Bedeutung der Proterosporen: Schlußfolgerungen aus Untersuchungen an *Ganoderma* (Basidiomycetes). *Plant Systematics and Evolution*, **141**, 53–79.

Paim, U. & Beckel, W. E. (1963). Seasonal oxygen and carbon dioxide content of decaying wood as a component of the microenvironment of *Orthosoma brunneum* (Foster) (Coleoptera: Cerambycidae). *Canadian Journal of Zoology*, **41**, 1133–47.

Park, D. (1976). Carbon and nitrogen levels as factors influencing fungal decomposers. In *The role of aquatic and terrestrial organisms in decomposition processes*, ed.

J. M. Anderson & A. Macfadyen, pp. 41–59. Oxford: Blackwell Scientific Publications.

Parr, J. F., Parkinson, D. & Norman, A. G. (1967). Growth and activity of soil micro-organisms in glass micro-beads. II. Oxygen uptake and direct observations. *Soil Science*, **103**, 303–10.

Raven, J. A. (1977). The evolution of vascular plants in relation to supracellular transport processes. *Advances in Botanical Research*, **5**, 153–82.

Rexrode, C. O. (1971). Insect damage to oaks. In *Oak Symposium Proceedings*, North Eastern Forest Service Experimental Station, pp. 129–34. Upper Derby, Pennsylvania: Department of Agriculture Forest service.

Rexrode, C. O. & Jones, T. W. (1970). Oak bark beetles – important vectors of oak wilt. *Journal of Forestry*, **68**, 294–7.

Ritter, F. J. & Coenen-Saraber, C. M. A. (1969). Food attractants and a pheromone as trail-following substances for the Saintoge termite. *Experimental and Applied Entomology*, **12**, 611–22.

Rodin, L. E. & Basilevic, N. I. (1967). *Production and mineral cycling in terrestrial vegetation*. Edinburgh and London: Oliver & Boyd.

Russell-Smith, A. (1979). A study of fungus flies (Diptera: Mycetophilidae) in beech woodland. *Ecological Entomology*, **4**, 355–64.

Schultze-Dewitz, G. & Unger, W. (1972). Das Verhalten von *Reticulitermes lucifugus var. santonensis* de Feyteaud gegenüber weibfaulem. *Beiträge zur Entomologie*, **22**, 487–90.

Striganova, B. R. (1967). Study of the role of wood lice and earth worms in the humification of decomposing wood. *Soviet Soil Science*, 1108–12.

Swift, M. J. (1976). Species diversity and the structure of microbial communities. In *The role of aquatic and terrestrial organisms in decomposition processes*, ed. J. M. Anderson & A. Macfadyen, pp. 185–222. Oxford: Blackwell Scientific Publications.

Swift, M. J. (1977). The role of fungi and animals in the immobilisation and release of nutrient elements from decomposing branch-wood. In *Soil Organisms as Components of Ecosystems*, ed. U. Lohm & T. Persson, pp. 193–202. Ecological Bulletin 25. Stockholm: Swedish Natural Science Research Council.

Swift, M. J. (1978). Growth of *Stereum hirsutum* during the long-term decomposition of oak branch-wood. *Soil Biology and Biochemistry*, **10**, 335–7.

Swift, M. J. (1982). Basidiomycetes as components of forest ecosystems. In *Decomposer Basidiomycetes*, ed. J. C. Frankland, J. N. Hedger & M. J. Swift, pp. 307–37. Cambridge University Press.

Swift, M. J., Boddy, L. & Healey, I. N. (1983). Wood decomposition in an abandoned beech and oak coppiced woodland in south-east England. II. The standing crop of wood on the forest floor with particular reference to *Tipula flavolineata* and other animals. *Holarctic Ecology* (in press).

Swift, M. J., Heal, O. W. & Anderson, J. M. (1979). *Decomposition in terrestrial ecosystems*. Oxford: Blackwell Scientific Publications.

Swift, M. J., Healey, I. N., Hibberd, J. K., Sykes, J. M., Bampoe, V. & Nesbitt, M. E. (1976). The decomposition of branch-wood in the canopy and floor of a mixed deciduous woodland. *Oecologia*, **26**, 139–49.

Talbot, P. H. B. (1952). Dispersal of fungus spores by small animals inhabiting wood and bark. *Transactions of the British Mycological Society*, **35**, 123–8.

Tattar, T. A., Shortle, D. & Rich, A. E. (1971). Sequence of microorganisms and changes in constituents associated with discoloration and decay of sugar maples infected with *Fomes connatus*. *Phytopathology*, **61**, 556–8.

Theden, G. (1961). Untersuchungen über die fanigkeit holzzerstorender pilze zur trockenstarre. *Angewandte Botanik*, **35**, 131–45.

Thomas, R. J. (1981). Ecological studies on the symbiosis of Termitomyces Heim with Nigerian Macrotermitinae. Unpublished Ph.D. thesis, University of London.

Thompson, W. & Boddy, L. (1983). Decomposition of suppressed oak trees in even-aged plantations. II. Colonisation of tree roots by cord and rhizomorph producing basidiomycete fungi. *New Phytologist*, **93**, 277–91.

Thompson, W. & Rayner, A. D. M. (1982). Structure and development of mycelial cord systems of *Phanerochaete laevis* in soil. *Transactions of the British Mycological Society*, **78**, 193–200.

Thompson, W. & Rayner, A. D. M. (1983). Extent, development and function of mycelial cord systems in soil. *Transactions of the British Mycological Society*, **81**, 333–45.

Watkinson, S. C. (1971). The mechanism of mycelial strand induction in *Serpula lacrimans:* A possible effect of nutrient distribution. *New Phytologist*, **70**, 1079–88.

Watling, R. (1982). Taxonomic status and ecological identity in the basidiomycetes. In *Decomposer basidiomycetes*, ed. J. C. Frankland, J. N. Hedger & M. J. Swift, pp. 1–32, Cambridge University Press.

Williams, R. M. C. (1965). Infestation of *Pinus caribaea* by the termite *Coptotermes niger* Snyder. *Twelfth International Congress of Entomology, London 1964*, pp. 675–6.

Wood, T. G. (1976). The role of termites (Isoptera) in decomposition processes. In *The role of terrestrial and aquatic organisms in decomposition processes*, ed. J. M. Anderson & A. Macfadyen, pp. 145–68. Oxford: Blackwell Scientific Publications.

5
The evolution of insect–fungus relationships in the primary invasion of forest timber

P. DOWDING
Botany School, Trinity College, Dublin 2, Eire

Key words: wood decomposition; primary colonization; pathogenesis; saprophagy; bark beetles; *Ceratocystis*

Introduction

Trees, and in particular the xylem of their trunk, branches and roots, form a long lasting, bulky resource for both fungi and insects in woodland ecosystems. As man has taken more and more wood for use in building, transport for paper and for fuel, insects and fungi have become increasingly important as his competitors for a dwindling and more costly resource. Where forestry activities have been restricted to selective felling and removal of timber, man has in fact removed potential habitat from the forest and may have *reduced* pest problems. The removed timber has then been used to recreate a rather specialized habitat in buildings, which has led to the selection of a cosmopolitan group of fungi and insects which attack structural timber and which are only rarely seen in natural forests. In Europe, in particular during this century, large areas of even-age single species communities have been planted, which have created further specialized habitats in which a few of the natural fauna and microflora have increased in number. Some of these organisms have become pests, either by an accelerated process of selection for aggressiveness, or by removal of natural control agents (Bevan, 1974) in such a uniform habitat.

Research effort has been concentrated towards the identification and control of pests in forest, sawmill and in buildings, without much regard to the natural context from which these organisms originally arose. In some instances, e.g. Dutch elm disease and sapstain, control measures were applied before the ecology and physiology of the causal organisms were properly understood, and with hindsight it is not surprising that the control achieved was at best partial and temporary.

Fragmentation of research effort has been brought about by the poor dissemination of information across the Atlantic, perhaps because of different names for the same species, as well as for different species, and because of different languages being used in the earlier work. Another communication problem which has hindered research effort is that existing between entomologists, mycologists and forest pathologists.

This review is meant to pull together the information about primary colonists (that is those dominant at early stages of community development) of xylem in standing trees and in freshly cut logs, in an attempt to illuminate a common pattern. This pattern is largely imposed upon the organisms by the structure and physiology of wood and, to a lesser extent, by the reproductive behaviour and nutritional preferences of those organisms which invade wood before it is fully dead. This group of organisms, which can be classified variously as weak parasites, necrotrophic pathogens, primary sugar saprotrophs, opportunistic pests, and some of their less obvious but ever-present associates (yeasts, nematodes and mites), constitutes the actors. Their stage is wooden, but has to be broken into before the act can begin, and most of the action goes on behind doors that are only partly ajar.

The substratum

The live standing tree is protected by a more or less impermeable bark which is often impregnated with anti-microbial substances. The bark surface is subject to wetting and drying regimes of a short periodicity and to great temperature extremes (Geiger, 1965). The combination of structure, chemistry and fluctuating environment means that only suitably adapted xeric and autotrophic organisms, such as algae, lichens and mosses together with a few fungi, such as *Athelia epiphylla* and *Dichaena rugosa*, can grow actively on or near the bark surface. Others depend on gaining access to phloem and xylem tissue via discontinuities in the bark surface. In natural ecosystems such discontinuities are provided by the feeding activities of bark penetrators such as bark beetles (Scolytids), woodwasps (Siricidae) and woodpeckers (Picidae), of bark strippers such as squirrels, deer and lagomorphs, and by the abscission of twigs and branches. In some genera (e.g. *Quercus*) small twigs are normally shed after the formation of an abscission layer at the junction between twig and branch, but in most trees the breakage of live twigs and branches only happens occasionally, as a result of large shear stresses from exceptional wind speeds and/or heavy snow or ice loads. The majority of twigs and branches die *in situ* and decay to some extent before they are broken off (see Swift & Boddy, Chapter 4). Exceptionally strong winds break whole trees, or on shallow soils and in wet weather uproot trees, creating wounds for entry and cutting off the water supply partially or completely.

In healthy trees with a good water supply a variety of mechanisms may help to seal these potential infection courts. For example, in abscission planes conducting elements in the xylem are blocked by tyloses, balloon-like outpushings into the lumina of the vessels and tracheids, which are impregnated with tannins. Provided that oxygen is present, all exposed living tissue which is damaged becomes impregnated with tannins by the mixing of polyphenoloxidase enzymes in the lysosomes, with phenol-like compounds in the vacuoles. Tannins impede the entry of potential invaders by 'locking up' plant proteins and by precipitating the exoenzymes of the invading organisms. Phloem-feeding insects, such as bark beetles and some weevils, ingest the broken-up live tissue too quickly for the tannins to form, and so they are able to digest the material for all of its available substrates. Tannins do form however in the material once it has been defaecated. In conifers wounded surfaces are sealed by the copious production of resin which contains anti-microbial and insecticidal compounds (Francke-Grossman, 1963) and sets hard after a few days' drying.

If a tree is below optimal water status, because of wind throw, root disease, drought or suppression by neighbours, or if the supply of photosynthate to the trunk is inadequate because of defoliation, water shortage, toxins, or shading out, then the mechanisms described above become more feeble or fail entirely (e.g. Vité, 1961; Courtois, Chararas & Debris, 1962; Gibbs, 1967; Coutts, 1969a, b) and successful invasion can occur. More importantly, under such conditions multiple wounds and potential inoculation sites are created in conifers by bark beetle attack (Unger, 1866; Vité, 1961; Bevan, 1974).

Forestry activities have increased the density of suppressed trees, because of dense planting and delayed thinning in even aged stands. Felling of trees during thinning or for the final crop creates large volumes of suitable material for attack, both because of its reduced water status (Münch, 1906; Lagerberg, Lundberg & Melin, 1927; Lekander & Rennerfelt, 1955; Pechmann, Graessle & Wutz, 1964; Henningsson, 1965) and because the crosscut ends and bark flaps caused by mechanical damage provide fresh surfaces for entry (Rumbold, 1929;

Verrall, 1941; Fystro & Bakke, 1962; Dowding, 1971). In addition the stumps which remain after felling, weakened by lack of photosynthate, fall prey to root-and collar-inhabiting insects and fungi (Krogerus, 1927; Wallace, 1953; Martin, 1965). One basidiomycete primary colonist, *Heterobasidion annosum*, has become of major importance in planted coniferous forests because of the new entry point created by the stump surface (Rishbeth, 1950; Meredith, 1959), but is not considered further in this review as it is primarily airborne to the stump surfaces. As stump surfaces are manifestly unnatural its mode of entry in wild conifer stands must be through some other entry point, possibly via root wounds, although Krebill & Patton (1962) failed to find *H. annosum* in wounded roots of *P. banksiana*.

Forestry activities thus increase the importance of primary colonists in the ecological sense by providing large amounts of novel substrata in high concentrations for relatively short periods. Short storage time of forest products, which is a management ideal, does not allow other colonists to develop and reproduce, except on waste wood (mostly of small diameter) and on stumps and roots, so that the balance of inocula of primary and secondary colonists is shifted in favour of the primary group. This may have contributed to the emergence of 'novel' disease problems by allowing one or several primary colonists to multiply. Under such circumstances evolutionary pressures towards pathogenicity and a short life cycle are stronger than those towards long-term saprotrophy, and in densely planted monocultures evolutionary pressures towards effective middle and long distance dispersal mechanisms are weaker than those towards local immediate transmission.

Primary animal invaders

The most important primary animal invaders of trees are the bark boring beetles, which feed mostly on living phloem, but also on outer xylem, both as adults and as larvae (Munro, 1926; Vité, 1952; Duffy, 1953; Dajos, 1980). Quarantine regulations for timber as tree trunks are designed specifically to prevent the impact of bark beetles.

The beetles carry with them mites (Krantz, 1965) and nematodes (Steiner & Buhrer, 1934). Their entry holes allow other winged insects to gain access to the surfaces of xylem and phloem exposed under the bark by their feeding and breeding activities (Séguy, 1950; Brauns, 1954; Johnsey, Nagel & Rudinsky, 1965; Dajos, 1980). These other insects are chiefly Diptera which are either mycophagous or predatory on the beetle larvae (Kleine, 1907; Tuomikoski, 1957; Johnsey, Nagel &

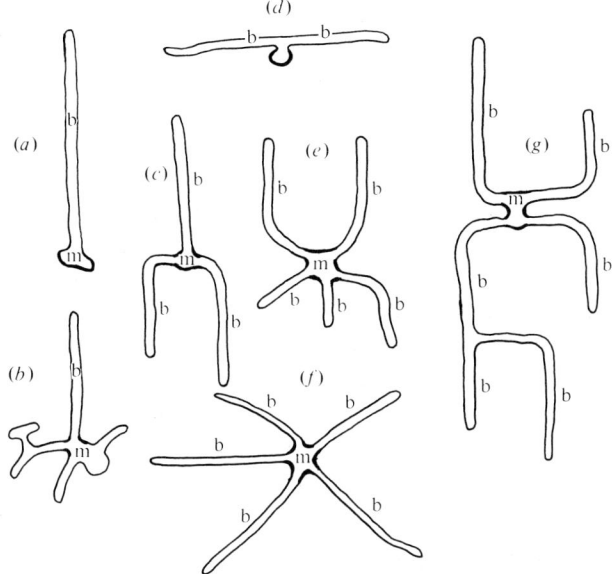

Fig. 5.1. Tunnel patterns of various bark beetles. m, mating chamber; b, brood tunnel; (a) *Myelophilus piniperda*, *Scolytus multistriatus*; (b) *Hylastes ater*; (c) *Ips typographus*, *Dendroctonus* spp.; (d) *Myelophilus minor*; (e) *Pityogenes chalcographus*; (f) *Pityogenes quadridens*; (g) *Ips acuminatus*, *Ips grandicollis*. Vertical orientation on the page represents longitudinal orientation in the phloem.

Rudinsky, 1965), although parastic Hymenoptera also occur (Elton *et al.*, 1964; Wallace, 1953; Savely, 1939).

Bark beetles excavate tunnels in the phloem under the bark. Feeding tunnels have no particular direction or pattern and do not have additional ventilation holes to the exterior (Munro, 1926). Breeding tunnels, however, have a particular shape and direction (Fig. 5.1) which is almost species specific (Duffy, 1953), and are made only during limited periods of the year – generally early and late summer (April/May and July/August). In more northerly latitudes the first breeding period is delayed, for instance *Scolytus scolytus* in most of the UK has only one brood in July/August, which overwinters as larvae to pupate and fly the following spring.

The breeding tunnel may be excavated by both male and female beetles, but more usually by the female, and the larger tunnels are ventilated by several extra holes to the exterior. During excavation sawdust and faecal material are pushed to the exterior, so keeping the mother tunnel clear. Eggs are laid singly or in groups at the side of the tunnel, and are not further tended by the female beetle. When she has finished a tunnel system the female beetle deserts it. The eggs hatch after 7–10 days and the larvae feed on the phloem, excavating progressively larger burrows as they grow (Fig. 5.2); however they do not move away from the freshly cut face for the entire growing period, and the tunnel behind them fills with frass (faecal material) and mycelium. Towards the end of their existence, the larvae may burrow into the sapwood or into the bark to excavate a pupation chamber, but more usually the pupation chamber is excavated in the phloem (see Webber & Brasier, Chapter 10). Larvae from the late summer brood in bivoltine species, or from the single midsummer brood, overwinter as larvae and pupate in the spring.

Pupation takes two weeks and after hardening off, the newly emerged adult either bores directly out of the bark, or feeds for a while on any remaining live phloem near its pupal chamber. The latter group usually fly to a similar habitat (e.g. log or weakened tree) on emergence for both feeding and breeding. The former group may fly to other logs or weakened trees to breed or feed, but as in the case of *Myelophilus piniperda, Scolytus scolytus*, and *S. multistriatus* may also fly to feed in young twigs of pines and elms respectively (see also Webber & Brasier, Chapter 10).

Other groups of beetles are primary invaders, but do not create as much devastation as the *Scolytid* bark beetles. Weevils (Curculionidae) attack roots and stem bases (Krogerus, 1927; Wallace, 1953; Elton *et al.*, 1964), sometimes of healthy trees, but more usually of weakened trees and stumps without root graft support (Johnsey *et al.*, 1965). Both adults and their solitary larvae feed on phloem and on the first annual ring of xylem. The larvae usually take ten months to mature over winter.

Fig. 5.2. A cross section of beetle infested phloem. b, brood tunnel; l, larval tunnel; p, pupal chamber; B, bark; Ph, phloem; W, wood. Entrance and exit tunnels are indicated by dotted lines and arrows.

Ambrosia beetles (Scolytidae) bore directly through the phloem into the sapwood, where they excavate radial, tangential and longitudinal tunnels, and inoculate them with specific ambrosia fungi (Leach, Hodson, Chilton & Christensen, 1940; Verrall, 1941; Bakshi, 1950; Bletchley & White, 1962). The larvae feed exclusively on the fungal mycelium and spores which coat the tunnels, and the adults have specialized structures to carry inoculum (Francke-Grossman, 1963). Logs and moribund trees are more at risk than weakened trees; healthy trees are not attacked.

Cerambycid beetles such as *Acanthocinus* and *Monochamus* oviposit beneath the bark, and their larvae burrow into the sapwood. They are associated with basidiomycetes (Soper & Olson, 1963) but are not implicated as vectors. They prefer to breed in recently dead trees and logs.

As the bark beetles do not defend their tunnels against intruders the spaces they create are almost immediately occupied by a variety of small insects and their larvae. The most frequent of these are saprophagous and predatory Diptera and predatory Coleoptera. Graham (1925), Savely (1939) and Wallace (1953) detailed this and later stages in their excellent accounts of animal successions on conifer logs and stumps. More recently Johnsey *et al.* (1965) described the association between *Dendroctonus frontalis* and two dipterans on Douglas fir. One of these was a predator in its third larval instar, and the other a fungivore. Dowding (1974) reported that Sciarid flies could be reared on *Ceratocystis* cultures through several generations, as both adults and larvae fed on the asexual spores of this ascomycete genus.

Graham (1967) considered that *Cecidomyid* Diptera were very common associates of bark beetles and of Ambrosia beetles, and were in turn associated with *Ceratocystis* spp. and their imperfect states. Krogerus (1927), in a carefully controlled study of *Picea* stumps over a three year period, observed adult Diptera emerging from beetle-infested stumps several weeks before the newly adult beetles. Similar findings were reported by Tuomikoski (1957) from fallen trees in Finland. Most mycologists have overlooked the presence of animals other than bark beetles.

Fungal primary colonists

As described by Hudson (1968), Käärik (1974) and Levy (1982) many primary fungal colonizers of timber do not cause decay but are able to invade weakened living tissue. The importance of decay fungi, particularly Basidiomycotina, as primary invaders has probably been considerably underestimated (see Swift & Boddy, Chapter 4), particularly in unmanaged woodland. However, the emphasis here will be on non-decay fungi, as the decay fungi are mostly airborne.

Perhaps the most frequently recorded non-decay genera belong to the Deuteromycotina and Ascomycotina, including *Ceratocystis* and its imperfect states, *Graphium, Leptographium, Endoconidiophora, Verticicladiella, Sporothrix*, and *Acremonium*, and surface colonizers such as *Cladosporium* spp. and *Aureobasidium pullulans*. The latter group are restricted to the surface by host-related mechanisms initially and by resource competition and/or intolerance of low oxygen and high CO_2 concentrations latterly. The *Ceratocystis* group can invade living or moribund tissue and can penetrate the sapwood radially via medullary rays and longitudinally via vessels and tracheids. In conifer wood at least, both radial and longitudinal mycelial growth rates are rapid – up to 3 mm radially and 10 mm longitudinally per day (Lagerberg, Lundberg & Melin, 1927), but tangential growth rates are much slower at 0.5 to 1 mm per day.

The hyphae of xylem-inhabiting *Ceratocystis* spp. and their allies are typically crowded in the ray cells (Fig. 5.3), penetrate walls via pits or by peg tubes (Levy, 1982), and occur sparsely in the lumina of the tracheids and vessels. The relatively slow tangential spread may be explained in part by nutrient gradients restricting the hyphae to the vicinity of rays, and partly to the relative mechanical ease of longitudinal growth, with long distances between cell walls. Whilst non-cellulolytic (Käärik, 1960), the fungi are able to utilize a wide variety of disaccharides, pectin

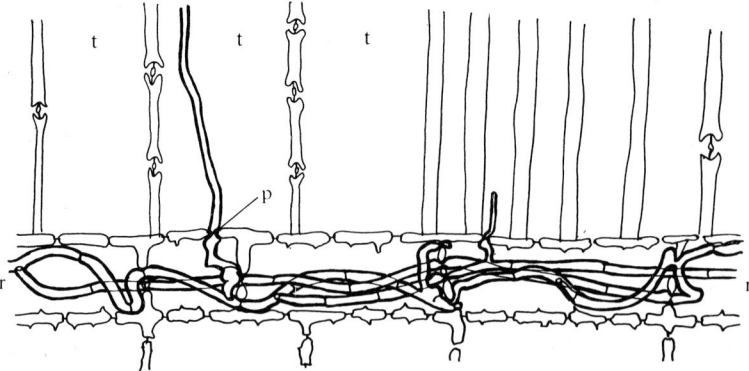

Fig. 5.3. The distribution of *Ceratocystis* hyphae in pine sapwood in radial longitudinal section. p, peg tube; r, ray cell; t, tracheid.

Fungal primary colonists 141

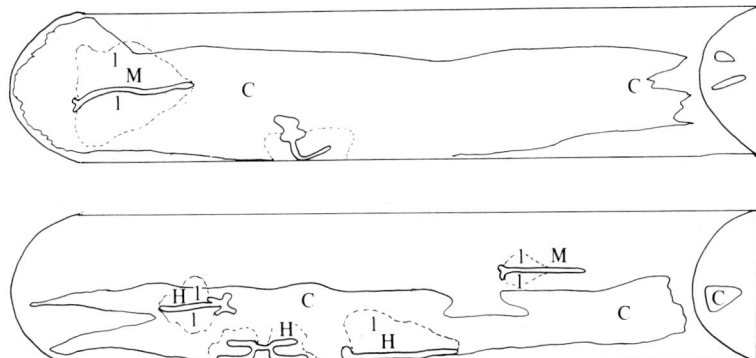

Fig. 5.4. The distribution of fungal colonies and bark beetle galleries in an experimental log exposed to colonization from February for three months. M, *Myelophilus piniperda*; H, *Hylastes ater*; C, *Ceratocystis* spp.; b, brood tunnel; l, larval tunnels.

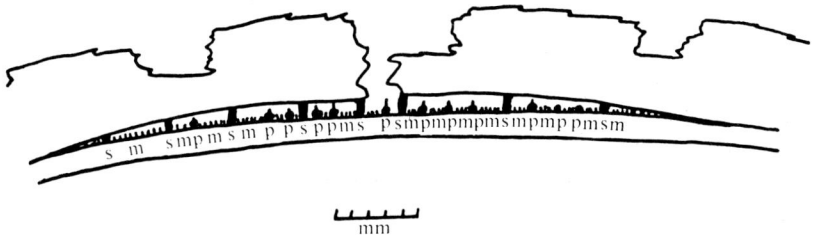

Fig. 5.5. A cross section of *Ceratocystis minor* infected wood showing sclerotia (s), perithecia (p) and *Sporothrix*-type sporing mycelium (m).

and starch, and most nitrogen sources. If inoculated under the bark in a felled tree or segment of trunk, their growth is stopped or severely restricted within 5–10 mm from the exposed end (Fig. 5.4). The simplest reason for this is that evaporation from the wood at the end reduces the water content of the 2–10 mm zone even if the 0–2 mm zone is subsequently re-wetted by rainfall. Under humid conditions some *Ceratocystis* spp., notably *C. pilifera* and *C. coerulescens*, will grow out to form a wedge-shaped grey/black patch of mycelium with fruiting structures on the sapwood surface.

In the majority of species fruiting structures are formed under the bark. Only *C. fagacearum* and *C. minor* can create a space for fruiting by their own activity: they both form pressure cushions (sclerotia?) which separate xylem from phloem initially, and then split the bark open (Fig. 5.5). Other species fruit in beetle excavations and under bark flaps created by mechanical wounding; the more complicated fruiting structures (*Graphium*, *Leptographium*, *Verticicladiella*, and perithecia) are located on the ends of rays and on the phloem, while the simpler structures (*Sporothrix*, *Endoconidiophora*, and *Cephalosporium*) occur on free mycelium.

The exposed fruiting structures of *C. minor* and *C. fagacearum* become feeding and breeding sites for a number of insects and mites. Where mycelia of *C. coerulescens* and *C. pilifera* grow out on to the crosscut ends the mats are frequently shredded by mites after a week's exposure. Under the bark, mites, nematodes and dipteran larvae are consistently found in spaces where *Ceratocystis* spp. and other fungi are fruiting. The gut contents of the larger animals of this group reveal that they cut down and consume *Leptographium* and *Graphium* conidiophores, as well as consuming mycelium. The large sporing heads of *Leptographium* and *Graphium* frequently contain one or more nematode worms. Old ascocarps are usually found broken open with their contents gone.

Two early Basidiomycotina colonists of conifer wood, *Heterobasidion annosum* and *Phlebia gigantea*, both produce asexual states in spaces beneath the bark, *Oedocephalum* and *Oidium* respectively. The former is sticky spored like the asexual states of *Ceratocystis* and may therefore perform the same ecological role – that of food for fungivorous arthropods which act as local dissemination agents for the fungus.

The characteristics of the spores of *Ceratocystis* spp. were investigated by Matheisen-Käärik (1960) and by Dowding (1969). The spores require a food source and water to germinate in the laboratory. Once separated from their conidiophore, they lose viability fairly rapidly. Dowding (1969) reported that separated spores of all asexual types lost 50% viability in under 2 hours at 50% RH in darkness. The rate of death of separated spores was further increased by exposure to sunlight or near UV light. Both authors observed that spore survival was much longer in the conidiophores, and the larger the sticky head of spores the longer the survival time. Even so asexual spores did not appear to be able to survive longer than 8 months under ideal laboratory conditions. Dowding (1969) found that asexual spores on their conidiophores were unable to survive more than 24 hours of direct sunlight, and were similarly sensitive to near UV light (except for the internally produced spores of *Endoconidiophora* which survived 32 hours near UV light). Cloud cover appeared to filter out the lethal agent in sky light, and spores survived as long under light from a cloudy sky as they did in darkness. Ascospores were somewhat different; their sticky matrix is hydrophobic and is not soluble in water (Cole & Fergus, 1956). Matheisen-Käärik (1960) reported increased germination rates after three months' storage, and survival for up to 14 months in three *Ceratocystis* species.

The physiological characteristics of the spores make it extremely unlikely that they are passively air dispersed, even though that assumption has been widely held (see Dowding, 1969). In their natural habitat – spaces under the bark – adaptations for air dispersal are unnecessary, and the stickiness cannot be an adaptation to dispersal by rainsplash either, as both the site of spore production and the infection court are usually protected from direct rain. It would appear therefore that the spores of these fungi are consistently adapted to arthropod carriage, and that the ascospores are uniquely so adapted as they are extruded in an undrying hydrophobic matrix 1–2 mm above the wood surface.

Ceratocystis spp. are poor competitors in wood and on agar. In pine logs columns of wood blued by *Ceratocystis* are either rapidly colonized by *Trichoderma* spp. which then fruit in a wedge-shaped conformation on the ends of the logs, or are overgrown by *Phlebia* and other Basidiomycotina (Dowding, 1974). *Trichoderma* spp. are unable to invade live wood even in logs, so depend on *Ceratocystis* to provide access, but several Basidiomycotina can invade as early as *Ceratocystis* provided that their spores are present. Where Basidiomycotina do get in first, as in logs felled in November, colonization by *Ceratocystis* (and by bark beetles) is prevented.

Pathogenic associations

The first important pathogenic association to be noticed was that between *Ceratocystis ulmi* and scolytid bark beetles on elms in Europe (Spierenberg, 1922). However, it was several years before the involvement of beetles in transmission was documented. This relationship is fully explored by Webber & Brasier in Chapter 10, and so will not be dealt with in detail here. At about the same time the associations between bark beetles and *Ceratocystis* spp. on pine were reported (Craighead, 1928; Bramble & Holst, 1935). These are explored in greater detail below, in relation to sap stains of conifer logs, which have been well documented since 1866 (Unger, 1866; von Schrenk, 1903; Münch, 1906, 1907).

About 20 years later (Norris, 1953; Bart & Griswold, 1953; Cole & Fergus, 1956), oak wilt caused by an association between *Ceratocystis fagacearum* and a variety of insects was the subject of many published investigations. This relationship is different in many respects from that between other *Ceratocystis* spp. and other fungi with bark boring beetles. The most important distinction is that the fungus produces pressure cushions (sclerotia) between the infected xylem and phloem, with which it bursts open the overlying bark (Figure 5.5). The ensuing split allows the entry of nitidulid beetles in particular – as they are dorsiventrally flattened – together with many Diptera. These insects feed and breed in the sporing fungus mat between bark and wood, depending on the biological pit-props provided by their food plant for space in which to move. Adult insects emerge after feeding or after pupation contaminated internally and externally with spores (Bart & Griswold, 1953; Jewell, 1954) and can then fly to contaminate wounds caused by other agencies, such as man and sapsuckers, on healthy trees (Norris, 1953; Dorsey, Leach, Jewell & True, 1953). Sapsuckers are specialized woodpeckers (*Sphyrapicus* spp.) which puncture holes in horizontal and vertical rows around the trunks of healthy trees (Farrand, 1981). The sap which exudes from these holes acts as an immediate food for the bird and also as an attractant to insects such as the nitidulids and small Diptera which feed and breed in *C. fagacearum* mats (Morris, Thompson, Hadley & Davis, 1955). The sapsuckers use these insects as food also, but not before spores have been inoculated into the fresh exuding wound, which are ideal infection sites (Cole & Fergus, 1956). It is possible, of course, that the bird itself might become contaminated with spores during feeding on the insects and then transfer them to fresh wounds, and so become a primary vector. It is more likely however that the behaviour of the bird in creating closely spaced wounds tangentially around the trunk is of much more importance in determining successful colonization by the fungus, as it grows much more rapidly longitudinally than radially or tangentially. As with Dutch elm disease only the outer ring of sapwood has to be invaded for disease expression to occur, and once sapwood has been killed the fungus can survive over winter from late summer inoculations (Gibbs, 1980) and as ascospores (Jewell, 1953). Oak-wilt is therefore spread primarily by fungivorous insects, some of which depend on the mycelium and spores of the fungus for larval and adult nutrition. The fungus does not depend on the activity of an insect to break the bark open. However the fungivore/fungus association is dependent on another agent to create infection sites, in this case an insectivorous bird, which uses the fungivorous insects associated with oak-wilt as a food source.

Another species of *Ceratocystis*, *C. fimbriata*, has been widely implicated in wilt and canker diseases of crop trees, for instance *Prunus* spp. (Moller & DeVay, 1966), Date palm (Bliss, 1941), *Hevea* sp. (Chevaugeon, 1957), *Platanus* spp. (Crandall, 1935), and *Ficus* sp. (Vacarini & Tokeshi, 1980). Wounds, caused by man, are the primary infection site of the first four hosts, and the vectors to these sites include Diptera (e.g. *Drosophila melanogaster*), and nitidulid beetles (e.g. *Carpophilus* spp.). *Xyleborus ferrugineus*, a scolytid, is implicated as the vector and inoculator of *C. fimbriata* in *Ficus*. The same fungus causes a wound-inoculated black rot in *Narcissus* (Limber, 1950), and in *Ipomoea* spp. (Chevaugeon, 1957) and *Crotalaria juncea* (Costa & Krug, 1935), the dissemination of which between hosts and within the host plant is dependent on a variety of animals including dipteran larvae, mites and nematodes.

Pathogenic associations affecting trees are not limited to *Ceratocystis* and beetles (see also Swift & Boddy, Chapter 4). Here two pertinent examples will be discussed. The first concerns a consistent association between bark-feeding aphids (woolly aphids) and apple canker (*Nectria* spp.) in America (Leach, 1940). The aphids are not the vectors of the disease, but feed preferentially on the callous tissue which develops around cankers each summer, stimulating the production of small raw galls. These raw galls burst on freezing in early winter, at which time the fungus is sporulating in the tissue close by, and reinfection of the canker occurs by transfer of spores to these galls by some unknown agent. Other insects and possibly also mites would be primary suspects. Apple canker in Europe is not associated with woolly aphids but with pruning wounds and, to a lesser extent, with premature leaf scar formation. Leach (1940) considered that pruning wounds are the primary avenue of infection for apple canker in the US, but that woolly aphids are required to make the cankers perennial. The disease in the US has been controlled by controlling the woolly aphids. In Europe, where the disease is air- and water-borne rather than insect-borne/associated, control has been by chemical or biological wound protection (Swinburne & Brown, 1976).

A pathogenic association between a woodwasp, *Sirex noctilio*, and *Amylostereum* sp. occurs in suppressed *Pinus radiata* grown in New Zealand (Rawlings, 1948). Coutts (1965) found that the wasp-inoculated fungus causes infected sapwood to dry out and become non-conducting (cf. Caird, 1935; Mathre, 1964), but that symptoms were observed in the crown before the fungus could affect enough sapwood to cut off water supply. Symptoms were also observed in healthy trees which had been recently attacked by *Sirex* and in which the fungus was unable to grow initially (Coutts, 1969a). Coutts (1969b) described the toxic action of mucus secreted by the wasp at the same time as it oviposited and inoculated the trees with *Amylostereum* spores. The toxin caused the translocation of photosynthate out of mature leaves to stop, with consequent starvation of the growing shoots, phloem, and most importantly of the roots, rendering the tree more susceptible to invasion by the fungus. *Sirex noctilio* oviposits in a series of closely spaced holes around the circumference of the trunk, thus ensuring that the tree is effectively girdled with fungus inoculum, and with toxic mucus. The mucus ensures successful colonization by *Amylostereum*, and *Amylostereum*-exploited wood forms the food and living quarters for the larvae of *Sirex*, which are unable to move into or derive sufficient nutrients from uninfected wood. *Amylostereum* infection also ensures that the tree does not recover to kill the larvae with resin.

Saprophagous associations

Münch (1907) was the first to suggest an association between bark beetles and primary fungal colonists of pine and spruce trees. He also reported that not all primary colonists were associated with beetles, and surmised that other insects were involved in carriage and inoculation of spores into beetle galleries and into wounds. Only one species, *C. minor*, was independent of wounds as fruiting spaces, as it produced pressure cushions. Leach, Orr & Christensen (1934) reported a set of elegant experiments which demonstrated that several American species of *Ceratocystis* were carried by bark beetles. Verrall (1939), by careful observation of small logs and pieces of timber exposed in a sawmill prior to incubation, was able to discard any visited by insects, and to demonstrate that the spores of *Ceratocystis* were not carried by air to new substrata. In spite of the earlier American work, European authors (Björkmann, 1946a, b; Rennerfelt, 1950; Matheisen-Käärik, 1953; Lekander & Rennerfelt, 1955; Pechmann, Graessle & Wutz, 1964; Savory, Pawsey & Lawrence, 1965) assumed that any *Ceratocystis* species not associated with a particular bark beetle was normally airborne. Rishbeth (1959) reported trapping *Ceratocystis* spp. in the air in Thetford Chase, in the course of enumerating spores of *Heterobasidion annosum*. Dowding (1969, 1971, 1974) demonstrated that airborne carriage of any spore of *Ceratocystis* was unlikely and that the survival of such spores was severely curtailed by UV light and by desiccation. By careful recording of associations in logs, and by observation of deflected successions caused by a sequence of felling times, Dowding (1974) demonstrated an association between insects and *Ceratocystis* spp., but none between primary basidiomycete colonists of logs and insects. Whitney & Cobb (1972) found that beetle tunnels were important avenues for entry by both Basidiomycotina and non-cellulolytic fungi into Ponderosa pine trees. Mercer (1982) considered that wounds are the primary avenues for entry by Basidiomycotina into timber. In standing trees, wounds created by bark beetle activity may not be as important as scars created by the abscission of dead twigs, branches and roots, which occur over a much longer time interval than the final lethal beetle attack. In logs and felled trees the crosscut ends and bark flaps and scrapes give a much larger surface area for colonization by Basidiomycotina than do beetle tunnel exits. The crosscut and tangentially exposed timber surfaces are also colonized by primary surface colonists (Dowding, 1971) such as *Cladosporium, Aureobasidium, Botrytis*, and *Penicillium* spp., which are all derived from airborne spores. These four genera were rarely isolated from wood or phloem surfaces in beetle tunnels.

The breeding behaviour of the beetle has a profound effect on the successful early colonization of the tunnels by other organisms. Brood tunnels of *Myelophilus piniperda* are longitudinal (Fig. 5.1) and are 'guarded' by the male beetle while the female is excavating and ovipositing. Entry of secondary insects is therefore prevented until the first laid eggs are hatched. Dowding (1974) observed that in short logs this delay was sufficient to allow Basidiomycotina (such as *Stereum sanguinolentum*) to grow into the wood under the beetle galleries. Primary colonists such as *C. piceae* which entered via the vacated brood tunnel were then restricted to a small volume of wood around the pairing chamber. The new adult beetles emerged uncontaminated with spores as *S. sanguinolentum* does not produce asexual fruiting structures in spaces under the bark. In standing trees this pre-emptive colonization by Basidiomycotina only occurred near the base of the tree. In the upper part of the trunk, primary colonists invaded the whole length of *M. piniperda* brood tunnels after the adult beetles had left, but, because of their slow tangential growth rate, did not fruit in the pupal chambers until after the new beetles had left.

Breeding in *Hylastes ater* is both preceded and followed by feeding on the phloem of the log or tree in which breeding takes place. Feeding tunnels do not follow any particular direction and are not guarded; the brood tunnel is longitudinal. While the female is laying eggs, the feeding galleries are occupied by various species of Diptera, and conidiophores of several *Ceratocystis* species appear on the wood and phloem surfaces. As the feeding tunnels of this species extend tangentially from the brood tunnel base, inoculation of *Ceratocystis* is over a wide band, and most pupal chambers have asexual fruiting bodies in them before the new adults emerge to feed under the bark. Most adults of *H. ater* were found to be contaminated with spores (Dowding, 1974), as were adults and larvae of Diptera associated with the beetles.

M. minor and most *Ips* and *Pityogenes* species have tangential components in their brood tunnels (Fig. 5.1), and therefore present a broader inoculation base to tangentially restricted fungi. At the same time, the pupal chambers, at the end of larval tunnels excavated at approximately 90° to the brood tunnel, are more or less longitudinally above and below the brood tunnel. This orientation ensures that the pupal chambers are excavated in fungus-infected wood, and therefore that the newly emerging adult is likely to be contaminated with spores (Fig. 5.6).

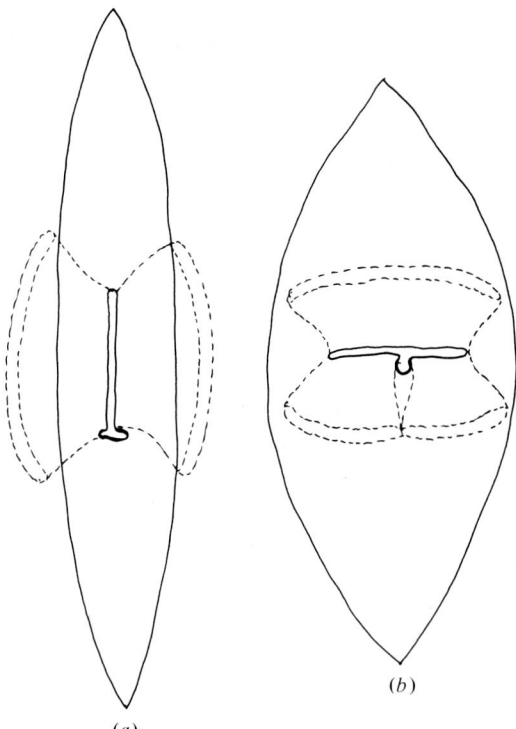

Fig. 5.6. The extent of larval tunnels and of bluestain colonization six weeks after initial attack by beetles with longitudinal brood tunnels (*a*) and with tangential brood tunnels (*b*).

Species which overwinter as larvae, pupae or adults in sub-bark spaces overlying infected wood will come into contact with asexual and sexual spores of *Ceratocystis* spp. As the ascospores of these fungi are particularly adherent and are long-lived (Matheisen-Käärik, 1960), vectors contaminated with these spores are likely to make more successful inoculations in the spring than are those contaminated with asexual spores.

Natural *Ceratocystis* colonies are invariably inhabited by fungivorous Coleoptera, Diptera, nematodes and mites, which are all dependent on the spores as food. Bark beetles on the other hand do not appear to depend on the fungi for food, but the presence of the fungus in the underlying wood may ensure that the tree does not recover its resin exudation pressure or other resistance mechanisms. It is plausible to suggest, therefore, that the primary fungus–insect relationship in the primary attack on trees and logs is between fungivorous insects and *Ceratocystis* and other sticky-spored fungi, and that both are dependent on bark beetles for creating an entry port, a niche for colonization, and an exit. Species of beetle which attack living trees may be benefited by a relationship with *Ceratocystis* because their chances of reproductive success are enhanced by the presence of the fungus. Beetles which attack logs, stumps and dead trees only become vectors of *Ceratocystis* if the patterns of their reproductive and feeding tunnels permit sporulation of the fungi in the pupal chambers, *or* if the new adults feed before emergence from their larval habitat *and* if they then fly directly to suitable habitats for the fungi.

References

Bakshi, B. K. (1950). Fungi associated with Ambrosia beetles in Great Britain. *Transactions of the British Mycological Society*, **33**, 111–20.

Bart, G. J. & Griswold, C. L. (1953). Recovery of viable spores of *Endoconidiophora fagacearum* from excrement of insects used in disease transmission studies. *Phytopathology*, **43**, 466.

Bevan, D. (1974). Control of forest insects: there is a porpoise close behind us. *Symposium of British Ecological Society*, **13**, 302–12.

Bjorkmann, E. (1946a). On the development of bluestain and storage decay in pine logs during floating. *Meddelanden Statens Skogsforskningsinstitut*, **35**, 1–56.

Bjorkmann, E. (1946b). On the conditions for the occurrence of timber yard bluestain, and methods for its control. *Meddelanden Statens Skogsforskningsinstitut*, **36**, 1–46.

Bletchley, J. D. & White, M. G. (1962). The significance and control of attack by the ambrosia beetle *Trypodendron lineatum* in Argyllshire forests. *Forestry*, **35**, 139–63.

Bliss, D. E. (1941). Relation of *Ceratostomella radicicola* to rhizosis of date palm. *Phytopathology*, **31**, 1123–9.

Bramble, W. C. & Holst, E. C. (1935). Microorganisms infecting pines attacked by *Dendroctonus frontalis*. *Phytopathology*, **25**, 7–10.

Brauns, A. (1954). *Die terricolen Dipteren larven*. Göttingen: Munster-Schmidt.

Caird, R. W. (1935). Physiology of pines infested with bark beetles. *Botanical Gazette*, **46**, 709–33.

Chevaugeon, G. J. (1957). *Ceratocystis fimbriata* Ellis & Halsted. *Revue Mycologique*, **22**, Supplément coloniale, 245–60.

Cole, H. & Fergus, C. L. (1956). Factors associated with the germination of oak wilt fungus spores in wounds. *Phytopathology*, **46**, 154–63.

Costa, A. S. & Krug, H. P. (1935). Eine durch *Ceratostomella* hervorgerufene Welkekrankheit der *Crotalaria juncea* in Brasilien. *Phytopathologische Zeitschrift*, **8**, 507–13.

Courtois, E. J., Chararas, C. & Debris, M-M. (1962). Etude de l'attaque enzymatique des glucides par un Coleoptère xylophage. *Comptes rendus hebdomadaires des séances de l'Academie des sciences*, **252**, 2608–9.

Coutts, M. P. (1965). *Sirex noctilio* and the physiology of *Pinus radiata*. *Bulletin of the Forest Timber Bureau, Australia*, **41**.

Coutts, M. P. (1969a). The mechanism of pathogenicity of *Sirex noctilio* on *Pinus radiata*. I. Effects of the symbiotic fungus, *Amylostereum*. *Australian Journal of Biological Sciences*, **22**, 915–24.

Coutts, M. P. (1969b). The mechanisms of pathogenecity of *Sirex noctilio* on *Pinus radiata*. II. Effects of *S. noctilio* mucus. *Australian Journal of Biological Sciences*, **22**, 1153–61.

Craighead, F. C. (1928). Interrelationships of tree-killing bark beetles (*Dendroctonus* spp.) and bluestains. *Journal of Forestry*, **26**, 886–7.

Crandall, B. S. (1935). *Endoconidiophora fimbriata* on Sycamore. *Plant Disease Reporter Supplement*, **90**, 98.

Dajos, R. (1980). *Ecologie des insectes forestiers*. Paris: Gauthier Villar.

Dorsey, Y. C. K., Leach, J. G., Jewell, F. F. & True, R. P. (1953). Experimental transmission of oak wilt by four species of Nitidulidae. *Plant Disease Reporter*, **37**, 419–20.

Dowding, P. (1969). The dispersal and survival of spores of fungi causing bluestain in pine. *Transactions of the British Mycological Society*, **52**, 125–37.

Dowding, P. (1971). Colonization of freshly bared pine sapwood surfaces by staining fungi. *Transactions of the British Mycological Society*, **55**, 399–412.

Dowding, P. (1974). Effects of felling time and insecticide treatment on the interrelationships of fungi and arthropods in pine logs. *Oikos*, **24**, 422–9.

Duffy, E. A. J. (1953). *Handbooks for the identification of British Insects*. V. Coleoptera. Part 15, Scolytidae and Platypodidae. London: Royal Entomological Society.

Elton, E. T. G., Blankwaardt, H. F., Burger, H. C., Steemers, W. F. & Tichelman, L. G. (1964). Insect communities in barked and unbarked pine stumps, with special reference to the large pine weevil. *Zeitschrift für angewandte Entomolgie*, **55**, 1–54.

Farrand, H. (1981). *World Atlas of Birds*. London: Paul Hamlyn.

Francke-Grossmann, H. (1963). Some new aspects in forest entomology. *Annual Reviews of Entomology*, **8**, 415–46.

Fystro, I. & Bakke, A. (1962). Damage on unbarked coniferous sawlogs and the effect of chemical spraying during summer storage in forests in different parts of Norway. *Norsk Skogbruk*, **8**, 272–8.

References

Geiger, R. (1965). *The Climate Near the Ground*. Cambridge, Massachusetts: Harvard University Press.
Gibbs, J. N. (1967). The role of host vigour in the susceptibility of pines to *Fomes annosus*. *Annals of Botany*, **31**, 801–15.
Gibbs, J. N. (1980) Survival of *Ceratocystis fagacearum* in branches of trees killed by oak wilt in Minnesota. *European Journal of Forest Pathology*, **10**, 218–24.
Graham, K. (1967). Fungus–insect mutualism in trees and timber. *Annual Review of Entomology*, **12**, 105–7.
Graham, S. A. (1925). The felled tree trunk as an ecological unit. *Ecology*, **6**, 397–411.
Henningsson, B. (1965). Undersökning av svampfloran in sommatavverkat sågtimmer. *Raporter Instituten Virkeslära Stockholm*, **50**, 1–18.
Hudson, H. J. (1968). The ecology of fungi on plant remains above the soil. *New Phytologist*, **67**, 578–604.
Jewell, F. F. (1953). Ascospore longevity in the oak wilt fungus. *Phytopathology*, **43**, 476.
Jewell, F. F. (1954). Viability of the conidia of *Endoconidiophora fagacearum* in the faecal material of various Nitidulidae. *US Dept. of Agriculture Plant Disease Reporter*, **38**, 53–4.
Johnsey, R. L., Nagel, W. P. & Rudinsky, J. A. (1965). The Diptera *Medetera alorichii* and *Lonchaea furnissi* with the Douglas Fir beetle in Western Oregon and Washington. *Canadian Entomologist*, **97**, 521–7.
Käärik, A. A. (1960). Growth and sporulation of *Ophiostoma* on synthetic media. *Symbolae Botanicae Uppsaliensis*, **16**, 1–159.
Käärik, A. A. (1974). Decomposition of wood. In *Biology of Plant Litter Decomposition*, ed. C. H. Dickinson & G. J. F. Pugh, pp. 129–74. London: Academic Press.
Kleine, R. (1907). Die Entwicklung von Dipteren in den Brutgängen von *Myelophilus piniperda*. *Berliner entomologische Zeitschrift*, **52**, 109–13.
Krantz, G. W. (1965). A new species of *Macrocheles* associated with bark beetles of the genera *Ips* and *Dendroctonus*. *Journal of the Kansas Entomological Society*, **38**, 145–53.
Krebill, R. G. & Patton, F. F. (1962). Wounds in Jack pine roots as entry points for a succession of fungi. *Phytopathology*, **52**, 739.
Krogerus, R. (1927). Beobachtungen über die Successionen einiger Insecten biocoenosen in Fichten stumpfen. *Suomi hyöntüki Aikak*, **7**, 121–6.
Lagerberg, T., Lundberg, G. & Melin, E. (1927). Biological and practical researches into blueing in Pine and Spruce. *Skogsvårdsforeningens Tidskrift*, **25**, 145–272.
Leach, J. G. (1940). *Insect transmission of plant diseases*. New York: McGraw Hill.
Leach, J. G., Hodson, A. C., Chilton, J. P. & Christensen, C. M. (1940). Observations on two ambrosia beetles and their associated fungi. *Phytopathology*, **30**, 227–36.
Leach, J. G., Orr, L. W. & Christensen, C. M. (1934). The interrelationships of bark beetles and bluestaining fungi in felled Norway pine timber. *Journal of Agricultural Research*, **49**, 315–42.
Lekander, B. & Rennerfelt, E. (1955). Undersökninger över insektsoch blänadsskador på sågtimmer. *Meddelanden Statens Skogsforskningsinstitut*, **45**, 5–36.
Levy, J. F. (1982). The place of basidiomycetes in the decay of wood in contact with the ground. *British Mycological Society Symposium*, **4**, 161–78.
Limber, D. P. (1950). *Ophiostoma* on *Narcissus*. *Phytopathology*, **40**, 493–6.
Martin, J. L. (1965). Living stumps and insect control. *Review of Applied Entomology*, **55**, 987–9.
Matheison-Käärik, A. (1953). Eine Ubersicht über die gewöhnlichsten mit Barkenkägern assozierten Bläuepilze in Schweden und einige für Schweden neue Bläuepilze. *Meddelanden Statens Skogforskningsinstitut*, **43**, 1–74.

Matheison-Käärik, A. (1960). Keimung und Lebensdauer der Sporen einiger Bläuepilze bei verscheidener Luftfeuchtigkeit. *Annales Societas Lit. Est.*, **3**, 57–81.
Mathre, D. E. (1964). Pathogenecity of *Ceratocystis ips* and *C. minor* to *Pinus ponderosa*. Contributions. *Boyce Thompson Institute for Plant Research*, **22**, 363–88.
Mercer, P. E. (1982). Basidiomycete decay of standing trees. *British Mycological Society Symposium*, **4**, 143–60.
Meredith, D. S. (1959). The infection of pine stumps by *Fomes annosus* and other fungi. *Annals of Botany*, **23**, 455–76.
Moller, W. J. & DeVay, J. E. (1966). Role of insect vectors in *Ceratocystis* canker disease of stone fruit trees. *Phytopathology*, **56**, 891.
Morris, C. L., Thompson, H. E., Hadley, B. L. & Davis, J. M. (1955). Use of radioactive tracer for investigation of the activity pattern of suspected insect vectors of the oak wilt fungus. *Plant Disease Reporter*, **39**, 61–3.
Münch, E. (1906). Die Blaufäule des Nadelholzes. I. *Naturwissenschaftliches Zeitschrift der Forst- und Landwirtschaft*, **5**, 531–73.
Münch, E. (1907). Die Blaufäule des Nadelholzes. II. *Naturwissenschaftliches Zeitschrift der Forst- und Landwirtschaft*, **6**, 32–47, 297–323.
Munro, J. W. (1926). *British Bark Beetles*. Bulletin No. 8, Forestry Commission, London.
Norris, D. M. (1953). Insect transmission of oak wilt in Iowa. *Plant Disease Reporter*, **37**, 417–18.
Pechmann, H., Graessle, E. & Wutz, A. (1964). Untersuchungen über Bläuepilze an Kiefernholze. *Forstwissenschaftlich Zentralblatt*, **83**, 183–93.
Rawlings, G. B. (1948). Recent observations on the *Sirex nociltio* population in *Pinus radiata* forests in New Zealand. *Journal of Forestry*, **5**, 1–11.
Rennerfelt, E. (1950). Uber den Zusammenhang zwischen den Verblauen des Holzes und den Insekten. *Oikos*, **2**, 1–14.
Rishbeth, J. (1950). Observations on the biology of *Fomes annosus*, with particular reference to East Anglia pine plantations. *Annals of Botany*, **14**, 365–83.
Rishbeth, J. (1959). Dispersal of *Fomes annosus* Fr. and *Peniophora gigantea* (Fr.) Massee. *Transactions of the British Mycological Society*, **42**, 243–60.
Rumbold, C. T. (1929). Bluestaining fungi found in the US. *Phytopathology*, **19**, 597–9.
Savely, H. E. (1939). Ecological relations of certain animals in dead pine and oak logs. *Ecological Monographs*, **9**, 321–85.
Savory, J. G., Pawsey, R. G. & Lawrence, J. S. (1965). Prevention of bluestain in unpeeled Scots pine. *Forestry*, **38**, 59–81.
Seguy, F. (1950). *La biologie des diptères*. Paris: Le Chevalier.
Soper, R. S. & Olsen, R. E. (1963). Survey of biota associated with *Monochamus* in Maine. *Canadian Entomologist*, **95**, 83–95.
Spierenberg, D. (1922). Ein Onbekende ziekte in den Iepen. II. *Verslag en Mededelingen Plantenziektenkundigen Dienst te Wageningen*, **24**.
Steiner, G. & Buhrer, E. M. (1934). *Aphelenchoides xylophilus* n. sp. a nematode associated with bluestain and other fungi in timber. *Journal of Agricultural Research*, **48**, 949–51.
Swinburne, T. R. & Brown, A. E. (1976). A comparison of the use of *Bacillus subilis* with conventional fungicides for the control of apple canker. *Annals of Applied Biology*, **82**, 365–8.
Tuomikoski, R. (1957). Beobachtungen über einige Sciariden, deren larven in faulen Holz oder unter den Rinde abgestorbener Bäume Leben. *Suomi Hyontüki Aikak*, **23**, 3–35, 210–11.
Unger, D. F. (1866). Uber einen in grosser Verbreitung an Nadelhölzern beobachteten Fadenpilze. *Botanische Zeitung*, **5**, 249–57.

Vacarini, P. J. & Tokeshi, H. (1980). *Ceratocystis fimbriata*, causal agent of fig die back and its control. *Summa Phytopathologica*, **6,** 102–6.
Verrall, A. F. (1939). Relative importance and seasonal prevalence of wood-staining fungi. *Phytopathology*, **29,** 1031–51.
Verrall, A. F. (1941). The dissemination of fungi that stain logs and lumber. *Journal of Agricultural Research*, **63,** 549–58.
Vité, J. P. (1952). *Die holzerstörenden Insekten Mitteleuropas*. Göttlingen: Munsterschmidt.
Vité, J. P. (1961). The influence of water supply on the oleo resin exudation pressure and resistance to bark beetle attack in *Pinus ponderosa*. *Contributions. Boyce Thompson Institute for Plant Research*, **21,** 37–66.
Von Schrenk, H. (1903). The blueing and red rot of western yellow pine with special reference to the Black Hills Forest Reserve. *Bulletin of the Bureau of Plant Industries*, **35,** US Dept. of Agriculture.
Wallace, M. R. (1953). The ecology of the insect fauna of pine stumps. *Journal of Animal Ecology*, **22,** 154–71.
Whitney, H. S. & Cobb, F. W. (1972). Non-staining fungi associated with the bark beetle *Dendroctonus brevicornis* on *Pinus ponderosa*. *Canadian Journal of Botany*, **50,** 1943–5.

6
The role of ingested enzymes in the digestive processes of insects

MICHAEL M. MARTIN

Division of Biological Sciences, University of Michigan, Ann Arbor, Michigan 48109, USA

Key words: fungal enzymes; termites; siricid woodwasps; polysaccharide digestion; attine ants

Introduction

It has been recognized since the classic investigations of Buchner (1921, 1928) and Cleveland (1924) that microbial symbionts can play an important role in the digestive processes of insects. Permanent populations of hindgut protozoa are responsible for cellulose digestion in the lower termites (Noirot & Noirot-Timothée, 1969; Honigberg, 1970; Lee & Wood, 1971; Breznak, 1975, 1982; LaFage & Nutting, 1978; O'Brien & Slaytor, 1982) and in the wood roach, *Cryptocercus punctulatus*, (Cleveland, Hall, Sanders & Collier, 1934), while permanent populations of hindgut bacteria contribute to cellulose digestion in the American cockroach, *Periplaneta americana* (Bignell, 1977; Cruden & Markovetz, 1979) as well as in the rhinocerus beetle, *Oryctes nasicornis* (Bayon & Mathelin, 1980). My co-workers and I have recently discovered a different mechanism by which insects accomplish the digestion of refractile dietary components, namely by the acquisition through ingestion of fungal enzymes active against plant cell wall polysaccharides (Martin & Martin, 1978, 1979; Martin, 1979). This review summarizes the results of our investigations of the significance of acquired fungal enzymes in the digestive processes of the fungus-growing termites, the siricid woodwasps, and the fungus-growing ants, and will conclude with a brief survey of other studies which suggest that the acquisition of a digestive capacity through the ingestion of a set of enzymes may be a fairly widespread phenomenon of considerable biological importance. Since acquired fungal enzymes have been shown to contribute to cellulose digestion in several species of wood-feeders, and seem likely to be implicated in other insects with diets high in cellulosic materials, it is appropriate to give a brief description of the enzymatic basis for cellulolysis in fungi.

The cellulolytic enzymes of fungi

In fungi, cellulose digestion is accomplished by a 'cellulase complex' that includes three major classes of hydrolytic enzymes (Table 6.1): endoglucanases (C_x-cellulases), cellobiohydrolases (C_1-cellulases), and cellobiases (Reese & Mandels, 1971; Bailey, Enari & Linko, 1975; Wood & McCrae, 1979; Ghose, Montenecourt & Eveleigh, 1981). The digestion of native cellulose is believed to be initiated when endoglucanases attack isolated amorphous regions of the predominantly crystalline cellulose, creating nicks in the linear cellulose chains. The cellobiohydrolases attack at these 'nick sites', liberating cellobiose, exposing additional potential sites for attack by the endoglucanases, and generally disrupting the highly ordered structure of the cellulose matrix. The continued combined action of the C_x- and C_1-cellulases results eventually in the complete degradation of the original cellulose and the production of cellobiose and a mixture of soluble linear oligosaccharides of varying chain lengths. The cellobiose, which is a potential inhibitor of both the C_1- and C_x-cellulases, is hydrolyzed to glucose by cellobiase, and the various oligosaccharides are further degraded, ultimately to glucose, by the action of both the C_x-cellulases and the cellobiases. Thus, the utilization of cellulose is dependent upon the concerted and synergistic action of three types of enzymes. If any one of the three is missing, cellulose digestion cannot proceed.

The fungus-growing termites

The fungus-growing termites (Termitidae, subfamily Macrotermitinae) forage for food in seasoned structural timber, fence posts, sound dead wood, stumps, grass, and herbivore dung (Coaton & Sheasby, 1972). They are higher termites which lack the assemblage of xylophagous protozoa that provide the lower termites with the cellulolytic enzymes they require in order to digest cellulose. These termites have long fascinated biologists because of their symbiotic association with fungi that grow in their nests on structures referred to as 'fungus combs' (Sands, 1969). The surface of the comb is covered with a sparse growth of mycelium and numerous white spheres or nodules, 0.5 to 2 mm in diameter (Fig. 6.1). These nodules are the conidia or conidiophores of a fungus, *Termitomyces* sp, believed to be restricted to the nests of the Macrotermitinae. The mycelium is a mixture of

Table 6.1. *Enzymes involved in cellulose depolymerization by fungi*

Enzyme (Alternate designations)	Mode of action and products	Substrates
The cellulase complex	A combination of the three categories of enzymes designated below, which brings about the complete digestion of native cellulose to glucose.	Microcrystalline cellulose powder, Avicell®, cotton, and filter paper.
1,4-β-D-Glucan 4-glucanohydrolase (EC 3.2.1.4) (Endo-β-1,4-glucanase) (Endoglucanase) (Carboxymethylcellulase) (CMCase) (C_x-Cellulase)	Random attack on β-1,4-glucosidic bonds, generating transient cellodextrins, cellobiose, and glucose	CMC and other soluble derivatives of cellulose, phosphoric acid swollen cellulose, and cellodextrins (increasing activity with increasing chain length). No activity toward crystalline cellulose. Hardly any activity toward cellobiose
1,4-β-D-Glucan cellobiohydrolase (EC 3.2.1.91) (Cellobiohydrolase) (C_1-cellulase)	Removal of cellobiose units from the non-reducing end of a β-1,4-glucan chain by attack on penultimate β-1,4-glucosidic bonds	Microcrystalline cellulose powder, Avicel®, cotton, swollen cellulose, and dextrins (increasing activity with increasing chain length). Limited activity toward CMC.
1,4-β-D-Glucoside 4-glucohydrolase (EC 3.2.1.21) (β-D-Glucosidase) (Cellobiase)	Hydrolysis of the β-1,4-glucosidic bond of cellobiose to generate glucose	Cellobiose, other β-linked disaccharides of glucose, and cellodextrins. No activity toward cellulose.

6 Role of ingested enzymes in digestion by insects 158

Fig. 6.1. *Macrotermes natalensis* worker on fungus comb bearing conidiophores of *Termitomyces* sp. (Photograph by Dr Robin Crewe, Dept. of Zoology, University of the Witwatersrand, Johannesburg, South Africa.)

Termitomyces and various xylariaceous fungi. The comb is constructed of small fragments of plant tissue in which intact cellular structure is still evident in scanning electron micrographs (Rohrmann, 1978; Wood, 1978). The termites rapidly starve to death on a diet composed exclusively of sound wood or filter paper, but survive for extended periods of time if they are also provided with a fragment of fungus comb complete with nodules (Sands, 1956; Ausat *et al.*, 1960). Survival is not prolonged if the fragment of fungus comb provided for the termites has been sterilized (Sands, 1956). These studies demonstrate that there is something in the live, intact fungus combs which is crucial to the nutritional well-being of the termites. The latter are known to consume both the *Termitomyces* nodules and the comb material (Grassé & Noirot, 1957; Alibert, 1964; Batra, 1975).

Martin & Martin (1978, 1979) showed that the midgut of the fungus-growing termite, *Macrotermes natalensis*, contains C_1-cellulases, C_x-cellulases, and β-glucosidases (presumably including cellobiases), whereas the paunch, which is the name given to the enlarged anterior portion of the hindgut where numerous bacteria reside, contains

Table 6.2. *Cellulase complex and C_x-cellulase activity of* Macrotermes natalensis *workers and their fungus gardens*

	Units of enzymatic activity (per termite gut section or per mg of fungus or comb material)	
	Cellulase complex	C_x-cellulase
Midgut (tissue + contents)	0.45 ± 0.05 (7)	1.19 ± 0.16 (6)
Paunch (tissue + contents)	0.04 ± 0.01 (5)	0.62 ± 0.08 (5)
Rectum (tissue + contents)	0.05 ± 0.02 (3)	0.50 ± 0.17 (3)
Midgut (tissue)	0.00 ± 0.00 (3)	0.33 ± 0.14 (3)
Salivary glands	0.01 ± 0.01 (5)	0.72 ± 0.14 (7)
Fungus nodules	0.53 ± 0.05 (4)	1.60 ± 0.46 (6)
Comb material	0.01 ± 0.01 (3)	0.04 ± 0.03 (3)

Note: Each value is the mean ± SEM for the number of replicates indicated in parenthesis. A unit of activity is the amount of enzyme required to liberate 1 micromole of maltose equivalents per hour under the conditions of the assay (37 °C, pH 5.0, incubation volume 1.0 ml). Substrates were microcrystalline cellulose and CMC.

Source: From Martin & Martin, 1978, 1979.

C_x-cellulases and β-glucosidases but only a barely detectable trace of activity attributable to the presence of C_1-cellulases (Table 6.2). These results clearly identify the midgut as a major site of cellulose digestion in this insect, especially the early stages of cellulose digestion when the C_1-enzymes are so critical. Extracts of the salivary glands and the midgut tissue actively degrade carboxymethylcellulose (CMC) but not microcrystalline cellulose, demonstrating that the insects are able to synthesize C_x-cellulases but not C_1-cellulases. However, an extract of the fungus nodules actively digests microcrystalline cellulose and contains high levels of C_x-cellulase activity (Table 6.3), demonstrating that C_1-cellulases are present in the *Termitomyces* nodules. In contrast to the fungus nodules, the comb itself is virtually devoid of enzymatic activity. These findings suggested to us the possibility that the *Termitomyces* nodules might be the source of the C_1-cellulases detected in the midgut fluids of the termite workers. Further evidence in support of this hypothesis was obtained when we demonstrated that after 72 hours the level of activity toward microcrystalline cellulose decreased by 40 to 50% in the midguts of termites provided only with water or with fungus comb from which the nodules had been removed, while the level of activity remained at 100% of the value measured in termites collected

Table 6.3. *Enzymatic activity toward various polysaccharides of extracts of midguts (tissue + contents) of Sirex cyaneus larvae and of extracts of cultures of their symbiotic fungus, Amylostereum chailletii*

	Units of enzymatic activity per milligram dry weight of dissolved solids in extract					
	Amylase	Cellulase complex	C_x-cellulase	Laminarinase	Pectinase	Xylanase
Larvae collected from natural galleries in a balsam fir log	6.06 ± 2.62 (5)	0.33 ± 0.13 (5)	8.86 ± 6.20 (5)	14.23 ± 8.91 (5)	8.47 ± 5.30 (5)	11.59 ± 7.21 (5)
Larvae cultured for one week on balsam fir chips permeated by *A. chailletii* mycelium	4.18 ± 1.30 (5)	0.48 ± 0.01 (5)	20.17 ± 0.59 (5)	12.91 ± 3.14 (5)	11.03 ± 1.11 (5)	33.41 ± 3.97 (5)
Larvae cultured for one week on sterile balsam fir chips	2.36 (1)	0.0 (1)	0.09 (1)	1.62 (1)	0.97 (1)	1.39 (1)
Balsam fir chips permeated by *A. chaelletii* mycelium	2.54 ± 0.00 (5)	0.37 ± 0.00 (5)	1.58 ± 0.65 (5)	5.65 ± 2.12 (5)	1.50 ± 0.74 (5)	6.78 ± 2.53 (5)
Sterile balsam fir chips	0.02 ± 0.01 (5)	0.0 (5)	0.0 (5)	0.02 ± 0.01 (5)	0.0 (5)	0.0 (5)
A. chailletii growing on microcrystalline cellulose suspended in a chemically defined liquid medium	9.48 ± 0.26 (5)	0.33 ± 0.03 (5)	7.45 ± 0.33 (5)	8.47 ± 0.38 (5)	9.76 ± 0.46 (5)	9.24 ± 0.33 (5)

Note: Each value is the mean ± SEM for the number of replicates indicated in parenthesis. A unit of activity is the amount of enzyme required to liberate 1 micromole of maltose equivalents per hour under the conditions of the assay (37°, pH 5.0, incubation volume 1.0 ml). Substrates were potato α-amylose, microcrystalline cellulose, CMC, seaweed laminarin, citrus pectin, and larchwood xylan.

Source: From Kukor & Martin, 1983.

from a nest if the termites were provided with a supply of nodules. C_x-activity remained at the normal level in termites which were not allowed to consume the *Termitomyces* nodules, suggesting that the decrease in activity toward microcrystalline cellulose in the termites deprived of access to the nodules was due to a decrease in the amount of C_1-cellulase present.

Finally, by the use of the technique of isoelectric focusing, we confirmed the identity of the C_1-cellulases in the nodules and the C_1-cellulases in the termites' midgut. Both the nodule and the midgut extracts contain two separate C_1-cellulases. The major enzyme from both sources has an isoelectric pH between 3.90 and 4.05, while the minor enzyme from both sources has an isoelectric pH in the range 4.20 to 4.35. Our isoelectric focusing studies also revealed that some of the C_x-cellulase activity detected in the midgut of the termite is due to enzymes originating from the ingested fungus nodules. These studies demonstrated that the ability to digest cellulose by *M. natalensis* is dependent upon the acquisition of critical enzymes, the C_1-cellulases, through the consumption by the termites of their fungal symbiont.

Abo-Khatwa (1978), studying the closely related termite *M. subhyalinus*, obtained comparable results and arrived at an identical interpretation.

The siricid woodwasps

More recently we have investigated the digestive processes in the larvae of the siricid woodwasp, *Sirex cyaneus* (Siricidae), another wood-feeder that is associated with a fungal symbiont (Francke-Grosmann, 1967; Kukor & Martin, 1983). The larvae develop in timber in galleries lined with the mycelium of a fungus introduced into the wood along with the egg at the time of oviposition. *Sirex* larvae can be grown normally on a pure culture of the symbiotic fungus, *Amylostereum chailletii*, but fail to survive in galleries devoid of fungal growth (Cartwright, 1929; Stillwell, 1966). The normal larval diet is a mixture of wood and the mycelium of their symbiotic fungus (Buchner, 1928, 1965).

An extract of whole guts dissected from larvae collected from natural galleries in a balsam fir log catalyzes the degradation of both microcrystalline cellulose and CMC (Table 6.3). Although activity toward microcrystalline cellulose is low compared to activity toward CMC and several other polysaccharides, the observed level is by no means a marginal one barely exceeding the lower limits of detection. The assay results provide an unambiguous demonstration that the gut fluids of this insect contain both C_1- and C_x-cellulases. Müller (1934) established the capacity of siricids to digest cellulose by determining that 22 and 31% of the cellulose in fir and spruce wood is assimilated during passage through the guts of *S. gigas* and *S. phantoma* respectively. These two species also assimilate significant quantities of ingested hemicellulose.

Experiments in which the diet of *S. cyaneus* is manipulated provide strong evidence that the C_1- and C_x-cellulases, as well as enzymes active against xylan, laminarin and pectin, are acquired from ingested fungal tissue (Table 6.3). The larvae can be maintained quite satisfactorily on a diet of balsam fir wood chips on which the symbiotic fungus has been allowed to grow for about a month. The larvae feed actively and appear to be normal. Their digestive fluids contain the same digestive enzymes as larvae collected from natural galleries, and at roughly comparable levels of activity. However, if the larvae are put on a diet of sterile balsam fir chips, they fare very badly. Mortality is high, and after a week enzymatic activity toward cellulose, CMC, xylan, laminarin, and pectin has declined to very low levels. Amylose is the only substrate toward which enzymatic activity remains high in larvae on a diet of sterile wood chips. Enzymes active against all of these substrates are present in the fungal symbiont of *S. cyaneus*, and are readily detectable in balsam wood chips on which the fungus is growing (Table 6.3).

The fungal origin of the enzymes responsible for C_x-cellulase and xylanase activity in the gut fluids of the woodwasp larvae was confirmed by establishing the identities of the gut and fungal enzymes using the techniques of isoelectric focusing and chromatofocusing. The C_x-cellulases and the xylanases present in a gut extract and in a liquid culture medium in which *A. chailletii* had been growing were purified by a protocol involving concentration, dialysis, ion exchange chromatography, gel permeation chromatography and chromatofocusing. C_x-cellulase and xylanase activity eluted in comparable fractions during the purification of the insect and fungal extracts. Both the larval midgut and the fungus culture extracts yielded two major C_x-cellulases and three major xylanases. The first cellulase from both sources eluted from the preparative chromatofocusing column at pH 4.9, and the second cellulase at a pH below 4.0. The first xylanase eluted at pH 5.7, the second at pH 5.3, and the third at pH 4.5. A further comparison of insect and fungal enzymes isolated from the chromatofocusing columns was made by subjecting the cellulase and xylanase fractions to analytical isofocusing on ultrathin polyacrylamide gels. This procedure revealed that several proteins were present in each of the cellulase and xylanase fractions. We did not determine whether the observed protein multiplicity reflected enzyme multiplicity or simply incomplete purification. However, the point of overriding importance is that the patterns seen in each of the cellulase and xylanase fractions from the fungal extract are identical with those seen in the comparable fractions derived from the insect extract. These experiments demonstrate that the C_x-cellulases and xylanases present in the gut of *S. cyaneus* larvae are identical to the enzymes produced by the fungus, *A. chailletii*, and demonstrate that the woodwasp larvae, like the fungus-growing termites, acquire essential digestive enzymes by the ingestion of their fungal symbiont.

The feeding experiments summarized in Table 6.3 also strongly suggest that enzymes active against pectin and β-1,3-glucans similar to laminarin are acquired fungal enzymes. That possibility has not yet been confirmed by direct comparisons of the gut and fungal enzymes, however.

In 1928, Prof. Paul Buchner, in discussing the siricid–fungus symbiosis, wrote the following: 'Man hat den Eindruck, dass die auf solche Weise aufgenommene Menge Pilzubstanz der Larve nicht genügen kann und ich mochte die Vermutung aussprechen, dass hier eine ganz neue Variante der Ambrosiazucht vorliegt, bei der es dem Insekt vielleicht weniger auf die Substanz des Pilzes ankommt, als auf die Bestandteile des Holzes lösenden Nahrungsenzyme, die es sich mit den Mycelstücken zuführt, und die ihre dem Tier abgehende Wirkung auf das Holz auch im Darmlumen geltend machen.'[1] This passage apparently went unnoticed by workers interested in symbiosis and insect–microbial interactions. We are pleased, honoured, and slightly humbled to find that our research conducted half a century later has verified the prescient speculation of this pioneer in the field of insect–microbial interactions.

The fungus-growing ants

The fungus-growing ants (Formicidae, subfamily Myrmicinae, tribe Attini) provide yet another variant to the role of acquired enzymes in the biology of an insect that is obligatorily associated with a fungal symbiont (Martin, Gieselmann & Martin, 1973; Boyd & Martin 1975*a*,

[1] 'One has the impression that the various fungal constituents are (dietically) insufficient for the larvae. I suggest that this represents an entirely new aspect of the Ambrosia cultivation in which the insect utilizes little of the fungus other than the range of wood-degrading enzymes. These are ingested with mycelial fragments and are active on wood in the gut lumen together with the insects' enzymes.'

b; Martin, Boyd, Gieselmann & Silver, 1975). In addition to being highly active against protein, the gut fluids of these ants are also active against a number of polysaccharides, including starch, CMC, pectin, xylan and chitin. We suspect that most, if not all, of these enzymes are acquired by the ants when they ingest the fluids present in the mycelium of their fungal symbiont (Fig. 6.2). We have established unequivocally that the proteolytic enzymes in the gut fluids of *Atta texana* are of fungal origin.

The proteolytic activity of the gut fluid is due to the presence of three enzymes, one serine proteinase and two metalloendoproteinases. A serine proteinase and two metalloendoproteinases are also present in an extract of the mycelial mat which results when the ants' fungal symbiont is cultured in a defined liquid medium. The fungal enzymes are not secreted into the culture medium to any significant extent. A comparison of molecular weights, substrate specificities, inhibition characteristics, pH-activity relationships and electrophoretic mobilities established the identity of the fungal enzymes and the enzymes present in the ants' gut fluids.

Unlike the fungus-growing termites and the siricid woodwasps, the food of the fungus-growing ants does not include highly refractile polysaccharides. Fungal tissue, or the fluids derived therefrom, is the exclusive food of the larvae and a minor component of the adults' diet, which consists largely of plant juices (Quinlan & Cherrett, 1979). Consequently, it does not seem likely that the ants would be dependent upon acquired fungal enzymes for the processing of ingested nutrients, and we are aware of no information which suggests a critical role for the fungal enzymes in the digestive processes occurring in the ants' digestive tracts. However, we have shown that the acquired fungal enzymes survive passage through the midgut, that they are concentrated in the rectum, and are still present and active in the faecal material, which plays a vital role in the fungus-culturing activities of these ants. The faecal fluid is regularly applied to the freshly cut leaf fragments which are used as the substrate on which *Atta* grows its fungal symbiont (Wheeler, 1907; Weber, 1972). We have shown that the fungal symbiont of *A. texana* grows very slowly in a culture medium in which the nitrogen is present in the form of polypeptides, that it grows rapidly if nitrogen is supplied as a mixture of amino acids, that it produces but does not secrete digestive proteinases, and that the growth rate of the fungus in a culture medium containing polypeptides as the sole nitrogen source is greatly increased by the addition of the ants' faecal material or

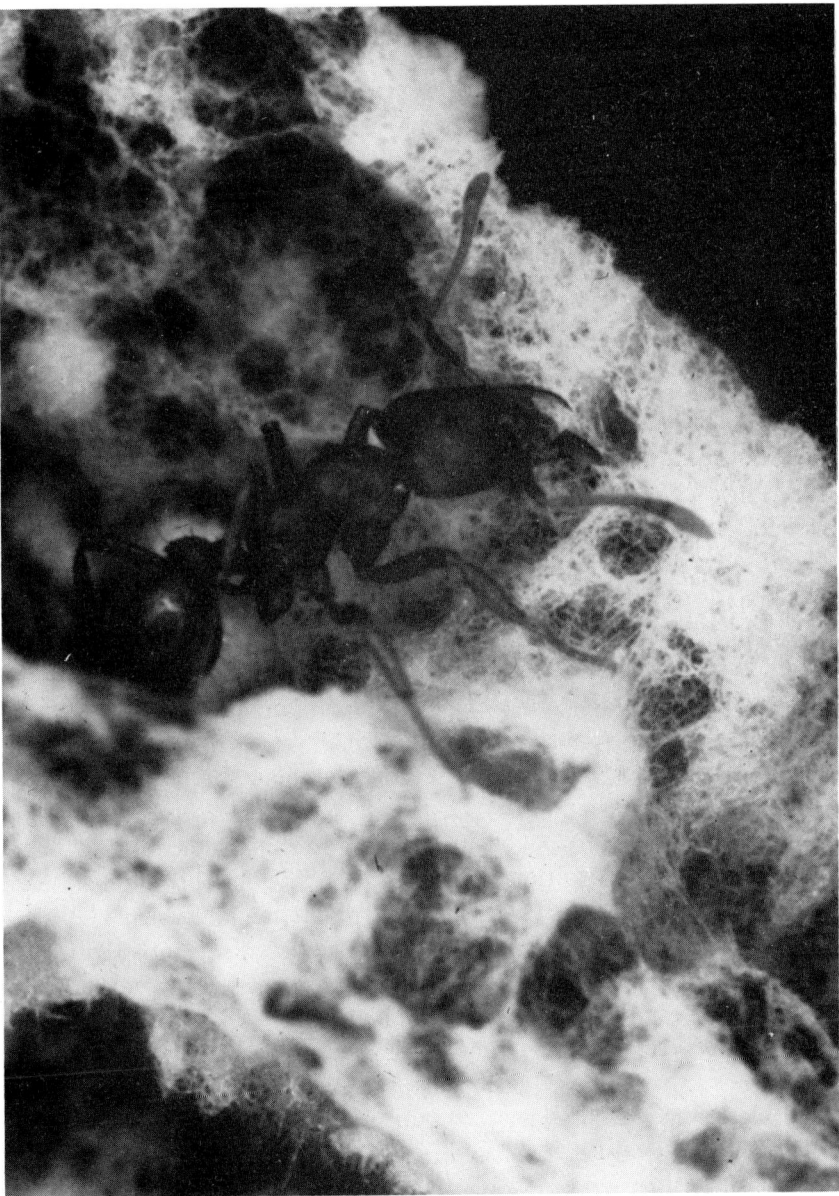

Fig. 6.2. *Atta colombica* worker on fungus garden. (Photograph by Mr David Bay, Division of Biological Sciences, University of Michigan, Ann Arbor, Michigan, USA.)

the proteolytic enzymes it contains (Boyd & Martin, 1975b). The application of the ants' faecal material to the freshly collected leaf fragments can, therefore, be viewed as a mechanism for increasing the initial growth rate of fungus by augmenting the enzymatic capacity of the new inoculum to degrade protein in the leaf tissue used as substrate. In addition, the pectinases present in the faecal material should foster rapid initial growth by degrading the intercellular middle lamella, thereby facilitating the penetration and ramification of the fungal hyphae in the leaf tissue.

Copious enzyme production and rapid hyphal penetration are two major factors which contribute to competitive ability in a fungus. Defined within the context of the adaptive strategies of the fungus, the feeding and defaecation behaviour of the adult ants can be viewed simply as a mechanism that has replaced the process of enzyme secretion. However, the ants achieve a distribution of enzymes more conducive to rapid initial growth than would result from the direct secretion of the enzymes by the fungus, since the ants feed in mature portions of the garden and transport enzymes produced there in abundance to the site of inoculation, where only small quantities of enzymes can be produced by the small mycelial fragment placed there by the ants. In summary then, in this truly remarkable symbiosis the ants acquire digestive enzymes from their fungal partner, only to return them in a manner which enhances fungal fitness.

Other examples of enzyme acquisition through ingestion

At the present time it is not clear whether the acquisition of enzymes through ingestion is a rare phenomenon which we have fortuitously encountered because of our interest in several spectacular, but highly atypical, coevolved symbiotic associations between insects and fungi, or whether it is a common phenomenon with broad biological significance. There is certainly no information which precludes the involvement of acquired enzymes in the digestive processes of other species of higher termites or of the wood-feeding buprestid (Schlottke, 1945), anobiid (Müller, 1934; Norman, 1936; Parkin, 1940) and cerambycid (Ripper, 1930; Mansour & Mansour-Bek, 1934; Müller, 1934; Parkin, 1940; Mishra & Singh, 1978) beetles, cellulose-digesting insects which are often presumed to produce their own cellulases, even though evidence in support of this presumption is usually permissive at best. There are scattered reports which suggest that the ingestion of enzymes may be important in organisms other than wood-feeders. This review will close with a summary of all of the other studies known to us which suggest that the ingestion of enzymes produced by another organism might be a widely occurring mechanism for acquiring a digestive capability.

Low levels of cellulase activity have been reported in many species of carnivorous marine and freshwater invertebrates (Araki & Giese, 1970; Kristensen, 1972; Monk, 1976), and Araki & Giese (1970) suggested that this activity may be due to prey enzymes. A similar origin has been proposed for low levels of cellulase activity in the gut fluids of several fish (Niederholzer & Hofer, 1979). It has not been suggested that these enzymes serve any adaptive function in the digestive processes of the predator species, but the observations at least underline the possibility that ingested enzymes can survive and remain active for extended periods of time following ingestion.

An adaptive role for ingested enzymes in several species of insects other than wood-feeders has been postulated. Applebaum (1964) reported the presence of a β-amylase of dietary origin along with an α-amylase of insect origin in the gut fluids of *Tenebrio* larvae reared on wheat. Since the insect's own α-amylase is inhibited by a substance in the wheat, starch digestion is presumably brought about entirely through the action of the acquired β-amylase. More recently Grogan & Hunt (1979) have suggested that protein digestion in adult honey bees is accomplished at least in part by proteolytic enzymes derived from pollen. Powning & Davidson (1979) have noted the bacteriolytic action of the gut fluids of larvae of the bean weevil *Acanthoscellides obsoletus*, and have proposed that this lytic activity is derived in part from the seeds they feed upon. The possibility that acquired enzymes might contribute to an organism's defenses against bacterial infections and aid in the detoxification of allelochemics is an important one deserving further consideration. In this context, the capacity of the terrestrial isopod *Oniscus asellus* to metabolize aromatic compounds has been attributed to ingested bacteria or bacterial enzymes (Kaplan & Hartenstein, 1978).

Detritus-feeding invertebrates would seem to be a group likely to include species that could benefit from the ingestion of microbial enzymes. Detritus consists of dead plant tissue of low nutritive value along with varying amounts of associated bacterial cells and fungal tissue. Numerous studies have demonstrated a preference on the part of detritivores for litter which supports a rich culture of microorganisms, especially fungi (Berrie, 1976; Anderson & Sedell, 1979; Cummins &

Klug, 1979). The preference for microbially colonized substrate has usually been interpreted as an ability of detritivores to select high quality over low quality food. Microbial tissue is more nutritious and more digestible than dead plant tissue, which is typically low in nitrogen and high in cell wall constituents. However, it is also possible that detritivores might benefit from selecting substrate supporting fungal growth because it provides fungal enzymes which are more effective than the invertebrate's own enzymes in degrading refractile components of the plant tissue. Low levels of C_x-cellulase activity have been detected in detritus-feeding stonefly nymphs (Plecoptera) (Martin, Martin, Kukor & Merritt, 1981; Sinsabaugh, Benfield & Linkins, 1981), caddisfly larvae (Trichoptera) (Bjarnov, 1972; Monk, 1976; Martin, Kukor, Martin, Lawson & Merritt, 1981), and crane fly (Sinsabaugh *et al.*, 1981), chironomid (Bjarnov, 1972) and sciarid (Terra, Ferreira & deBianchi, 1979) larvae (Diptera). On the basis of a correlation between the level of C_x-activity in the gut and the presence or level in the food, it has been proposed that acquired microbial enzymes are responsible at least in part for activity in the terrestrial isopod *Philoscia muscorum* (Hassal & Jennings, 1975), the aquatic amphipod *Gammarus fossorum* (Bärlocher, 1982), and the aquatic nymphs of the stonefly *Pteronarcys proteus* (Benfield, personal communication). The role of ingested microbial enzymes in the nutrition of detritus-feeding invertebrates is a subject deserving further investigation at this time.

Acknowledgments. I thank my co-workers, Drs N. D. Boyd, J. J. Kukor and J. S. Martin, and Ms M. J. Gieselmann, for their many excellent contributions over the years to the research described in this review, and to the National Institutes of Health (Grant AI-07386), the National Science Foundation (Grants GB-31581, PCM-78-22733, PCM-82-03537, and a Faculty Fellowship in Science), CSIRO, Pretoria (Grant 398-77-21-REI), the Sloan Foundation, and the Horace Rackham School of Graduate Studies of the University of Michigan for their generous and timely financial support.

References

Abo-Khatwa, N. (1978). Cellulase of fungus-growing termites: a new hypothesis on its origin. *Experientia*, **34**, 559–60.
Alibert, J. (1964). L'evolution dans le temps des meule à champignons construites par les termites. *Comptes rendus hebdomadaire des séances de l'Académie des sciences*, **258**, 5260–3.

Anderson, N. H. & Sedell, J. R. (1979). Detritus processing by macroinvertebrates in stream ecosystems. *Annual Review of Entomology*, **24**, 351–77.

Applebaum, S. W. (1964). The action pattern and physiological role of *Tenebrio* larval amylase. *Journal of Insect Physiology*, **10**, 897–906.

Araki, G. S. & Giese, A. C. (1970). Carbohydrases in sea stars. *Physiological Zoology*, **43**, 296–305.

Ausat, A., Cheema, P. S., Koskhi, T., Petri, S. L. & Ranganathan, S. K. (1960). Laboratory culturing of termites. In *Termites in the Humid Tropics, Proceedings of the New Delhi Symposium*, Oct. 4–12, 1960, pp. 121–5. Paris: UNESCO.

Bailey, M., Enari, T. M. & Linko, M. (eds.) (1975). *Symposium on Enzymatic Hydrolysis of Cellulose*. Helsinki: SITRA.

Bärlocher, F. (1982). The contribution of fungal enzymes to the digestion of leaves by *Gammarus fossarum* Koch. *Oecologia*, **52**, 1–4.

Batra, S. W. T. (1975). Termites (Isoptera) eat and manipulate symbiotic fungi. *Journal of the Kansas Entomological Society*, **48**, 89–92.

Bayon, C. & Mathelin, J. (1980). Carbohydrate fermentation and by-product absorption studied with labelled cellulose in *Oryctes nasicornis* larvae (Coleoptera: Scarabaeidae). *Journal of Insect Physiology*, **26**, 833–40.

Berrie, A. D. (1976). Detritus, microorganisms and animals in fresh water. In *The Role of Terrestrial and Aquatic Organisms in Decomposition Processes*, ed. J. M. Anderson & A. Macfadyen, pp. 328–38. Oxford: Blackwells.

Bignell, D. E. (1977). An experimental study of cellulose and hemicellulose degradation in the alimentary canal of the American cockroach. *Canadian Journal of Zoology*, **55**, 579–89.

Bjarnov, N. (1972). Carbohydrases in *Chironomus, Gammarus* and some Trichopteran larvae. *Oikos*, **23**, 261–3.

Boyd, N. D. & Martin, M. M. (1975a). Faecal proteinases of the fungus-growing ant, *Atta texana*: properties, significance and possible origin. *Insect Biochemistry*, **5**, 619–35.

Boyd, N. D. & Martin, M. M. (1975b). Faecal proteinases of the fungus-growing ant, *Atta texana*: their fungal origin and ecological significance. *Journal of Insect Physiology*, **21**, 1815–20.

Breznak, J. A. (1975). Symbiotic relationships between termites and their intestinal microbiota. *Symposium of the Society for Experimental Biology*, **29**, 559–80.

Breznak, J. A. (1982). Intestinal microbiota of termites and other xylophagous insects. *Annual Review of Microbiology*, **36**, 323–43.

Buchner, P. (1921). Studien an intracellularen Symbionten. III. Die Symbiose der Anobiinen mit Hefepilzen. *Archiv für Protistenkunde*, **42**, 317–36.

Buchner, P. (1928). *Holznahrung und Symbiose*. Berlin: Springer Verlag.

Buchner, P. (1965). *Endosymbiosis of Animals with Plant Microorganisms, Revised English Version*. New York: Interscience.

Cartwright, K. St. G. (1929). Notes on a fungus associated with *Sirex cyaneus*. *Annals of Applied Biology*, **16**, 182–7.

Cleveland, L. R. (1924). The physiological and symbiotic relationship between the intestinal protozoa of termites and their host, with special reference to *Reticulitermes flavipes* Kollar. *Biological Bulletin*, **46**, 117–227.

Cleveland, L. R., Hall, S. R., Sanders, E. P. & Collier, J. (1934). The wood-feeding roach *Cryptocerus*, its protozoa, and the symbiosis between protozoa and roach. *Memoirs of the American Academy of Arts and Sciences*, **17**, 185–342.

Coaton, W. G. H. & Sheasby, J. L. (1972). Preliminary report on a survey of the termites (Isoptera) of South West Africa. *Cimbebasia, Series A*, **2**, 1–129.

Cruden, D. L. & Markovetz, A. J. (1979). Carboxymethylcellulose decomposition by intestinal bacteria of cockroaches. *Applied and Environmental Microbiology*, **38**, 369–72.

Cummins, K. W. & Klug, M. J. (1979). Feeding ecology of stream invertebrates. *Annual Review of Ecology and Systematics*, **11**, 147–72.

Francke-Grosmann, H. (1967). Ectosymbiosis in wood-inhabiting insects. In *Symbiosis*, Vol. 2, ed. S. M. Henry, pp. 142–205. New York: Academic Press.

Ghose, T., Montenecourt, B. S. & Eveleigh, D. E. (1981). *Measure of Cellulase Activity*. International Union of Pure and Applied Chemistry, Commission on Biotechnology.

Grassé, P. P. & Noirot, C. (1957). Le signification meules à champignons des Macrotermitinae (Ins., Isoptera). *Comptes rendus hebdomadaire des séances de l'Académie des sciences*, **244**, 1845–50.

Grogan, D. E. & Hunt, J. H. (1979). Pollen proteases: their potential role in insect digestion. *Insect Biochemistry*, **9**, 309–13.

Hassall, M. & Jennings, J. B. (1975). Adaptive features of gut structure and digestive physiology in the terrestrial isopod *Philoscia muscorum* (Scopoli) 1763. *Biological Bulletin*, **49**, 348–64.

Honigberg, B. M. (1970). Protozoa associated with termites and their role in digestion. In *Biology of Termites*, Vol. 2, ed. K. Krishna & F. M. Weesner, pp. 1–36. New York: Academic Press.

Kaplan, D. L. & Hartenstein, R. (1978). Studies on monooxygenases and dioxygenases in soil macroinvertebrates and bacterial isolates from the gut of the terrestrial isopod, *Oniscus asellus* L. *Comparative Biochemistry and Physiology*, **60B**, 47–50.

Kristensen, J. H. (1972). Carbohydrases of some marine invertebrates with notes on their food and on the natural occurrence of the carbohydrate studied. *Marine Biology*, **14**, 130–42.

Kukor, J. J. & Martin, M. M. (1983). Siricid wood wasps acquire digestive enzymes from their fungal symbiont. *Science*, **220**, 1161–3.

LaFage, J. P. & Nutting, W. L. (1978). Nutrient dynamics of termites. In *Production Ecology of Ants and Termites*, ed. M. V. Brian, pp. 165–232. Cambridge University Press.

Lee, K. E. & Wood, T. G. (1971). *Termites and Soils*. New York: Academic Press.

Mansour, K. & Mansour-Bek, J. J. (1934). The digestion of wood by insects and the supposed role of micro-organisms. *Biological Reviews*, **9**, 363–82.

Martin, M. M. (1979). Biochemical implications of insect mycophagy. *Biological Reviews*, **54**, 1–21.

Martin, M. M., Boyd, N. D., Gieselmann, M. J. & Silver, R. G. (1975). Activity of faecal fluid of a leaf-cutting ant toward plant cell wall polysaccharides. *Journal of Insect Physiology*, **21**, 1887–92.

Martin, M. M., Gieselmann, M. J. & Martin, J. S. (1973). Rectal enzymes of attine ants. α-Amylase and chitinase. *Journal of Insect Physiology*, **19**, 1409–16.

Martin, M. M., Kukor, J. J., Martin, J. S., Lawson, D. L. & Merritt, R. W. (1981). Digestive enzymes of larvae of three species of caddisflies (Trichoptera). *Insect Biochemistry*, **5**, 501–5.

Martin, M. M. & Martin, J. S. (1978). Cellulose digestion in the midgut of the fungus-growing termite *Macrotermes natalensis*: the role of acquired digestive enzymes. *Science*, **199**, 1453–5.

Martin, M. M., & Martin, J. S. (1979). The distribution and origins of the cellulolytic enzymes of the higher termite *Macrotermes natalensis*. *Physiological Zoology*, **52**, 1–11.

Martin, M. M., Martin, J. S., Kukor, J. J. & Merritt, R. W. (1981). The digestive enzymes of detritus-feeding stonefly nymphs (Plecoptera: Pteronarcyidae). *Canadian Journal of Zoology*, **59**, 1947–51.

Mishra, S. C., & Singh, P. (1978). Polysaccharide digestive enzymes in the larvae of *Stromatium barbatum* (Fabr.), a dry wood borer (Coleoptera: Ceramycidae). *Materiel und Organismen*, **13**, 115–22.

Monk, D. C. (1976). The distribution of cellulase in freshwater invertebrates of different feeding habits. *Freshwater Biology*, **6**, 471–5.

Müller, W. (1934). Untersuchungen über die Symbiose von Tieren mit Pilzen und Bakterien. *Archiv für Mikrobiologie*, **5**, 84–147.

Niederholzer, R. & Hofer, R. (1979). The adaptation of digestive enzymes to temperature, season and diet in roach *Rutilus rutilus* L. and rudd *Scardinius erythrophthalmus* L. Cellulase. *Journal of Fish Biology*, **15**, 411–16.

Noirot, C. & Noirot-Timothée, C. (1969). The digestive system. In *Biology of Termites*, Vol. 1, ed. K. Krishna & F. M. Weesner, pp. 49–88. New York: Academic Press.

Norman, A. G. (1936). The destruction of oak by the death watch beetle. *Biochemical Journal*, **30**, 1135–7.

O'Brien, R. W. & Slaytor, M. (1982). Role of microorganisms in the metabolism of termites. *Australian Journal of Biological Sciences*, **35**, 239–62.

Parkin, E. A. (1940). The digestive enzymes of some wood-boring beetle larvae. *Journal of Experimental Biology*, **17**, 364–77.

Powning, R. F. & Davidson, W. J. (1979). Studies on insect bacteriolytic enzymes – III. Lytic activities in some plant materials of possible benefit to insects. *Comparative Biochemistry and Physiology*, **63B**, 199–206.

Quinlan, R. J. & Cherrett, J. M. (1979). The role of fungus in the diet of the leaf-cutting ant *Atta cephalotes* (L.). *Ecological Entomology*, **4**, 151–60.

Reese, E. T. & Mandels, M. (1971). Degradation of cellulose and its derivatives; enzymic degradation. In *High Polymers*, 2nd edn, vol. 5, ed. N. M. Bikales & L. Segal, pp. 1079–94. New York: Wiley-Interscience.

Ripper, W. (1930). Zur Frage des Celluloseabbaus der Holzverdauung xylophager Insektenlarven. *Zeitschrift für vergleichende Physiologie*, **13**, 314–33.

Rohrmann, G. F. (1978). The origin, structure and nutritional significance of the comb in two species of Macrotermitinae (Isoptera: Termitidae). *Pedobiologia*, **18**, 89–98.

Sands, W. S. (1956). Some factors affecting the survival of *Odontotermes badius*. *Insectes Sociaux*, **3**, 531–6.

Sands, W. S. (1969). The association of termites and fungi. In *Biology of Termites*, Vol. 1, ed. K. Krishna & F. M. Weesner, pp. 495–524. New York: Academic Press.

Schlottke, E. (1945). Über die Verdauungsfermente im Holzfressender Käferlarven. *Zoologische Jahrbücher, Zoologie und Physiologie*, **61**, 88–140.

Sinsabaugh, R. L., Benfield, E. F. & Linkins, A. E. (1981). Cellulose digestion and assimilation by a stream shredder. *Abstract, 29th Annual Meeting of the North American Benthological Society*, Provo, Utah, April 27–30, 1981.

Stillwell, M. A. (1966). Woodwasps (Siricidae) in conifers and the associated fungus, *Stereum chailletii* in eastern Canada. *Forest Science*, **12**, 121–8.

Terra, W. R., Ferreira, C. & deBianchi, A. G. (1979). Distribution of digestive enzymes among the endo- and ectoperitrophic spaces and midgut cells of *Rhynchosciara* and its physiological significance. *Journal of Insect Physiology*, **25**, 487–94.

Weber, N. A. (1972). Gardening ants, the attines. *Memories of the American Philosophical Society*, **92**, 1–146.

Wheeler, W. M. (1907). The fungus-growing ants of North America. *Bulletin of the American Museum of Natural History*, **23**, 669–807.

Wood, T. G. (1978). Food and feeding habits of termites. In *Production Ecology of Ants and Termites*, ed. M. V. Brian, pp. 55–80. Cambridge University Press.

Wood, T. M. & McCrae, S. I. (1979). Synergism between enzymes involved in the solubilization of native cellulose. In *Hydrolysis of Cellulose: Mechanisms of Enzymatic and Acid Hydrolysis, Advanced in Chemistry Series, No. 181*, ed. R. D. Brown, & L. Jurasek, pp. 179–209. Washington, DC: American Chemical Society.

7
Biochemical aspects of symbiosis between termites and their intestinal microbiota

JOHN A. BREZNAK
Department of Microbiology and Public Health, Michigan State University, East Lansing, Michigan 48824-1101, USA

Key words: termites; invertebrate gut microbiota; enzyme activity; nitrogen metabolism; wood digestion

Introduction
A fascinating and frequently cited example of nutritional symbiosis is the association between certain xylophagous termites and their intestinal protozoa. The classical early studies by Cleveland (1923, 1924, 1925a, b), Trager (1932, 1934), and Hungate (1938, 1939, 1943) indicated that the protozoa were key cellulolytic agents in certain groups and that their presence in the gut was critical to the survival of termites on a diet of sound wood or cellulose. However, this is an oversimplification of the biochemical details of the symbiosis, since only fragmentary information has been available on the importance of the bacterial component of the termite's intestinal microbiota. Further, most extant termite species harbour only bacteria in their intestinal tract.

Several factors have undoubtedly contributed to the slow pace at which research in this area has progressed. First, the relatively small size of most termites hampers *in situ* studies of the gut microbiota, as well as biochemical studies on termite tissues. Second, the diversity and complexity of the gut microbiota (discussed below) makes the microbial community inherently difficult to analyse from a functional standpoint. Third, the refractility of many of the gut organisms to *in vitro* cultivation, coupled with the lack of an available system to rear germ-free termites, has made it difficult to determine the quantitative importance of individual gut microbes, versus the biochemical capabilities of the insect, in meeting the termite's nutritional requirements. Nevertheless, some of these difficulties have been overcome in recent years, so that we now have a better understanding of the role of gut microbes in termite digestive processes and nutrition. Accordingly, the purpose of this review is to summarize our current knowledge on this topic with particular reference to research conducted in the last decade.

In this review the term symbiosis, and derivatives thereof, will be used in the broad sense of de Bary (1879), i.e. the permanent or semi-permanent association of two dissimilar organisms (symbionts). More specific terms will be used where appropriate and where possible. For other perspectives on this topic, as well as more detailed information on termite nutrition, behaviour, and ecology, several additional reviews are available (Breznak, 1975, 1982; Brian, 1978; Hungate, 1946a; Krishna & Weesner, 1969–1970; LaFage & Nutting, 1978; Lee & Wood, 1971; O'Brien & Slaytor, 1982; Potrikus, 1983).

Nutritional biology of termites

Of all the intriguing activities and properties of termites, none seems as widely recognized (or as often quoted) as the ability of termites to utilize wood as a food resource. Indeed, many species of termites thrive on sound, decay-free wood which contains as little as 0.03–0.1% nitrogen (dry weight basis; Cowling & Merrill, 1966). However, it is also important to recognize that there are nearly 2000 species of termites, and their biology and behaviour can be quite varied (Krishna, 1969). For example many species prefer wood that is partially decayed by fungi (Lee & Wood, 1971), whereas some termites (subfamily Macrotermitinae) actually cultivate fungi in elaborate gardens for use as a nutrient resource (Sands, 1969). Still others, depending on the species, feed on leaves, roots, grasses, dung of herbivorous animals, humus, or soil (Lee & Wood, 1971). Clearly, the diet of termites as a group is quite diverse, but is basically one rich in cellulose, hemicellulose, and lignin or lignin derivatives. Since their diet is also relatively poor in combined nitrogen, termites may be thought of as oligonitrotrophic saprovores. This trait places termites in an important position ecologically, particularly in tropical regions where their activities can dominate the processes of decomposition and nutrient cycling. Further, the biomass density of termites can be so large ($10-20\,g\,m^{-2}$) that their impact is similar to, and may surpass, that of grazing mammals (Wood & Sands, 1978).

The termite gut as a microbial habitat

All termites that have been examined possess a dense and diverse population of microorganisms in their alimentary tract. In the phylogenetically 'lower' termites (families Masto-, Kalo-, Hodo-, Rhino-, and Serritermitidae) the intestinal microbiota includes bacteria, as well as unique genera and species of oxymonad, trichomonad, and hypermastigote protozoa found almost nowhere else in nature (Honigberg, 1970; Yamin, 1979). Many of these protozoa are capable of ingesting wood particles and are cellulolytic. In the phylogenetically 'higher' termites (family Termitidae), which constitute roughly 75% of all termite species, the gut microbiota consists essentially of bacteria alone. In those instances where protozoa have been observed in higher termites, their populations were generally low and the protozoa were not of the cellulolytic type (Honigberg, 1970). At present, there is no evidence to indicate that any termite harbours a prominent and stable population of fungi in the gut.

Anatomy

The gross anatomy, histology, and cytology of the termite alimentary tract, as well as its phylogenetic variations, have been described in detail by Noirot & Noirot-Timothée (1969). Basically, the termite gut consists of three main divisions: the foregut or stomodeum (which includes the crop and muscular gizzard); the tubular midgut or mesenteron; and the voluminous hindgut or proctodeum (which includes the paunch, colon and rectum). For the present discussion it is the bulbous 'paunch' region of the hindgut which is particularly important, because it is almost invariably colonized by a dense microbiota. The paunch region of the gut of a lower termite, *Reticulitermes flavipes*, is shown in Fig. 7.1. The digestive tube of higher termites shows greater anatomical variability and may also include a mixed segment. The latter is formed by a posterior elongation of part of the mesenteron, such that the lumen of the digestive tube is bounded on one side by the mesenteron and on the other side by the proctodeum.

Studies of intestinal transit of food have been few, but indicate that passage of food through the gut takes about 24 hours with the food being retained longest in the hindgut (Kovoor, 1967a; Noirot & Noirot-Timothée, 1969). Recent work by Odelson and Breznak (unpublished), who tracked food movement through the gut of *Reticulitermes flavipes* by using sawdust amended with ^{14}C-labelled lignin, showed that the retention time of particulate food in the hindgut was 26 hours.

An enteric valve prevents refluxing of hindgut contents to the midgut, which is the main site of nutrient absorption in most insects. Consequently, microbial fermentation products, which are produced primarily in the hindgut, are absorbed from the hindgut. Unless proctodeal

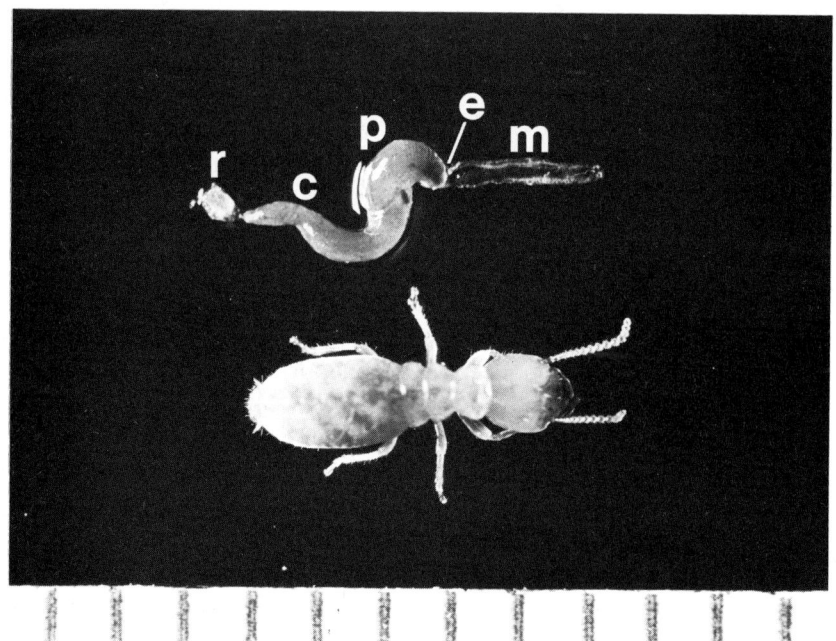

Fig. 7.1. A gut from a *Reticulitermes flavipes* worker (top), extracted from the posterior end of the insect with fine-tipped forceps, is positioned next to a separate, intact worker. Note the bulbous paunch region of the hindgut. Malpighian tubules are not readily seen in this preparation, because they are very closely apposed to the hindgut. Individual marks on the rule are 1 mm apart. Abbreviations: c, colon; e, enteric valve region; m, midgut; p, paunch; r, rectum.

trophallaxis occurs, absorption in the midgut is limited to soluble nutrients present in the food, or those liberated from the food by termite-secreted enzymes or by bacteria which may colonize the midgut (see below).

Another key feature of the alimentary tract is the presence of Malpighian tubules which empty at the precise juncture of the midgut and hindgut. Consequently, urine and urinary metabolites must traverse the hindgut before either being voided to the exterior with faeces, or resorbed from the hindgut, or metabolized by hindgut microorganisms.

Physico-chemical characteristics of the gut

The physico-chemical characteristics of the gut are similar in both higher and lower termites (reviewed by O'Brien & Slaytor, 1982).

The pH in regions where microbes are found (hindgut and midgut) usually ranges from about 6.0 to 7.5. Exceptions exist, however, in the mixed segment of higher termites (pH 7.0 to >9.6), as well as in the anterior hindgut of some soil-feeding termites where pH values as high as 10.4 have been recorded (Bignell & Anderson, 1980). In addition, sites such as the paunch which are heavily colonized by microorganisms appear anaerobic with an E'_o usually in the range of -230 to -270 mV. By contrast, midgut regions appear to be aerobic. Such determinations have been made by feeding the insects redox dyes and observing the resulting colour in the gut (Veivers, O'Brien & Slaytor, 1980), or by physiological measurements on gut homogenates (Bignell & Anderson, 1980). Anaerobicity of the hindgut is consistent with the oxygen sensitivity of the hindgut protozoa of lower termites (Cleveland, 1925a; Hungate, 1939), with the demonstration of strict anaerobic bacteria in termite hindguts (Schultz & Breznak, 1978; Potrikus & Breznak, 1980b), with the demonstration of *in situ* methanogenesis by the hindgut microbiota (Breznak, 1975), and with the isolation of methanogenic bacteria from homogenates of termite hindguts (Odelson and Breznak, unpublished).

Gut microbes and sites of colonization

As mentioned, the hindgut is heavily colonized by microorganisms. This is true for both lower and higher termites. By using light and electron microscopy, Breznak (1975) and Breznak & Pankratz (1977) found the hindgut of *Reticulitermes flavipes* and *Coptotermes formosanus* to be colonized by a morphologically heterogeneous assemblage of bacteria and protozoa. Most of the bacteria were situated close to the paunch epithelium, and many possessed holdfast elements that appeared to secure them to the epithelial tissue and to other bacterial cells. Lesser numbers of bacteria were observed free in the lumen of the paunch, which was occupied mainly by protozoa. However, some of the protozoa (*Pyrsonympha vertens*) also possessed a holdfast organelle that enabled them to attach by their anterior end to the paunch epithelium. The midgut was more sparsely colonized, but did possess distinctive cuboidal-shaped, endospore forming bacteria situated between the microvilli of the epithelium. Similar observations have been made with other lower (Bloodgood, 1975; To, Margulis, Chase & Nutting, 1980) and higher termites (Kovoor, 1959, 1968; Noirot & Noirot-Timothée, 1967). In the soil-feeding higher termites *Procubitermes aburiensis* and *Cubitermes severus*, the proctodeum is heavily colonized by bacteria, and many of these bacteria are attached to cuticular spines emanating from the wall of the posterior colon (Bignell, Oskarsson & Anderson, 1979, 1980a, b). Some of the bacteria are filamentous, actinomycete-like forms. The crop, mesenteron, and mixed segment also harboured bacteria (Bignell et al., 1980c). Interestingly, when a mixed segment is present the bacteria therein tend to be so similar in morphology as to suggest a pure culture (Noirot & Noirot-Timothée, 1969).

As determined by direct microscopic count, population levels of bacteria range from 10^6–10^7 bacteria gut^{-1} (Krasil'nikov & Satdykov, 1969; Breznak, Brill, Mertins, and Coppel, 1973; Schultz & Breznak, 1978), whereas those of protozoa are about 3–4 × 10^4 gut^{-1} for *Reticulitermes flavipes* (Mauldin & Rich, 1980). Considering that the gut of some termites such as *R. flavipes* is 1 µl or less in volume, these data translate to about 10^9–10^{10} bacteria ml^{-1} and 10^7 protozoa ml^{-1} gut contents, i.e. the gut (particularly the hindgut) is almost a solid, packed mass of microbes. Indeed, Odelson & Breznak (1983) determined that 61% of the hindgut contents of *R. flavipes* was microbial cells, the remainder being extracellular hindgut fluid. It is noteworthy that the ratio of protozoan cells to bacteria in the hindgut of lower termites such as *R. flavipes* (1:100) is quite large compared to a gastrointestinal ecosystem such as the bovine rumen (1:100 000; Hungate, 1975). This undoubtedly has a bearing on the pattern of fermentation in the hindgut and is discussed further below.

Attempts at enumeration and identification of termite gut bacteria by means of microbiological culture techniques have been reviewed by Breznak (1975, 1982) and O'Brien & Slaytor (1982). Most of the bacteria isolated from lower and higher termites have proven to be facultative or strict anaerobes including strains of *Streptococcus*, *Bacteroides*, various Enterobacteriaceae, *Staphylococcus*, and *Bacillus*. The biochemical characteristics of these bacteria, as they relate to symbiotic interaction with the termites, are discussed in the appropriate sections below.

Spirochaetes are of constant and abundant occurrence in the hindgut of all termites examined and exhibit considerable diversity in terms of their size range (0.2 × 3.0 to 1.0 × 100 µm) and morphology (Breznak, 1973, 1983; To et al., 1980). They generally occur free in the gut fluid, as well as attached to the surface of polymastigote and hypermastigote protozoa in hindguts of lower termites (Breznak, 1983). In some of the latter associations they have evolved spectacular 'motility symbioses', whereby their coordinated undulations can actually serve to propel the protozoans through the hindgut fluid. This has been described for spirochaetes adherent to *Mixotricha paradoxa* from the hindgut of the primitive Australian termite *Mastotermes darwiniensis* (Cleveland & Grimstone, 1964), as well as for spirochaetes adherent to unidentified flagellates from *Kalotermes schwarzi, Pterotermes occidentis*, and *Marginitermes hubbardi* (To, Margulis & Cheung, 1978). Unfortunately, almost nothing is known of their biochemical role in the gut system, as termite hindgut spirochaetes have defied all attempts at *in vitro* cultivation (Breznak, 1973). However, they may be important to the vitality of termites. Eutick, Veivers, O'Brien & Slaytor (1978) observed decreased survival of the higher termite *Nasutitermes exitiosus* when spirochaetes were eliminated from the gut by means of antibiotics.

Digestive processes in termites and the role of gut microorganisms

Cellulose and hemicelluloses (xylans, mannans, etc.) undergo substantial dissimilation on passage through the termite gut (65–99%), and the assimilation efficiency of wood-feeders is quite high (54–93%) (Esenther & Kirk, 1974; Wood, 1978). Values for lignin digestion vary considerably and will be discussed later. Nevertheless, a key question has been, and continues to be, the contribution of specific gut organisms, versus the termite's own enzymes, to the digestive process.

Lower termites

The bulk of wood polysaccharide digestion occurs in the hindgut of lower termites, and all available evidence indicates that the hindgut microbiota, principally the anaerobic flagellate protozoa, is the driving force of such dissimilatory activity (reviewed by Breznak, 1975, 1982 and O'Brien & Slaytor, 1982). The early work of Hungate (1938, 1939, 1943) in particular led to the following model for wood-cellulose digestion. Wood particles consumed by termites are passed to the hindgut where they are endocytosed by protozoa. The cellulose is fermented anaerobically within the protozoa to CO_2, H_2 and acetate, which are liberated from the cells. Acetate is subsequently absorbed by the termites and used as their major oxidizable energy source. This model is conceptually appealing, because it permits acquisition of energy (ATP) by both the protozoa (via anaerobic fermentation of cellulose) and the termites (via aerobic oxidation of acetate).

Acetate is indeed the major volatile fatty acid (VFA) in hindgut fluid of lower termites examined. Odelson & Breznak (1983) used a 'micro'

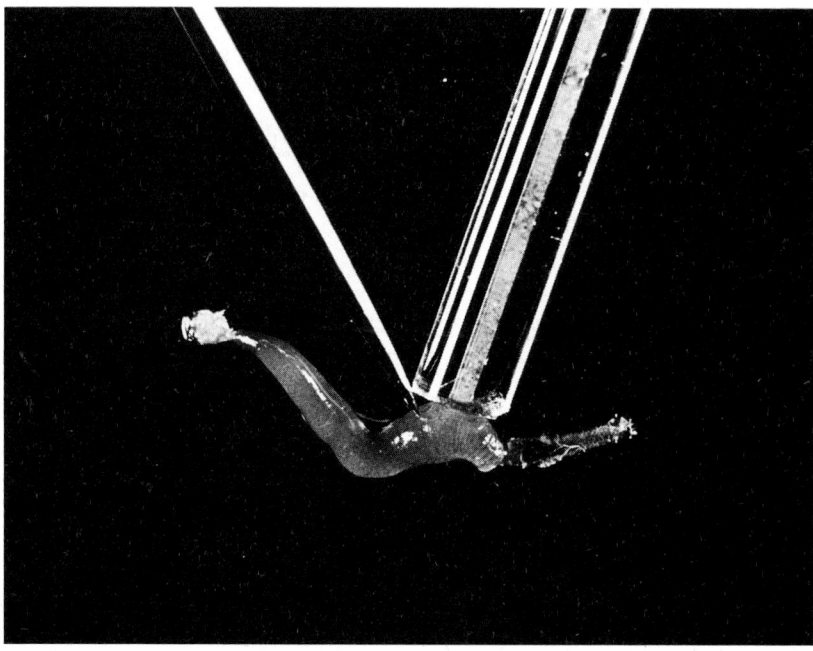

Fig. 7.2. A 'micro' sampling method is used to sample paunch contents from (in this case) the gut of *Reticulitermes flavipes*. After puncturing the paunch with a fine dissecting needle (left), the contents which issue from the puncture site are immediately aspirated into a 1 μl capacity glass capillary tube (right).

technique to sample hindgut fluid from termites for subsequent analysis by gas chromatography and mass spectroscopy (Fig. 7.2). These investigators found that acetate was present in extracellular hindgut fluid of worker termites (*Reticulitermes flavipes*, *Zootermopsis angusticollis*, and *Incistermes schwarzi*) at concentrations of 58–81 mM, and it accounted for 94–99 mole % of all VFAs. Small amounts of propionate and butyrate were also detected. By feeding *R. flavipes* sawdust amended with ^{14}C-labelled polymers, they also found that about 80% of the acetate was derived from cellulose, whereas about 20% was derived from hemicellulose.

If the main role of protozoa is to degrade wood-cellulose to acetate, one would predict that defaunated termites (i.e. with protozoa removed) should be able to survive on a cellulose-free diet if supplied with acetate. While this has been tried, negative results were obtained with

Z. angusticollis (Cook, 1943; Hungate, 1946a) and with *R. flavipes* (Odelson & Breznak, 1983). However, the lack of success might be due to toxicity to termites of the cation moiety (Ca^{2+} or Na^+) of the acetate salt that was added to the diet. It should be remembered that in the faunated termites there is a continuous and gradual production of free acetic acid – a situation difficult to mimic in feeding studies.

A major contribution to our understanding of termite hindgut protozoa is the recent work of Yamin (1978, 1981), who obtained axenic cultures of major cellulolytic species *Trichomitopsis termopsidis* and *Trichonympha sphaerica* from the termite *Zootermopsis*. The medium used to cultivate the protozoa was prepared under anaerobic conditions and included: cellulose, serum, yeast extract, glutathione, and dead (heat-killed) rumen bacteria. Initial cultures also contained antibiotics to inhibit bacterial growth, and final axenic cultures appeared to be free of live extracellular bacteria as well as endosymbiotic bacteria. Both species of protozoa were cellulolytic and fermented cellulose to CO_2, H_2, and acetate as major products (Yamin, 1980, 1981). These products were identical to those formed from cellulose by mixed suspensions of hindgut protozoa from *Zootermopsis* (Hungate 1939, 1943). For *T. termopsidis*, cellulose in the culture medium could not be replaced by glucose, cellobiose, carboxymethylcellulose, glycogen, inulin, or rice starch (Yamin, 1978). Moreover, the dead rumen bacteria (which were endocytosed and digested within food vacuoles) could not be replaced by cells of *Escherichia coli*, *Bacillus subtilis*, *Pseudomonas maltophilia*, or a *Clostridium* sp. Presumably, the rumen bacteria were the *in vitro* correlate of termite hindgut bacteria (on which the protozoa graze *in vivo*) and one or more species in the mixture provided a growth factor necessary for the protozoa. Crude extracts of *T. termopsidis* possessed carboxymethylcellulase (Yamin & Trager, 1979), as well as cellobiase, coenzyme-A-dependent pyruvate:ferredoxin oxidoreductase, and hydrogenase (Yamin, 1980).

If defaunated *Zootermopsis* was reinfected with axenic cultures of *T. termopsidis*, the termites survived appreciably longer than did non-reinfected controls and as well as those reinfected with a mixed population of microbes derived from hindgut contents (Yamin & Trager, 1979).

These experiments buttressed the conclusion that hindgut protozoa are key agents of cellulose digestion in lower termites, and they showed that cellulase activity in the protozoa does not necessarily require the presence of endosymbiotic bacteria (Yamin, 1981). The availability of pure cultures should make these protozoa important subjects for further biochemical studies.

Considering the extent to which wood polysaccharides are dissimilated in termites, it appears that the protozoa are quite efficient at stripping these polymers from their complex with lignin. Accordingly, it would be of intrinsic interest, and perhaps also practical value, to study the nature of their cellulases and other carbohydrases in greater detail. Unfortunately, the growth yields of the protozoa *in vitro* are relatively poor. For example, cell yields of *T. termopsidis* are about 2000 cells ml^{-1} after 37 days (Yamin, 1980), whereas those of the larger *T. sphaerica* are only 150 cells ml^{-1} (Yamin, 1981). Accordingly, in order to facilitate biochemical studies, efforts were made in our laboratory to increase the cell yields of these protozoa. Since hydrogen was the only reduced product formed during growth of these protozoa, it seemed likely that some of the H_2 was derived from $NADH + H^+$. However, H_2 production from $NADH + H^+$ is thermodynamically unfavourable even at partial pressures of H_2 as low as 10^{-3} atmospheres (Wolin, 1974). This suggested that accumulation of H_2 in protozoan cultures might actually be suppressing cell yields. Accordingly, attempts were made to co-culture *T. termopsidis* with avid hydrogen-consuming bacteria. It was reasoned that if the pH_2 in protozoan cultures could be kept very low, by employing hydrogen-consuming bacteria as an 'electron sink' (Wolin, 1974), better growth of the protozoa might result. Preliminary experiments (Table 7.1; Odelson and Breznak, unpublished) revealed that yields of *T. termopsidis* were increased in the presence of a methane-forming *Methanospirillum*. Additionally, it was found that replacement of dead rumen bacteria in the culture medium with heat-killed cells of *Bacterioides* sp. strain JW20, a bacterium isolated from the hindgut of *Reticulitermes flavipes* (Schultz & Breznak, 1978, 1979), boosted cell yields considerably. In the presence of both the *Methanospirillum* and *Bacterioides*, the cell yield of *T. termopsidis* was increased roughly 15-fold. These preliminary data suggest that termite bacteria may be a better food source for cellulolytic protozoa than mixed rumen bacteria, and that the hydrogen-consuming activities of some bacteria (e.g. methanogens) might help to 'pull' anaerobic cellulose decomposition in the termite hindgut. The importance of methanogens as terminal organisms in anaerobic food webs has been pointed out by Wolfe (1979).

Methane is emitted from live termites, and methane bacteria have been isolated from hindguts of *R. flavipes*. The bacteria are rod-shaped

Table 7.1. *Effects of* Methanospirillum hungatii *and heat-killed termite gut bacteria on growth of* Trichomitopsis termopsidis *in vitro*[a]

Omitted from basal medium[b]	Added to basal medium[b]	Growth yield of T. termopsidis (cells ml^{-1})[c]	Presence of H$_2$ in headspace
—	—	3800 (23)	+
—	Msp	7800 (39)	–
AWRB	JW20	31 900 (31)	+
AWRB	Msp + JW20	58 000 (26)	–

Note:
[a] *T. termopsidis* strain 6057 was obtained from Dr M. A. Yamin.
[b] The basal medium and culture conditions were as described by Yamin (1978). Abbreviations: AWRB, autoclaved, washed rumen bacteria (Yamin, 1978); Msp, viable cells of *Methanospirillum hungatii* strain JF 1 (obtained from Dr M. P. Bryant); JW20, heat-killed cells of *Bacteroides* sp. strain JW20 isolated from hindguts of *Reticulitermes flavipes* termites (Schultz & Breznak, 1978).
[c] Determined by direct microscopic count. Numbers in parenthesis indicate the number of days required to reach maximum cell density. Initial cell densities were approximately 500 cells ml^{-1}.

and await complete characterization (Odelson and Breznak, unpublished). Rates of methane emission by termites vary, however, depending on the termite species and their diet prior to assay. Rates of emission as low as 0.3 nmole CH_4 hr^{-1} g fresh wt^{-1} were observed for *Coptotermes formosanus* worker larvae feeding on nest wood, whereas rates up to 73.1 nmole CH_4 hr^{-1} g fresh wt^{-1} were seen with *R. flavipes* under the same conditions (Breznak, 1975). When feeding on cellulose filter paper, rates of methane emission by *R. flavipes* increased to 1340 nmole hr^{-1} g fresh wt^{-1} (Breznak, 1975). Low amounts of methane emission were reported for *Marginitermes hubbardi* (LaFage & Nutting, 1979). Zimmerman, Greenberg, Wandiga & Crutzen (1982) found that methane was emitted from *Reticulitermes tibialis* (approximately 277 nmole hr^{-1} g fresh wt^{-1}) and *Gnathamitermes perplexus* (approximately 258 nmole hr^{-1} g fresh wt^{-1}) (values calculated from Zimmerman's data by assuming a fresh weight of 4 mg per termite). Based on these determinations, Zimmerman *et al.* (1982) suggested that of the total annual global production of methane ($3-12 \times 10^{14}$ g), termites could contribute as much as 1.5×10^{14} g. This estimate seems high, and it is based on analyses of only a few termite species as well as on termite population density data which are fairly limited (Wood & Sands, 1978). However, it is not unreasonable to assume that termites could make a significant contribution to local methane emissions, particularly at sites where they abound.

If polysaccharide dissimilation in the hindgut of lower termites is basically considered as the anaerobic fermentation of glycosyl units by the protozoa ($nC_6H_{12}O_6 + 2nH_2O \rightarrow 2nC_2H_4O_2 + 2nCO_2 + 4nH_2$), then a maximum of nCH_4 could be formed by methanogens from H_2 and CO_2 according to the reaction $nCO_2 + 4nH_2 \rightarrow nCH_4 + 2nH_2O$. Assuming all the acetate produced in the hindgut was then oxidized to CO_2 and H_2O, CH_4 emission by termites should be 20% that of total CO_2 emission on a molar basis. However, recent studies indicate that it is less than 1% for *R. flavipes* (Odelson & Breznak, 1983) and *R. tibialis* (Zimmerman et al., 1982). Hydrogen not used in methanogenesis is not merely evolved, however. On a molar basis H_2 emission is only 10% that of CO_2 for *Zootermopsis nevadensis* (Gilmour, 1940), and it is usually 0.7–1.5% that of CO_2 for *R. flavipes* (Odelson & Breznak, 1983). Consequently, although some hydrogen escapes from the hindgut system, most of the reducing equivalents derived from cellulose fermentation are somehow consumed. Either free H_2 is not formed appreciably by protozoa *in situ*, or it is formed and rapidly used in reactions other than methanogenesis. One possible explanation is that hydrogen evolved from protozoa (and perhaps from other hindgut bacteria) is also consumed by *Acetobacterium*-type bacteria that are able to reduce CO_2 to acetate according to the reaction $2CO_2 + 4H_2 \rightarrow C_2H_4O_2 + 2H_2O$ (Balch, Schoberth, Tanner & Wolfe, 1977). Such bacteria could function along with methanogens as electron sink organisms and could also serve as an additional source of acetate for the termite. Moreover, use of H_2 and CO_2 for acetogenesis in the gut, with subsequent oxidation of acetate by the termite, would not alter the respiratory quotient of termites from the normal value of approximately 1.0 (Peakin & Josens, 1978). Acetogenesis from CO_2 appears to occur in other gastrointestinal ecosystems such as the caecum of guinea pigs, rats, and rabbits (Prins & Lankhorst, 1977), although it must be emphasized that such a reaction has not yet been confirmed for termite hindguts.

A number of bacterial isolates from lower termites have been found to possess C_x-type cellulase activity with carboxymethylcellulose as substrate (Krelinova, Kirku & Skoda, 1977; Thayer, 1976, 1978). However, at present there is no convincing evidence that bacteria are quantitatively important to the hydrolysis of crystalline cellulose *in vivo*; at least not in lower termites which possess cellulolytic protozoa (Breznak, 1975, 1982). Considering that many of the bacteria isolated from hindguts are fairly common heterotrophs (see above), one wonders what substrate (or substrates) supports their growth *in situ*. A clue may be found in the studies of Yamin (1980) and Hungate (1943) who observed that 10–30% of cellulose carbon fermented by hindgut protozoa *in vitro* could not be recovered as acetate and CO_2. Yamin suggested that some of the missing carbon might be present as soluble intermediates (oligomers) of cellulose hydrolysis released from the protozoa. It may be advantageous for protozoa to release hydrolysis intermediates *in situ*, because, if these could be used by the bacteria, the protozoa could essentially 'culture' a population of bacterial cells on which they might subsequently feed. The requirement for killed bacteria by axenic protozoan cultures (Yamin, 1978, 1981; Table 7.1), the early mixed culture studies conducted by Trager (1934), and the common occurrence of bacteria attached to (or within) hindgut protozoa (Ball, 1969), all support such speculation. Furthermore, Veivers, O'Brien & Slaytor (1982) found that metabolic activities of bacteria are important in maintaining anaerobic conditions in the gut of *Mastotermes darwiniensis*, and this may be an additional benefit to the protozoa, which are strict anaerobes.

Based on considerations discussed above, a working model is proposed for wood polysaccharide fermentation by the hindgut microbiota of lower termites (Fig. 7.3). The model is based on the original scheme

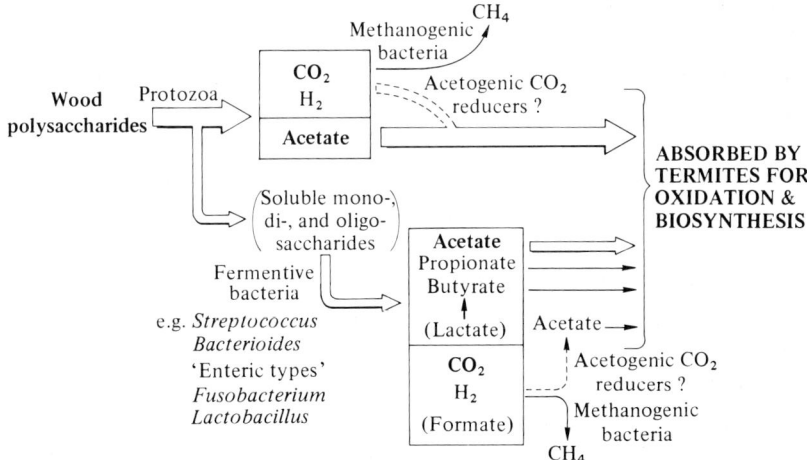

Fig. 7.3. Proposed working model for wood polysaccharide degradation in the hindgut of lower termites. Thickness of arrows represents approximate relative quantitative contribution of the respective reactions. Major products of the hindgut fermentation are indicated in boldface letters. See text for details.

postulated by Hungate (1939, 1943), but is modified to accommodate some of our current information on termite hindgut bacteria and metabolites present in the hindgut. According to this scheme, VFAs (principally acetate) are the major microbial metabolites supplied to the termites: these are readily oxidizable substrates for termites (Odelson & Breznak, 1983), and they are also known to be important precursors for synthesis of termite cuticular hydrocarbons (Blomquist, Howard & McDaniel, 1979), amino acids (Mauldin, Rich, & Cook, 1978), and terpenes (Prestwich, Jones & Collins, 1981). Most of the acetate is derived from fermentation by the protozoa. This is undoubtedly because the protozoa are very abundant in the hindgut and have the ability to endocytose, and thereby sequester, wood particles during the degradation. Consequently, most of the fermentation products formed should be those liberated by the protozoa and be dominated by acetate, CO_2 and H_2. Smaller amounts of products are thought to be derived from the bacteria, via the fermentation of mono-, di-, and soluble oligosaccharides released from protozoan cells. Fermentative bacteria that have been isolated from hindguts are known to ferment glucose and cellobiose (Schultz & Breznak, 1978), although none have been tested on soluble oligosaccharides of C_1 or greater. This is an important question which should be examined. Metabolites such as formate and lactate, which are known to be produced by pure cultures of termite hindgut bacteria (Schultz & Breznak, 1978), do not accumulate in the hindgut. In the case of lactate, *in vitro* co-culture studies by Schultz & Breznak (1979) have shown that lactate production by termite hindgut streptococci can support the growth of bacterioides, which ferment the compound to propionate, acetate, and CO_2. A similar type of cross-feeding of lactate is believed to occur *in situ*. If formate is produced in the gut, it is probably used as a methanogenic substrate.

The present scheme is derived mainly from studies of cellulose decomposition. However, virtually nothing is known about the mechanisms of the degradation of the other major polysaccharides of wood, i.e. the hemicelluloses. This is another important area which must be examined in greater detail.

Role of termite enzymes. The working model depicted in Fig. 7.3 is intended to describe the microbial decomposition of wood polysaccharides which reach the hindgut. It is not meant to imply that termite enzymes have no role in wood polysaccharide degradation. Indeed, many years ago Hungate (1938) found that *Zootermopsis* could digest approximately one third of the polysaccharides in wood without the aid of their hindgut protozoa. Lower termites have been shown to possess a variety of carbohydrases in their gut, including cellulases and β-glucosidases (LaFage & Nutting, 1978; McEwen, Slaytor & O'Brien, 1980; Mishra, 1980; O'Brien et al., 1980; Veivers, Musca, O'Brien & Slaytor, 1982; Yamaoka & Nagatani, 1975; Yokoe, 1964), and there is some evidence that cellulase is synthesized by the termites themselves. However, most of such claims have been made by using carboxymethylcellulose or reprecipitated cellulose as the assay substrate. These substrates can be used to demonstrate the presence of C_x-type cellulases (active against noncrystalline cellulose and soluble derivatives or degradation products of cellulose), but not C_1-type cellulases (active against crystalline cellulose) (Wood & McCrae, 1979). On the other hand, Veivers, Musca, O'Brien & Slaytor (1982) showed the presence of C_1-cellulase in salivary glands and hindguts of *Mastotermes darwiniensis* where salivary enzyme activity appeared to be entirely of termite origin. Thus, it may well be that some cellulose hydrolysis can be initiated by enzymes of lower termites in the fore- and midgut. However, the importance of the hindgut microbiota (particularly the protozoa) to the survival of lower termites on sound wood or cellulose is undeniable.

Higher termites

Surprisingly little work has been done on the role of gut organisms in digestive processes of higher termites. Since the higher termites lack protozoa in their hindguts, it has been generally assumed that bacteria are primarily responsible for cellulose digestion in these insects. However, at present there is no good evidence to support that notion. Hungate (1946b) isolated an anaerobic, cellulolytic actinomycete (*Micromonospora propionici*) from the gut of *Amitermes minimus*, but he regarded the organism to be insignificant *in situ* because of its low population density in the gut and its slow growth rate *in vitro*. We have been unable to demonstrate cellulolytic bacteria in the gut of *Nasutitermes corniger*, using strict anaerobic techniques and ball-milled Whatman filter paper as a substrate (Breznak, unpublished). By contrast, higher termites appear to synthesize carbohydrases (McEwen et al., 1980; O'Brien et al., 1980; Kovoor, 1970; Potts & Hewitt, 1973), including cellulase active on crystalline cellulose (Potts & Hewitt, 1974). It may be that phylogenetically 'higher' termites are independent of their hindgut bacteria for cellulose hydrolysis. Nevertheless, a prominent fermentation appears to occur in the hindgut, and acetate and other VFAs have been identified in hindguts of *Microcerotermes edentatus* (Kovoor, 1967b) and *Nasutitermes corniger* (Odelson & Breznak, 1983). Perhaps fermentative activities of the gut bacteria, rather than cellulose hydrolysis *per se*, are important to the nutrition of higher termites. Clearly, more research on this group of termites is sorely needed.

The importance of mycophagy, as a source of nutrients and digestive enzymes for fungus-cultivating Macrotermitinae, is dealt with by Martin elsewhere in this volume (Chapter 6).

Role of gut microbes in termite nitrogen economy

The oligonitrotrophic habit of termites is most conspicuous for those species that thrive on sound, decay-free wood containing as little as 0.03–0.1% nitrogen on a dry weight basis (Cowling & Merrill, 1966) and exhibiting a C:N ratio of about 1000:1. Since termite tissues contain nitrogen in amounts similar to those of other animals (approximately 11% nitrogen on a dry weight basis; LaFage & Nutting, 1978; Potrikus & Breznak, 1980a), it would appear that termites have evolved efficient mechanisms for acquiring and/or conserving combined nitrogen. This seems to be true, and it appears that gut microbes can aid termites in both of these processes.

Nitrogen fixation

Nitrogen fixation by termites would constitute a means for acquisition of combined nitrogen, and it is a process that has been suspected ever since Cleveland's (1925b) claim that termites could live '... perhaps indefinitely on a diet of pure cellulose.' Only recently, however, has this suspicion received experimental support. By using the acetylene reduction assay, a reliable indicator of nitrogen fixation (Hardy, Burns & Holsten, 1973), nitrogen fixation has been inferred for a variety of termites (Table 7.2). Since it is a uniquely prokaryotic phenomenon, the fixation has generally been assumed to be mediated by bacteria, probably gut bacteria. This assumption appears to be valid, however, inasmuch as nitrogen fixation activity can be abolished by feeding termites antibacterial drugs (Breznak *et al.*, 1973). Moreover, nitrogen fixing bacteria have been isolated from termite guts. French, Turner & Bradbury (1976) isolated nitrogen fixing *Citrobacter freundii* strains from Australian termite species (*Coptotermes lacteus*, *Mastotermes darwiniensis*, and *Nasutitermes exitiosus*), whereas Potrikus &

Breznak (1977) isolated another member of the Enterobacteriaceae (*Enterobacter agglomerans*) from hindguts of *Coptotermes formosanus*.

In Table 7.2, the rates of nitrogen fixation predicted from acetylene reduction have been converted to a TDN equivalent. The latter represents 'time to double nitrogen' and is the amount of time it would take for a given termite species to double its nitrogen content, assuming that the measured fixation rates remained constant. A striking feature of these data is the wide variation of nitrogen fixation rates in termites, both interspecifically as well as intraspecifically. Considering these rates in *R. flavipes* workers for example, it would be difficult to argue that nitrogen fixation is important to this termite's nitrogen economy: particularly since the TDN equivalent is nearly 1000 years. A similar difficulty would be encountered with data from other species such as *Cubitermes* sp., *Labiotermes* sp., and *Zootermopsis* sp. By contrast, nitrogen fixation might be more important to termites such as *Cornitermes* sp., *I. minor*, or *T. trinervoides*, whose TDN values are far lower. In fact, for whole colonies of *N. corniger* the maximum rates of nitrogen fixation reported could allow for complete doubling of colony nitrogen in 0.5 years (Table 7.2). This certainly seems significant, although it would help greatly to know how fast termite populations increase *in the field*, and such data are scarce. Nevertheless, J.P.E.C. Darlington (cited by Prestwich & Bentley, 1981) estimates that *Macrotermes* species in East Africa turnover four populations annually. If *N. corniger* population growth is similar, then nitrogen fixation could conceivably contribute 50% of all the termite's nitrogen needs.

Several factors might account for the large variations in nitrogen fixation rates seen in Table 7.2. One of these is the nitrogen content of the food being eaten. For example, Breznak *et al.* (1973) found that acetylene reduction rates in *C. formosanus* varied inversely with the amount of combined nitrogen (e.g. KNO_3, $(NH_4)_2SO_4$, aspartic acid, or killed cells of *Escherichia coli*) added to a Whatman filter paper food disc. A significant change in fixation rates could be detected within five hours of a dietary shift, and variations in fixation rates up to 200-fold were observed. Inasmuch as nitrogenase synthesis by bacteria is repressed by readily utilizable sources of combined nitrogen (Brill, 1975), the modulation of nitrogen-fixing activity in *C. formosanus* presumably reflected a response of nitrogen-fixing gut bacteria to dietary nitrogen levels. Similarly, the difference in fixation rates between *R. perarmatus* and *N. corniger* was attributed to the fact that the former feeds on

Table 7.2. *Nitrogen fixation (acetylene reduction) rates of live termites*

Termites	Caste[a]	g of N fixed g fresh wt^{-1} day^{-1}[b]	TDN (yr)[c]	Reference
Amitermes sp.	W	0.36	126	Sylvester-Bradley, Bandeira & de Oliveira, 1978
	S	0.45	100	Sylvester-Bradley, Bandeira & de Oliveira, 1978
Armitermes sp.	W + S	1.44	31	Sylvester-Bradley, Bandeira & de Oliveira, 1978
Coptotermes formosanus	W	0.16–49.39	238–1	Breznak, 1975; Breznak, Brill, Mertins & Coppel, 1973
	S	0.03	1507	Breznak, Brill, Mertins & Coppel, 1973
Coptotermes lacteus	W	0.37–1.87	122–24	French, Turner & Bradbury, 1976
Cornitermes sp.	W + S	1.06	43	Sylvester-Bradley, Bandeira & de Oliveira, 1978
Cryptotermes brevis	BL	0.38	119	Breznak, Brill, Mertins & Coppel, 1973
Cubitermes sp.	W	0.17	265	Rohrmann & Rossman, 1980
	R	0.00	∞	Rohrmann & Rossman, 1980
Heterotermes sp.	W	0.94	48	Sylvester-Bradley, Bandeira & de Oliveira, 1978
	S	0.00	∞	Sylvester-Bradley, Bandeira & de Oliveira, 1978
Incisitermes minor	W	1.00–18.87	45–2	Benemann, 1973
	S	0.33	137	Benemann, 1973
	R	0.66	68	Benemann, 1973
Labiotermes sp.	W	0.16	282	Sylvester-Bradley, Bandeira & de Oliveira, 1978
Macrotermes ukuzii	W, S, R	0.00	∞	Rohrmann & Rossman, 1980
Mastotermes darwiniensis	W	0.00–23.47	∞–2	French, Turner & Bradbury, 1976
Nasutitermes corniger[d]	W	6.00–8.00	8–6	Prestwich & Bentley, 1981; Prestwich, Bentley & Carpenter, 1980
	S	0.90–28.40	50–2	Prestwich & Bentley, 1981; Prestwich, Bentley & Carpenter, 1980
	Whole colony	27.40–81.68	2–0.5	Prestwich & Bentley, 1981
Nasutitermes exitiosus	W	0.00–5.60	∞–8	French, Turner & Bradbury, 1976

Species	Caste	TDN	Reference
Nautitermes sp.	W	0.20–13.28	226–3	Sylvester-Bradley, Bandeira & de Oliveira, 1978
	S	0.87–7.44	52–6	Sylvester-Bradley, Bandeira & de Oliveira, 1978
	W + S	1.11–18.31	41–2	Sylvester-Bradley, Bandeira & de Oliveira, 1978
Neocapritermes sp.	W	0.00	∞	Sylvester-Bradley, Bandeira & de Oliveira, 1978
Reticulitermes flavipes	W	0.05	904	Breznak, Brill, Mertins & Coppel, 1973
	S	0.02	2260	Breznak, Brill, Mertins & Coppel, 1973
Rhynchotermes perarmatus	W	3.5	13	Prestwich, Bentley & Carpenter, 1980
	S	0.5	90	Prestwich, Bentley & Carpenter, 1980
Trinervitermes trinervoides	W	6.86	7	Rohrmann & Rossman, 1980
	S	4.48–4.73	10–9	Rohrmann & Rossman, 1980
	R	0.00	∞	Rohrmann & Rossman, 1980
Zootermopsis sp.	W + BL	0.06	753	Breznak, Brill, Mertins & Coppel, 1973

Note:
[a] W, Workers; S, soldiers; BL, brachypterous larvae; R, reproductives.
[b] Values calculated assuming that N_2 fixation rates are one third that of acetylene reduction (Hardy, Burns & Holsten, 1973), and that fresh weight of termites = 6.7 × dry weight.
[c] TDN, time required for termites to double their nitrogen content (see text). It is assumed that the nitrogen content of all termites is 11% (dry wt basis; Potrikus & Breznak, 1980a).
[d] Mistakenly identified as *Nasutitermes ephratae* (Prestwich, Bentley & Carpenter, 1980) and corrected in a subsequent paper (Prestwich & Bentley, 1981).

material higher in combined nitrogen (leaf litter) than does the latter (which feeds on wood litter) (Prestwich, Bentley & Carpenter, 1980).

The feeding preferences of some termites may in fact obviate the need for nitrogen fixation, simply because their diets contain adequate, albeit low, amounts of combined nitrogen. For example, many of the Hodo-, Rhino-, and Kalotermitidae actually prefer wood that has undergone some decay by fungi (Lee & Wood, 1971). Experiments by Hungate (1941, 1944) suggested that fungi have the effect of increasing the relative amount of combined nitrogen in wood, by degrading the polysaccharide components and by translocating nitrogen into the resource from the surroundings (e.g. soil). Thus, Hungate's (1941, 1944) nitrogen balance studies on *Zootermopsis*, *Incisitermes* (*Kalotermes*), and *Reticulitermes* species, which fed preferentially on decayed portions of wood, showed no evidence for nitrogen fixation. Perhaps it is not surprising that these species also show only low to moderate rates of nitrogen fixation when assayed by the acetylene reduction method (Table 7.2). It is interesting that termites are not the only animals whose nitrogen fixation rates appear to vary inversely with dietary nitrogen: a similar response has been observed with sea urchins (Guerinot, Fong & Patriquin, 1977) and even man (Bergersen & Hipsley, 1970).

Age or developmental stage of termites may also have a bearing on rates of nitrogen fixation and may account for some of the intraspecific variations seen in Table 7.2. For example, Breznak (1975) found that small worker larvae of *C. formosanus* had rates 300-fold greater than larger, more fully developed workers. Presumably, the nitrogen needs of the smaller workers were greater, and this was reflected in their elevated nitrogen fixation rates. An inverse relationship between nitrogen fixation rate and body size has also been observed with marine shipworms (Carpenter & Culliney, 1975). Other factors that could contribute to intraspecific variations in fixation rates are cyclic demands for nitrogen placed on colony members during bursts of reproductive activity (Prestwich & Bentley, 1981), as well as physical disturbances to the termites during their manipulation for assay (Potrikus & Breznak, 1977; Prestwich & Bentley, 1981; Prestwich *et al.*, 1980).

Nitrogen conservation

The complementary side to nitrogen acquisition is nitrogen conservation and there are three main ways this could be done: storage/recycling of nitrogenous metabolic wastes; recycling of termite tissues (e.g. exoskeletons); and digestion and assimilation of gut microbes or lytic or secretory products thereof. The contribution of these processes to termite nitrogen economy has been largely speculative. However, recent evidence suggests that recycling of nitrogenous wastes by termite gut bacteria is an important conservation mechanism.

Uric acid (UA) is a common, and well-suited, nitrogenous excretory product of a variety of terrestrial insects (Cochran, 1975). Because of its poor solubility in aqueous solution, UA can be voided with faeces as a non-toxic solid, thereby minimizing water loss to the insect. However, Leach & Granovsky (1938) speculated that in termites UA might be degraded by hindgut organisms to a form of nitrogen reusable by the insects. For more than 40 years this hypothesis remained untested, until Potrikus & Breznak (1980a, b, c; 1981) critically examined its validity. Working mainly with *Reticulitermes flavipes*, we found that termite tissues contain UA (Potrikus & Breznak, 1980a), as well as key enzymes for UA biosynthesis (purine nucleoside phosphorylase, EC 2.4.2.1, and xanthine dehydrogenase, EC 1.2.1.37; Potrikus & Breznak, 1981). However, little or no UA was voided in faeces (Potrikus & Breznak, 1980a), despite the fact that termite tissues did not possess a uricase or any other enzyme to degrade UA (Potrikus & Breznak, 1980a; 1981). This apparent paradox was reconciled by the demonstration that uricolysis did, in fact, occur in *R. flavipes*, but as an anaerobic process mediated by hindgut bacteria (Potrikus & Breznak, 1980b, c; 1981). Major uricolytic isolates were *Streptococcus* sp., *Bacteroides termitidis* and *Citrobacter* sp. All isolates used UA as an energy source for growth, and major products of UA fermentation were NH_3, CO_2, and acetate. Uric acid decomposition by the bacteria was a strictly anaerobic processes, not only for *B. termitidis* which is a strict anaerobe, but for uricolytic streptococci and citrobacters as well. As expected, uricolytic activity of gut homogenates was also inhibited by exposure to air and was markedly reduced in termites that had been fed antibacterial drugs.

By using ^{14}C- and ^{15}N-labelled UA, Potrikus & Breznak (1981) were able to show that UA was transported from its site of synthesis and storage (fat body tissue) to the gut by Malpighian tubules, and that microbial uricolysis *in situ* liberates nitrogen that is reused by the termites for biosynthesis.

Based on these results a model for microbial-mediated UA-nitrogen (UA-N) recycling in *R. flavipes* can be proposed (Fig 7.4). The precise form of UA-N taken up by termites is unknown, however. Ammonia is one possibility since it is the major nitrogenous product of uricolysis by the gut bacteria in pure culture (Potrikus & Breznak, 1980a), and termite tissues possess glutamine synthetase (EC 6.3.1.2; Potrikus & Breznak, 1981). Alternatively, UA-N might first be assimilated by gut microbes to become available to termites as organic nitrogen, either by secretion from gut microbes, or by lysis of cells in the gut, or through coprophagy or trophallaxis. However, at present virtually nothing is known about the dynamics of UA synthesis and mobilization in *R. flavipes* under natural conditions, although it seems safe to assume that UA biosynthesis will be favoured when the intake of dietary nitrogen by termites exceeds that needed for biosynthesis, whereas UA mobilization will be favoured when colony demand for nitrogen is high (e.g. during peaks of reproductive activity). Nevertheless, UA-N recycling is almost certainly significant to the nitrogen economy of *R. flavipes*. It was

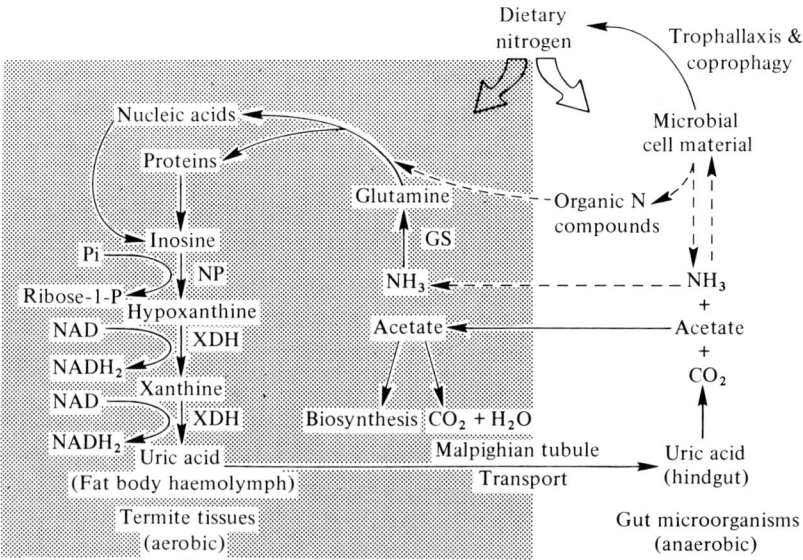

Fig. 7.4. Proposed working model for recycling of uric acid nitrogen in *Reticulitermes flavipes*. Stippled portion depicts reactions mediated by termite enzymes and tissues, whereas the clear portion depicts reactions mediated by hindgut bacteria under anaerobic conditions. Solid arrows indicate reactions for which there is experimental proof or which are extremely likely, whereas dashed arrows depict reactions which are still speculative. Abbreviations: GS, glutamine synthetase; NP, purine nucleoside phosphorylase; Pi, phosphate ion; XDH, xanthine dehydrogenase. See text for details. (Modified from Potrikus, 1980.)

calculated that biosynthesis of UA was an energetically sound investment for *R. flavipes*, and that bacterial uricolysis could liberate enough nitrogen annually to support biosynthesis of termites equivalent to 30% of the colony biomass (Potrikus & Breznak, 1981). Thus, in termites UA may not function as a waste product *per se*, but could be an important storage form of nitrogen. Such a situation appears to hold for cockroaches: in an elegant study with *Periplaneta americana*, Mullins & Cochran (1975a, b) showed that UA accumulated by the insects on high nitrogen diets was mobilized when shifted to a low nitrogen diet. Part of the UA-N mobilized under such conditions was used for oothecal production by females. However, microbial involvement in cockroach uricolysis is unclear (see Cochran, 1975 for review of this issue). By contrast, Nazarczuk, O'Brien & Slaytor (1981) found that *Nasutitermes exitiosus* and *Coptotermes lacteus* failed to mobilize UA over a seven week period when starved, or when their gut bacteria were killed with tetracycline. However, if fully developed workers were used in these experiments, their nitrogen needs may have been very low and limited to that needed for protein turnover. Consequently, it may not be surprising that UA was not mobilized under these conditions. It would be interesting to start examining nitrogen metabolism in termites under laboratory conditions that mimic nutritional stresses expected in nature, e.g. by using test groups of termites containing young, nutrient-dependent larvae.

Another potentially important strategy for nitrogen conservation in termite colonies would be the consumption of exuviae, as well as dead, dying, or supernumerary individuals (via cannibalism). The incidence of such behaviour has been reviewed by LaFage & Nutting (1978). A rich source of nitrogen in such food is chitin, and chitinase has in fact been demonstrated in termites (Tracy & Youatt, 1958; Waterhouse, Hackman & McKellar, 1961). However, since whole insects were used to prepare the enzyme extracts, the anatomical origin of such activity is unknown. By contrast, Rohrmann & Rossman (1980) found that most of the chitinase of *Macrotermes ukuzii* workers and nymphs was gut-associated. Since *M. ukuzii* cultivates and consumes a symbiotic *Termitomyces* fungus, a digestive chitinase may make fungal cell wall chitin available to this termite as a nitrogen source. These investigators also isolated a chitinolytic bacterium from the gut of *M. ukuzii*, but they were uncertain of the contribution of the bacterium to chitinase activity of gut homogenates. It would be interesting to examine termite gut microbes further as a source of chitinase.

It is not yet known if termites can conserve nitrogen by actively digesting some of their gut microbiota. While lysis of protozoa in the hindgut of lower termites appears insignificant as a means of nitrogen conservation (Hungate, 1955), proctodeal trophallaxis could conserve microbial nitrogen if cells could be digested in the gut of recipients. Rates of growth, lysis, and turnover of bacteria in the hindgut are unknown, but would add greatly to our understanding of the nutritive potential of bacterial cell material. On the other hand, the ability of termite hindgut bacteria to supply amino acids to their host is suggested by two studies. Mauldin *et al.* (1978) found that normal synthesis of amino acids by *Coptotermes formosanus* fed ^{14}C-acetate was dependent on the combined presence of normal gut bacteria and two protozoa (*Holomastigotoides hartmanni* and *Spirotrichonympha leidyi*). A similar role for bacteria was suggested by Speck, Becker & Lenz (1971), who used *Reticulitermes santonensis* fed ^{14}C-glucose.

Lignin degradation in termites

The question of lignin degradation by termites is intriguing, since much of the termite gut is anaerobic and natural anaerobic mechanisms for lignin degradation are unknown. Conclusions based on analysis of termite faeces are conflicting, with some authors reporting as much as 83% lignin degradation (reviewed by LaFage & Nutting, 1978; Lee & Wood, 1971), and others reporting virtually none (Esenther & Kirk, 1974). The controversy concerning lignin degradation has recently been discussed (Breznak, 1982; O'Brien & Slaytor, 1982) and need not be repeated in detail here. However, the work of Butler & Buckerfield (1979) must be cited as one of the most convincing arguments for lignin digestion in the termite gut. These investigators found that the higher termite *Nasutitermes exitiosus* readily respired ^{14}C-labelled lignins. Fourteen to 32% and 15–63% of the ^{14}C-label was evolved from synthetic and maize lignins respectively, which were labelled in various positions in the polymer (methoxy; C_2; ring). An important control was their demonstration that maximum $^{14}CO_2$ emission required the presence of live termites in the incubation vessels; little or no $^{14}CO_2$ was evolved when termites were removed. This indicated that lignin degradation was occurring *within* the termites' bodies, and not from voided faecal pellets. The specific site of degradation in the gut was not determined, nor was the involvement of gut microbes in the process.

However, Butler & Buckerfield (1979) speculated that the polymer might be degraded in the gut to smaller molecular weight derivatives which might be taken up through the gut epithelium and oxidized aerobically by termite tissues.

Concluding remarks

A significant amount of information has been obtained in the last decade or so on the biochemical activities of the termite gut microbiota and the symbiotic interaction between the microbiota and its host. We are now in a position to refine pre-existing models for such symbioses (e.g. Figs. 7.3, 7.4) and to begin to test the generality of these models. Particularly satisfying is the availability in pure culture of some of the dominant and seemingly important gut microbes of termites, including the anaerobic cellulolytic protozoa of lower termites. The biochemistry of such organisms will be an important topic for future research. Nevertheless, some quantitatively significant gut symbionts of termites remain to be isolated and studied, notably the spirochaetes. It seems certain that a better understanding of the physiology, biochemistry, and nutrition of such forms will greatly increase our understanding of the gut ecosystem from which they came and their importance to termite nutrition.

We now also have a clearer notion of the termites' own biochemical capabilities, and it is beginning to look as though higher termites might be independent of their gut microbes, at least with regard to the initial steps of cellulose degradation. Further studies must continue to discriminate carefully the termites' capabilities from those provided by the gut microbiota. This is of utmost importance as we probe more deeply into the biochemistry of polysaccharide and lignin degradation, and nitrogen metabolism, in these ecologically important insects.

Acknowledgments. I am grateful to R. W. O'Brien and P. R. Zimmerman for providing me with preprints of unpublished papers. Some of the research reported in this review was conducted by the writer with support from the National Science Foundation (BMS 75-05850, PCM 77-05732, and PCM 80-12026) and the Michigan Agricultural Experiment Station of Michigan State University, and for which he is grateful. This is journal article no. 10609 from the Michigan Agricultural Experiment Station.

References

Balch, W. E., Schoberth, S., Tanner, R. S. & Wolfe, R. S. (1977). *Acetobacterium*, a new genus of hydrogen-oxidizing, carbon dioxide-reducing, anaerobic bacteria. *International Journal of Systematic Bacteriology*, **27**, 355–61.

Ball, G. H. (1969). Organisms living on and in protozoa. In *Research in Protozoology*, vol. 3, ed. T.-T. Chen, pp. 565–718. New York: Pergamon Press.

de Bary, A. (1879). *Die Erscheinung der Symbiose*. Strasburg: Trubner.

Benemann, J. R. (1973). Nitrogen fixation in termites. *Science*, **181**, 164–5.

Bergerson, F. J. & Hipsley, E. H. (1970). The presence of N_2-fixing bacteria in the intestines of man and animals. *Journal of General Microbiology*, **60**, 61–5.

Bignell, D. E. & Anderson, J. M. (1980). Determination of pH and oxygen status in the guts of lower and higher termites. *Journal of Insect Physiology*, **26**, 183–8.

Bignell, D. E., Oskarsson, H. & Anderson, J. M. (1979). Association of actinomycete-like bacteria with soil-feeding termites (Termitidae, Termitinae). *Applied and Environmental Microbiology*, **37**, 339–42.

Bignell, D. E., Oskarsson, H. & Anderson, J. M. (1980*a*). Distribution and abundance of bacteria in the gut of a soil-feeding termite *Procubitermes aburiensis* (Termitidae, Termitinae). *Journal of General Microbiology*, **117**, 393–403.

Bignell, D. E., Oskarsson, H. & Anderson, J. M. (1980*b*). Specialization of the hindgut wall for the attachment of symbiotic micro-organisms in a termite *Procubitermes aburiensis* (Isoptera, Termitidae, Termitinae). *Zoomorphology*, **96**, 103–12.

Bignell, D. E., Oskarsson, H. & Anderson, J. M. (1980*c*). Colonization of the epithelial face of the peritrophic membrane and the ectoperitrophic space by actinomycetes in a soil-feeding termite. *Journal of Invertebrate Pathology*, **36**, 426–8.

Blomquist, G. J., Howard, R. W. & McDaniel, C. A. (1979). Biosynthesis of the cuticular hydrocarbons of the termite *Zootermopsis angusticollis* (Hagen). Incorporation of propionate into dimethylalkanes. *Insect Biochemistry*, **9**, 371–4.

Bloodgood, R. A. (1975). Ultrastructure of the attachment of *Pyrsonympha* to the hindgut wall of *Reticulitermes tibialis*. *Journal of Insect Physiology*, **21**, 391–9.

Breznak, J. A. (1973). Biology of nonpathogenic, host-associated spirochetes. *CRC Critical Reviews in Microbiology*, **2**, 457–89.

Breznak, J. A. (1975). Symbiotic relationships between termites and their intestinal microbiota. In *Symbiosis: 29th Symposium of the Society for Experimental Biology*, ed. D. H. Jennings & D. L. Lee, pp. 559–80. Cambridge University Press.

Breznak, J. A. (1982). Intestinal microbiota of termites and other xylophagous insects. *Annual Review of Microbiology*, **36**, 323–43.

Breznak, J. A. (1983). Hindgut spirochetes of termites and *Cryptocercus punctulatus*. In *Bergey's Manual of Systematic Bacteriology*, ed. N. R. Krieg (in press). Baltimore: Williams & Wilkins.

Breznak, J. A., Brill, W. J., Mertins, J. W. & Coppel, H. C. (1973). Nitrogen fixation in termites. *Nature, London*, **244**, 577–80.

Breznak, J. A. & Pankratz, H. S. (1977). In situ morphology of the gut microbiota of wood-eating termites [*Reticulitermes flavipes* (Kollar) and *Coptotermes formosanus* Shiraki]. *Applied and Environmental Microbiology*, **33**, 406–26.

Brian, M. V. (ed.) (1978). *Production Ecology of Ants and Termites*. Cambridge University Press.

Brill, W. J. (1975). Regulation and genetics of bacterial nitrogen fixation. *Annual Review of Microbiology*, **29**, 109–29.

Butler, J. H. A. & Buckerfield, J. C. (1979). Digestion of lignin by termites. *Soil Biology and Biochemistry*, **11**, 507–11.

Carpenter, E. J. & Culliney, J. L. (1975). Nitrogen fixation in marine shipworms. *Science*, **187**, 551–2.

Cleveland, L. R. (1923). Symbiosis between termites and their intestinal protozoa. *Proceedings of the National Academy of Sciences of the USA*, **9**, 424–8.

Cleveland, L. R. (1924). The physiological and symbiotic relationships between the intestinal protozoa of termites and their host, with special reference to *Reticulitermes flavipes* Kollar. *Biological Bulletin*, **46**, 178–227.

Cleveland, L. R. (1925*a*). The effects of oxygenation and starvation on the symbiosis between the termite, *Termopsis*, and its intestinal flagellates. *Biological Bulletin*, **48**, 309–26.

Cleveland, L. R. (1925*b*). The ability of termites to live perhaps indefinitely on a diet of pure cellulose. *Biological Bulletin*, **48**, 289–93.

Cleveland, L. R. & Grimstone, A. V. (1964). The fine structure of the flagellate *Mixotricha paradoxa* and its associated micro-organisms. *Proceedings of the Royal Society of London, Series B*, **159**, 668–86.

Cochran, D. G. (1975). Excretion in insects. In *Insect Biochemistry and Function*, ed. D. J. Candy & B. A. Kilby, pp. 178–281. London: Chapman & Hall.

Cook, S. F. (1943). Nonsymbiotic utilization of carbohydrates by the termite *Zootermopsis angusticollis*. *Physiological Zoology*, **16**, 123–8.

Cowling, E. B. & Merrill, W. (1966). Nitrogen in wood and its role in wood deterioration. *Canadian Journal of Botany*, **44**, 1539–54.

Esenther, G. R. & Kirk, T. K. (1974). Catabolism of aspen sapwood in *Reticulitermes flavipes* (Isoptera: Rhinotermitidae). *Annals of the Entomological Society of America*, **67**, 989–91.

Eutick, M. L., Veivers, P., O'Brien, R. W. & Slaytor, M. (1978). Dependence of the higher termite, *Nasutitermes exitiosus* and the lower termite, *Coptotermes lacteus* on their gut flora. *Journal of Insect Physiology*, **24**, 363–8.

French, J. R. J., Turner, G. L. & Bradbury, J. F. (1976). Nitrogen fixation by bacteria from the hindgut of termites. *Journal of General Microbiology*, **95**, 202–6.

Gilmour, D. (1940). The anaerobic gaseous metabolism of the termite, *Zootermopsis nevadensis* Hagen. *Journal of Cellular and Comparative Physiology*, **15**, 331–42.

Guerinot, M. L., Fong, W. & Patriquin, D. G. (1977). Nitrogen fixation (acetylene reduction) associated with sea urchins (*Strongylocentrotus droebachiensis*) feeding on seaweeds and eelgrass. *Journal of the Fisheries Research Board of Canada*, **34**, 416–20.

Hardy, R. W. F., Burns, R. C. & Holsten, R. D. (1973). Applications of the acetylene–ethylene assay for measurement of nitrogen fixation. *Soil Biology and Biochemistry*, **5**, 47–81.

Honigberg, B. M. (1970). Protozoa associated with termites and their role in digestion. In *Biology of Termites*, vol. 2, ed. K. Krishna & F. M. Weesner, pp. 1–36. New York: Academic Press.

Hungate, R. E. (1938). Studies on the nutrition of *Zootermopsis*. II. The relative importance of the termite and the protozoa in wood digestion. *Ecology*, **19**, 1–25.

Hungate, R. E. (1939). Experiments on the nutrition of *Zootermopsis*. III. The anaerobic carbohydrate dissimilation by the intestinal protozoa. *Ecology*, **20**, 230–45.

Hungate, R. E. (1941). Experiments on the nitrogen economy of termites. *Annals of the Entomological Society of America*, **34**, 467–89.

Hungate, R. E. (1943). Quantitative analysis on the cellulose fermentation by termite protozoa. *Annals of the Entomological Society of America*, **36**, 730–9.

Hungate, R. E. (1944). Termite growth and nitrogen utilization in laboratory cultures. *Texas Academy of Sciences Proceedings and Transactions*, **27**, 1–7.
Hungate, R. E. (1946a). The symbiotic utilization of cellulose. *Journal of Elisha Mitchell Scientific Society*, **62**, 9–24.
Hungate, R. E. (1946b). Studies on cellulose fermentation. II. An anaerobic cellulose-decomposing actinomycete, *Micromonospora propionici*, n. sp. *Journal of Bacteriology*, **51**, 51–6.
Hungate, R. E. (1955). Mutualistic intestinal protozoa. In *Biochemistry and Physiology of Protozoa*, vol. 2, ed. S. H. Hutner & A. Lwoff, pp. 159–99. New York: Academic Press.
Hungate, R. E. (1975). The rumen microbial ecosystem. *Annual Review of Ecology and Systematics*, **6**, 39–66.
Kovoor, J. (1959). Anatomie du tractus intestinal dans le genre *Microcerotermes* (Silvestri), (Isoptera, Termitidae). *Bulletin de la Société zooligique de France*, **84**, 445–57.
Kovoor, J. (1967a). Etude radiographique du transit intestinal chez un termite supérieur. *Experientia*, **23**, 820–1.
Kovoor, J. (1967b). Présence d'acides gras volatils dans la panse d'un termite supérieur (*Microcerotermes edentatus* Was., Amitermitinae). *Comptes rendus hebdomadaire des séances de L'Académie des sciences*, **264D**, 486–8.
Kovoor, J. (1968). L'intestin d'un termite supérieur (*Microcerotermes edentatus*, Wasman, Amitermitinae). Histophysiologie et flore bactérienne symbiotique. *Bulletin biologique de la France et de la Belgique*, **102**, 45–84.
Kovoor, J. (1970). Présence d'enzymes cellulolytiques dans l'intestin d'un termite supérieur, *Microcerotermes edentatus* (Was.). *Annales des sciences naturelles* (*Zoologie*), **12**, 65–71.
Krasil'nikov, N. A. & Satdykov, S. I. (1969). Estimation of the total bacteria in the intestines of termites. *Microbiology*, **38**, 289–92.
Krelinova, D., Kirku, V. & Skoda, J. (1977). The cellulolytic activity of some intestinal bacteria of termites. *International Biodeterioration Bulletin*, **13**, 81–7.
Krishna, K. (1969). Introduction. In *Biology of Termites*, vol. 1, ed. K. Krishna & F. M. Weesner, pp. 1–17. New York: Academic Press.
Krishna, K. & Weesner, F. M. (1969–1970). *Biology of Termites*, vols. 1 and 2. New York: Academic Press.
LaFage, J. P. & Nutting, W. L. (1978). Nutrient dynamics of termites. In *Production Ecology of Ants and Termites*, ed. M. V. Brian, pp. 165–232. Cambridge University Press.
LaFage, J. P. & Nutting, W. L. (1979). Respiratory gas exchange in the dry wood termite, *Marginitermes hubbardi* (Banks) (Isoptera: Kalotermitidae). *Sociobiology*, **4**, 257–67.
Leach, J. G. & Granovsky, A. A. (1938). Nitrogen in the nutrition of termites. *Science*, **87**, 66–7.
Lee, K. E. & Wood, J. G. (1971). *Termites and Soils*. New York: Academic Press.
McEwen, S. E., Slaytor, M. & O'Brien, R. W. (1980). Cellobiase activity in three species of Australian termites. *Insect Biochemistry*, **10**, 563–7.
McMahan, E. A. (1969). Feeding relationships and radioisotope techniques. In *Biology of Termites*, vol. 1, ed. K. Krishna & F. M. Weesner, pp. 387–406. New York: Academic Press.
Martin, M. M. & Martin, J. S. (1979). The distribution and origins of the cellulolytic enzymes of the higher termite, *Macrotermes natalensis*. *Physiological Zoology*, **52**, 11–21.

Mauldin, J. K. & Rich, N. M. (1980). Effect of chlortetracycline and other antibiotics on protozoan numbers in the eastern subterranean termite. *Journal of Economic Entomology*, **73**, 123–8.

Mauldin, J. K., Rich, N. M. & Cook, D. W. (1978). Amino acid synthesis from ^{14}C-acetate by normally and abnormally faunated termites, *Coptotermes formosanus*. *Insect Biochemistry*, **8**, 105–9.

Mishra, S. C. (1980). Carbohydrases in *Neotermes basei* Snyder (Isoptera: Kalotermitidae). *Material und Organismen*, **15**, 253–61.

Mullins, D. E. & Cochran, D. G. (1975a). Nitrogen metabolism in the American cockroach. I. An examination of positive nitrogen balance with respect to uric acid stores. *Comparative Biochemistry and Physiology*, **50A**, 489–500.

Mullins, D. E. & Cochran, D. G. (1975b). Nitrogen metabolism in the American cockroach. II. An examination of negative nitrogen balance with respect to mobilization of uric acid stores. *Comparative Biochemistry and Physiology*, **50A**, 501–10.

Nazarczuk, R. A., O'Brien, R. W. & Slaytor, M. (1981). Alterations of the gut microbiota and its effect on nitrogen metabolism in termites. *Insect Biochemistry*, **11**, 267–75.

Noirot, C. & Noirot-Timothée, C. (1967). L'epithélium absorbant de la panse d'un termite supérieur. Ultrastructures et rapport avec la symbiose bactérienne. *Annales de la Société entomologique de France*, **3**, 577–92.

Noirot, C. & Noirot-Timothée, C. (1969). The digestive system. In *Biology of Termites*, vol. 1, ed. K. Krishna & F. M. Weesner, pp. 49–88. New York: Academic Press.

O'Brien, G. W., Veivers, P. C., McEwen, S. E., Slaytor, M. & O'Brien, R. W. (1979). The origin and distribution of cellulase in the termites, *Nasutitermes exitiosus* and *Coptotermes lacteus*. *Insect Biochemistry*, **9**, 619–25.

O'Brien, R. W. & Slaytor, M. (1982). Role of microorganisms in the metabolism of termites. *Australian Journal of Biological Sciences*, **35**, 239–62.

Odelson, D. A. & Breznak, J. A. (1983). Volatile fatty acid production by the hindgut microbiota of xylophagous termites. *Applied and Environmental Microbiology*, **45**, 1602–13.

Peakin, G. J. & Josens, G. (1978). Respiration and energy flow. In *Production Ecology of Ants and Termites*, ed. M. V. Brian, pp. 111–63. Cambridge University Press.

Potrikus, C. J. (1980). 'Uric acid and uricolytic gut bacteria in wood-eating termites: a strategy for nitrogen conservation.' Unpublished Ph.D. dissertation, Michigan State University.

Potrikus, C. J. (1983). Role of microorganisms in the nitrogen economy of termites. *Bioscience* (in press).

Potrikus, C. J. & Breznak, J. A. (1977). Nitrogen-fixing *Enterobacter agglomerans* isolated from guts of wood-eating termites. *Applied and Environmental Microbiology*, **33**, 392–9.

Potrikus, C. J. & Breznak, J. A. (1980a). Uric acid in wood-eating termites. *Insect Biochemistry*, **10**, 19–27.

Potrikus, C. J. & Breznak, J. A. (1980b). Uric acid-degrading bacteria in guts of termites [*Reticulitermes flavipes* (Kollar)]. *Applied and Environmental Microbiology*, **40**, 117–24.

Potrikus, C. J. & Breznak, J. A. (1980c). Anaerobic degradation of uric acid by gut bacteria of termites. *Applied and Environmental Microbiology*, **40**, 125–32.

Potrikus, C. J. & Breznak, J. A. (1981). Gut bacteria recycle uric acid nitrogen in termites: a strategy for nutrient conservation. *Proceedings of the National Academy of Sciences of the USA*, **78**, 4601–5.

Potts, R. C. & Hewitt, P. H. (1973). The distribution of intestinal bacteria and cellulase activity in the harvester termite *Trinervitermes trinervoides* (Nasutitermitinae). *Insectes sociaux*, **20**, 215–20.
Potts, R. C. & Hewitt, P. H. (1974). Some properties and reaction characteristics of the partially purified cellulase from the termite *Trinervitermes trinervoides* (Nasutitermitinae). *Comparative Biochemistry and Physiology*, **47B**, 327–37.
Prestwich, G. D. & Bentley, B. L. (1981). Nitrogen fixation by intact colonies of the termite *Nasutitermes corniger*. *Oecologia*, **49**, 249–51.
Prestwich, G. D., Bentley, B. L. & Carpenter, E. J. (1980). Nitrogen sources for neotropical nasute termites: fixation and selective foraging. *Oecologia*, **46**, 397–401.
Prestwich, G. D., Jones, R. W. & Collins, M. S. (1981). Terpene biosynthesis by nasute termite soldiers (Isoptera: Nasutitermitinae). *Insect Biochemistry*, **11**, 331–6.
Prins, R. A. & Lankhorst, A. (1977). Synthesis of acetate from CO_2 in the cecum of some rodents. *FEMS Microbiology Letters*, **1**, 255–8.
Rohrmann, G. F. & Rossman, A. Y. (1980). Nutrient strategies of *Macrotermes ukuzii* (Isoptera: Termitidae). *Pedobiologia*, **20**, 61–73.
Sands, W. A. (1969). The association of termites and fungi. In *Biology of Termites*, vol. 1, ed. K. Krishna & F. M. Weesner, pp. 495–524. New York: Academic Press.
Schultz, J. E. & Breznak, J. A. (1978). Heterotrophic bacteria present in hindguts of wood-eating termites [*Reticulitermes flavipes* (Kollar)]. *Applied and Environmental Microbiology*, **35**, 930–6.
Schultz, J. E. & Breznak, J. A. (1979). Cross-feeding of lactate between *Streptococcus lactis* and *Bacteroides* sp. isolated from termite hindguts. *Applied and Environmental Microbiology*, **37**, 1206–10.
Speck, U., Becker, G. & Lenz, M. (1971). Ernährungsphysiologische Untersuchungen an termiten nach selektiver medikamentöser Ausschaltung der Darmsymbionten. *Zeitschrift für angewandte Zoologie*, **58**, 475–91.
Sylvester-Bradley, R., Bandeira, A. G. & de Oliveira, L. A. (1978). Fixacão de nitrogênio (reducão de acetileno) em cupins (Insecta: Isoptera) da Amazônia Central. *Acta Amazonica*, **8**, 621–7.
Thayer, D. W. (1976). Facultative wood-digesting bacteria from the hindgut of the termite *Reticulitermes hesperus*. *Journal of General Microbiology*, **95**, 287–96.
Thayer, D. W. (1978). Carboxymethylcellulase produced by facultative bacteria from the hind-gut of the termite *Reticulitermes hesperus*. *Journal of General Microbiology*, **106**, 13–18.
To, L. P., Margulis, L., Chase, D. & Nutting, W. L. (1980). The symbiotic microbial community of the Sonoran Desert termite: *Pterotermes occidentis*. *Biosystems*, **13**, 109–37.
To, L. P., Margulis, L. & Cheung, A. T. W. (1978). Pillotinas and hollandinas: distribution and behaviour of large spirochetes symbiotic in termites. *Microbios*, **22**, 103–33.
Tracy, M. V. & Youatt, G. (1958). Cellulase and chitinase in two species of Australian termites. *Enzymologia*, **19**, 70–2.
Trager, W. (1932). A cellulase from the symbiotic intestinal flagellates of termites and of the roach, *Cryptocercus punctulatus*. *Biochemical Journal*, **26**, 1762–71.
Trager, W. (1934). The cultivation of a cellulose-digesting flagellate, *Trichomonas termopsidis* and of certain other termite protozoa. *Biological Bulletin*, **66**, 182–90
Veivers, P. C., Musca, A. N., O'Brien, R. W. & Slaytor, M. (1982). Digestive enzymes of the salivary glands and gut of *Mastotermes darwiniensis*. *Insect Biochemistry*, **12**, 35–40.

Veivers, P. C., O'Brien, R. W. & Slaytor, M. (1980). The redox state of the gut of termites. *Journal of Insect Physiology*, **26**, 75–7.
Veivers, P. C., O'Brien, R. W. & Slaytor, M. (1982). Role of bacteria in maintaining the redox potential in the hindgut of termites and preventing entry of foreign bacteria. *Journal of Insect Physiology*, **28**, 947–51.
Waterhouse, D. F., Hackman, R. H. & McKellar, J. W. (1961). An investigation of chitinase activity in cockroach and termite extracts. *Journal of Insect Physiology*, **6**, 96–112.
Wolfe, R. S. (1979). Methanogenesis. In *International Review of Biochemistry, Microbial Biochemistry*, vol. 21, ed. J. R. Quayle, pp. 270–300. Baltimore: University Park Press.
Wolin, M. J. (1974). Metabolic interactions among intestinal microorganisms. In *Intestinal Microecology*, ed. M. H. Floch & D. J. Hentges, *American Journal of Clinical Nutrition*, **27**, 1320–8.
Wood, T. G. (1978). Food and feeding habits of termites. In *Production Ecology of Ants and Termites*, ed. M. V. Brian, pp. 55–80. Cambridge University Press.
Wood, T. G. & Sands, W. A. (1978). The role of termites in ecosystems. In *Production Ecology of Ants and Termites*, ed. M. V. Brian, pp. 245–92. Cambridge University Press.
Wood, T. M. & McCrae, S. I. (1979). Synergism between enzymes involved in the solubilization of native cellulose. In *Hydrolysis of Cellulose: Mechanisms of Enzymatic and Acid Catalysis*, ed. R. D. Brown & L. Jurasick, pp. 181–209. Washington: American Chemical Society.
Yamaoka, I. & Nagatani, Y. (1975). Cellulose digestion system in the termite, *Reticulitermes speratus* (Kolbe). I. Producing sites and physiological significance of two kinds of cellulase in the worker. *Zoological Magazine*, **84**, 23–9.
Yamin, M. A. (1978). Axenic cultivation of the cellulolytic flagellate *Trichomitopsis termopsidis* (Cleveland) from the termite *Zootermopsis*. *Journal of Protozoology*, **25**, 535–8.
Yamin, M. A. (1979). Flagellates of the orders Trichomonadida Kirby, Oxymonadida Grassé, and Hypermastigida Grassi and Foà reported from lower termites (Isoptera families Mastotermitidae, Kalotermitidae, Hodotermitidae, Termopsidae, Rhinotermitidae, and Serritermitidae) and from the wood-feeding roach *Cryptocercus* (Dictyoptera: Cryptocercidae). *Sociobiology*, **4**, 1–120.
Yamin, M. A. (1980). Cellulose metabolism by the termite flagellate *Trichomitopsis termopsidis*. *Applied and Environmental Microbiology*, **39**, 859–63.
Yamin, M. A. (1981). Cellulose metabolism by the flagellate *Trichonympha* from a termite is independent of endosymbiotic bacteria. *Science*, **211**, 58–9.
Yamin, M. A. & Trager, W. (1979). Cellulolytic activity of an axcnically-cultivated termite flagellate, *Trichomitopsis termopsidis*. *Journal of General Microbiology*, **113**, 417–20.
Yokoe, Y. (1964). Cellulase activity in the termite, *Leucotermes speratus*, with new evidence in support of a cellulase produced by the termite itself. *Scientific Papers of the College of General Education, University of Tokyo*, **14**, 115–20.
Zimmerman, P. R., Greenberg, J. P., Wandiga, S. O. & Crutzen, P. J. (1982). Termites: a potentially large source of atmospheric methane, carbon dioxide, and molecular hydrogen. *Science*, **218**, 563–5.

8
The arthropod gut as an environment for microorganisms

D. E. BIGNELL

Department of Zoology, Westfield College, University of London, Hampstead, London NW3 7ST, UK

Key words: gut cuticle; microbial adherence; peritrophic membrane; Actinomycetes; fluid fluxes; pH

Introduction

The objective of this contribution is to review the general structural and physiological organization of the arthropod gut for a readership of microbiologists. Emphasis will be placed on the variations in physical and chemical conditions that occur from one part of the alimentary tract to another and on the functional differences shown by the gut in two of the major taxa, the insects and crustaceans. While acknowledging that the arthropod gut is in general a rich source of microorganisms, it is proposed that the suitability of individual sites for colonization by commensal organisms depends as much on the host physiological activities taking place at the site as on adaptations of the microorganisms themselves. Hence a challenge is laid against some of the assumptions frequently made in work on gut floras, for example that whole-gut homogenates are adequate sources of isolates or that microbes which proliferate in the intestinal environment face predominantly anaerobic conditions. With a knowledge of gut functions it is possible in many cases to predict the sites where colonization and growth will be most intensive and to identify those arthropods in which proliferation is likely to be most marked. This has obvious importance where the animals are vectors of disease, or are candidates for this status, or where they contribute significantly to decomposition processes.

The vast majority of arthropods are accommodated in four major classes: Insecta, Crustacea, Myriapoda (centipedes and millipedes), and Arachnida (spiders, scorpions, ticks, and mites). Other taxa are recognized but are so poorly researched that nothing useful can be said about them in the context of this discussion. Amongst the four major groupings the literature is biased heavily towards the insects and one must of necessity draw much of the physiological information from this source, with progressively reduced consideration of crustaceans, myriapods and arachnids. I have attempted to redress the balance to some extent by drawing on millipede material for some of the micrographs, but a shortage of physiological information about this group is a handicap. In order to keep the text free of obstructive detail the figure legends are extended so that each becomes a self-contained treatment of a particular aspect of intestinal structure or organization.

Structural considerations

Although enormous variations exist in the general morphology of arthropod guts, a tripartite division into foregut (stomodaeum), midgut (mesenteron) and hindgut (proctodaeum) can be universally recognized. The foregut and hindgut are lined by a cuticle which

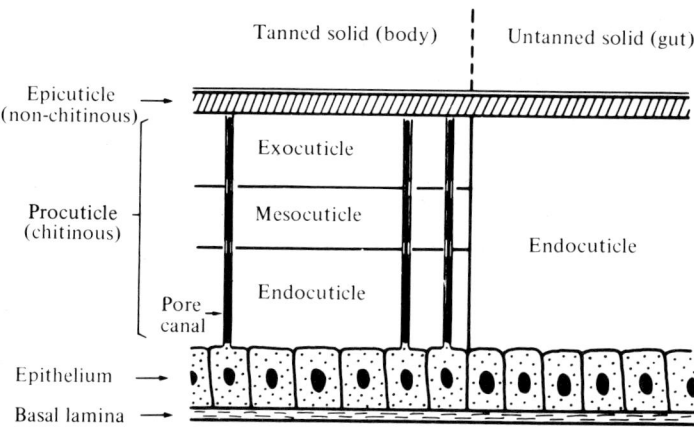

Fig. 8.1. Generalized structure of the body wall and gut cuticles. Although the bulk of cuticle consists of various layers containing chitin, the physiological properties are largely determined by the nature and prominence of the non-chitinous epicuticle which may contain a number of lipid and phenolic materials in addition to protein. Lipid also occurs in exocuticle and mesocuticle, but not in endocuticle. Rigidity, where required, is achieved by aromatic tanning of the epicuticle and exocuticle (plus calcification of the exocuticle in crustaceans), but most gut cuticle is untanned. Pore canals are present in integumentary cuticle as extensions of the epidermal cells and apparently function in the transport of impregnating waxes and protein-bound polyphenols. Rubber-like cuticle and arthrodial membrane (non-solid cuticles) also occur in the body wall, but are absent from the gut. (Modified from Neville, 1975.)

resembles that of the body wall in some respects, and which is also secreted by an underlying epithelium (Fig. 8.1). Despite the existence of a large literature on the structure and chemistry of integumentary cuticle (reviews by Neville, 1975; Richards, 1978), little information is available about the cuticle lining the gut and comparisons must therefore be made with caution. A common feature is that both are largely composed of a chitin–protein composite (procuticle) overlain by a non-chitinous epicuticle. The latter layer confronts the gut lumen and therefore provides a surface to which microorganisms can attach (Fig. 8.2). It has long been recognized that the epicuticle of the body wall is itself multilayered, two layers being distinguishable when the cuticle is newly formed at ecdysis, with additional wax and cement layers forming subsequently from constituents which reach the surface through pore canals and dermal gland ducts respectively. These secondarily secreted layers contain labile, straight chain hydrocarbons which confer substantial waterproofing to the integument (Lockey, 1980; Neville, 1975). Pore canals are absent in gut cuticle, the obvious inference being that the wax layer is not secreted. This would be consistent with a requirement for partial or selective permeability, and in most gut cuticles examined microscopically only two layers of epicuticle can be distinguished (e.g. Figs. 8.1 and 8.7e). Sclerotized (tanned) cuticle does occur in guts where it is used to construct triturating surfaces and valves, but little or nothing is known of its structure. In general the epicuticle is more prominent in foregut than hindgut cuticle, reflecting the lower permeability of the former to aqueous fluids and small, neutrally charged hydrophilic molecules (Maddrell & Gardiner, 1980). Hindgut epicuticle may show discontinuous but regular thinning, producing dome-like protrusions of the cuticular surface (e.g. Fig. 8.7e), simple pitting (Noirot & Noirot-Timothée, 1969, 1976; Bayon, 1971) or deep invaginations (e.g. Fig. 8.7f). Permeability is probably facilitated at these sites, which seem to be best developed in animals such as termites and cockroaches where the uptake of organic materials from the hindgut is of nutritional significance. Maddrell & Gardiner (1980) showed that the hindgut cuticle of a number of insects was more permeable to cations than anions, an observation which suggests that this cuticle may carry a net negative charge. Such charging might militate against bacterial adherence (Ofek & Beachey, 1980), but it does not necessarily follow that the charges are located at the cuticular surface.

The midgut is the digestive surface *sensu strictu* and in consequence does not secrete a cuticle. However it does elaborate a chitinous

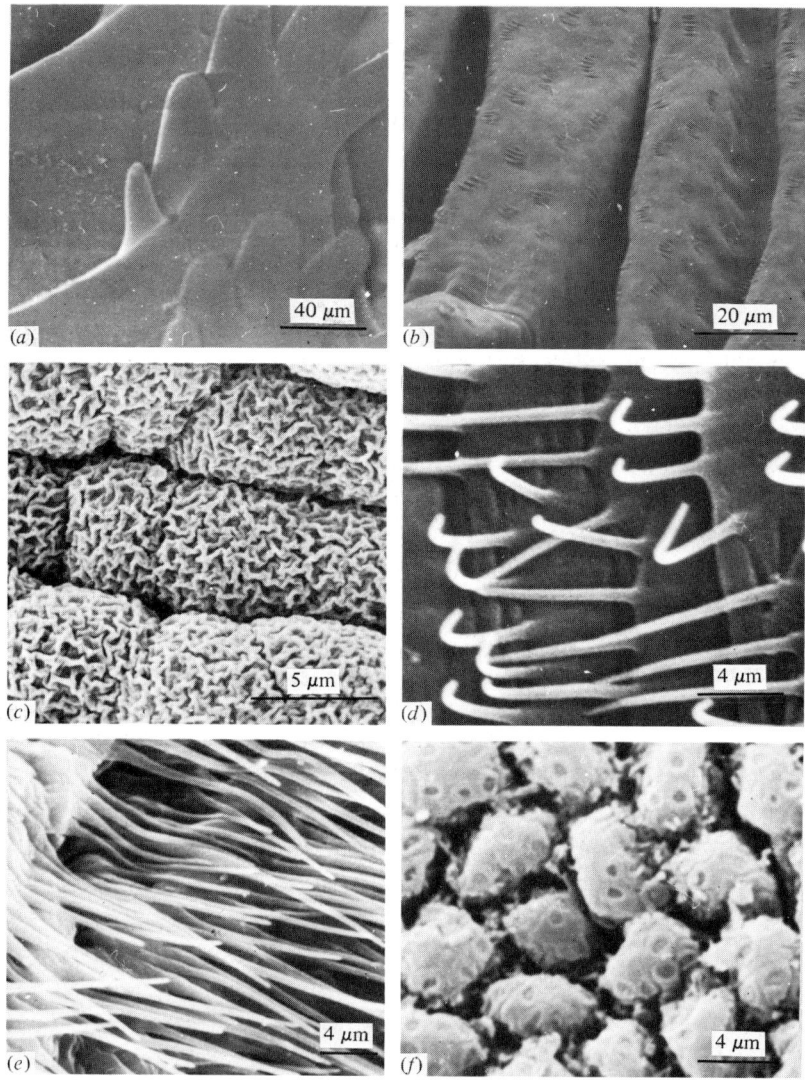

Fig. 8.2. Surface architecture of arthropod gut cuticles. (a) Flat surfaces are rare because of the requirement that most of the gut wall be extensible for peristalsis, but are sometimes present as components of rigid gizzard structures; *Acheta domestica*. (b) Simple longitudinal folding allows periodic expansion of the gut and provides microniche crypts in which microorganisms often proliferate; colon of *Schistocerca gregaria*. (c) More complex patterns of folding seem to occur frequently in millipedes and isopods, but in the latter group cuticular surfaces are not usually colonized; anterior hindgut of *Oniscus asellus*. (d) Spines of various shapes and sizes are common features, although their functions

peritrophic membrane which encloses solid food material and prevents direct contact with the epithelium (Fig. 8.6*f–h*). The functions of the peritrophic membrane are not fully known, but its most obvious purpose is to protect the absorptive surfaces in the absence of mucus secretion (Richards & Richards, 1977; Bignell, 1981*a*). Whereas the cuticular linings of the foregut and hindgut are stable surfaces for the duration of each instar, the peritrophic membrane is continuously secreted, and moves rearwards at rates of up to 1 cm hr^{-1} in some cases. Where the intestinal transit time is short (e.g. in many herbivores), the peritrophic membrane passes out of the gut largely intact as a wrapping around the faecal pellets, but more commonly the membrane is broken up in the hindgut by muscular contractions or spinous projections of the cuticle. The latter are often associated with an enteric valve (e.g. Bignell, 1981*b*). Disruption of the peritrophic membrane may be of importance in allowing ingested organisms access to colonization sites on the hindgut cuticle, and also permits some parasites to reach the mesenteric epithelium by forward migration along the ectoperitrophic space (Le Berre, 1967). Whether the peritrophic membrane is intended to be a barrier to microorganisms is unclear: that it is in general effective as one may be deduced from the relative rarity of non-pathogenic associations with mesenteric tissue (Bignell, Oskarsson & Anderson, 1980).

Systematic examination of the gut wall in a number of arthropods has shown that the cuticle is unevenly colonized (compare Figs. 8.2 and 8.3). Foreguts are, in general, poorly utilized by microorganisms while the hindguts of groups such as insects and millipedes may usually be relied on to reveal a rich and varied flora (e.g. Boyle & Mitchell, 1978; Bracke, Cruden & Markovetz, 1979; Bignell, unpublished observations). While some adherent organisms are certain to be removed during processing for microscopy, the consistency of these observations suggests that they are not an artefact of specimen preparation, but represent genuine differences in composition or environment which are sufficient to determine whether attachment occurs or not. In isopods much of the hindgut cuticle is devoid of adherent organisms *in situ*

>Caption for fig. 8.2. (*cont.*).
>are not always evident; hindgut of *Polydesmus* sp. (*e*) Where spines are too slender for secure attachment their function may be to prevent microbial colonization or to assist in shredding the peritrophic membrane; posterior hindgut of *Cylindroiulus* sp. (*f*) Sites of facilitated permeability are sometimes indicated by pitting of the cuticle surface; posterior hindgut of *Tachypodiulus* sp.

8 Arthropod gut as an environment for microorganisms   210

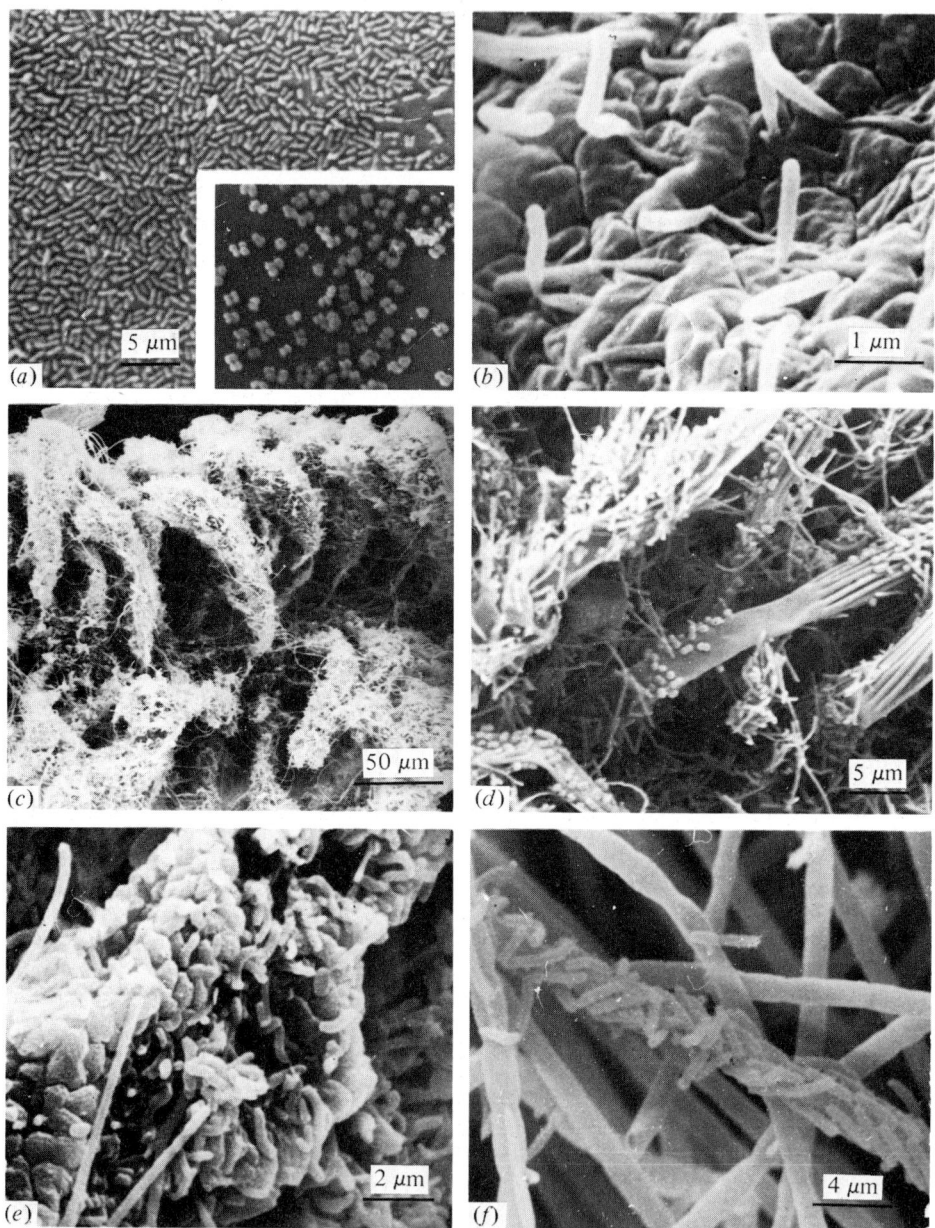

Fig. 8.3. Patterns of microbial attachment to cuticle. (a) On flat, relatively permeable surfaces a monolayer of rods or coccobacilli is most commonly encountered; *Schistocerca gregaria*, colon. The inset shows coccoid forms which are occasionally encountered in the fore-

(Boyle & Mitchell, 1978; Fig. 8.2c) (an observation which has led some authors to the erroneous conclusion that the guts of crustaceans are sterile), but bacteria can be made to adhere to the cuticle *in vitro* while the attendant muscle layers are untouched (Bignell, unpublished observations). If this is accepted as evidence that cuticle composition has no adverse effect on microbial adherence, it seems most probable that permeability is the determining factor, and organisms colonizing cuticular surfaces through which a significant fluid flux is occurring will be least affected by the adjacent unstirred layers.

The chemical environment

It is obvious that the chemical milieu of the gut lumen will be influenced both by the nature of ingested food material and the secretions of the gut wall and its associated glands. In arthropods, as in all the more highly evolved animals, digestive processes are sequenced such that optimum conditions can be achieved in turn for the enzymatic degradation and absorption of each major type of nutrient. This involves not only differentiation of the gut wall (in particular the restriction of absorptive surfaces to parts of the midgut), but also regulation of the rates of transit to and from these surfaces and the sites of digestive enzyme secretion, control of the rates of enzyme synthesis and discharge, manipulation of the pH and, possibly, the redox potential of digestive sites, and restriction of the physical access of enzyme molecules to their substrates (Dadd, 1970; House, 1974; Bignell, 1981*b*). The presence of the semi-permeable peritrophic membrane makes it probable that digestion is also sequenced radially, with partial

Caption for fig. 8.3. (*cont.*).
gut. (*b*) End-on attachment is sometimes employed on less permeable cuticles, for example crustacean hindgut. It would be a suitable orientation for spore-formers or serve to remove parts of the cell from non-stirred layers adjacent to the surface; *Oniscus asellus*, hindgut. (*c*) Elongated spines may be furnished by the host as attachment surfaces for symbiont organisms, ensuring that the microbes come into intimate contact with food material in transit; posterior colon of *Procubitermes aburiensis* showing filamentous procaryotic organisms adhering to spines. (*d*) An alternative form of spine, branched distally, serving as an attachment site for long rods and coccoid couples; ileum of *Acheta domestica*. (*e*) filaments may also attach to flat cuticular surfaces although a clearly differentiated holdfast is not apparent; posterior hindgut of *Cylindroiulus* sp. (*f*) Epibionts may be found in some locations, suggesting secondary symbiotic relationships amongst the constituents of a gut flora; hindgut of *Tachypodiulus* sp.

degradation of macromolecules only taking place in the endoperitrophic space, the hydrolysis to dimers and oligomers occurring principally in the ectoperitrophic space separating the membrane and the midgut epithelium, and some terminal digestion occurring intracellularly or mediated by enzymes bound to the apical cell membrane (Terra, Ferreira & de Bianchi, 1979; Terra & Ferreira, 1981).

Some mention has already been made of the difference between insects and some crustaceans in respect of microbial colonization of the hindgut wall. This may be related to a fundamental distinction between the two groups, namely that the guts of insect (and also myriapods) accommodate Malpighian tubules as an additional secretory structure (Figs. 8.4 and 8.5). Their function is to deliver to both the midgut and the hindgut a fluid containing many of the haemolymph constituents at concentrations proportional to (though not necessarily equal to) their concentrations in the haemolymph. The hindgut reabsorbs those constituents required by the animal, rejecting others; in this way the composition of the haemolymph can be regulated while nitrogenous excretion and detoxification are effected at the same time (Maddrell, 1977; Wall, 1977). Although Malpighian tubules have the ability to transport organic and inorganic anions from haemolymph to the lumen, the principal active process is the secretion of K^+. Fluid movement accompanies this influx, with anions and low molecular weight organic compounds such as sugars and amino acids diffusing passively and appearing in the secreted fluid at relative concentrations proportional to molecular size. For reasons not fully understood, phosphate ions become concentrated in the tubule fluid so that the emergent secretion constitutes not only a nutrient broth of balanced composition but also an effective buffer solution, stabilizing hydrogen ion concentration in the hindgut and the ectoperitrophic space into both of which the fluid is directed. Maddrell (1981) has pointed out that whereas most insects produce a primary urine that is usually more dilute than the haemolymph, the Malpighian tubules of millipedes generate a fluid in which the solutes achieve a concentration roughly equal to that of the bathing medium. In physiological terms this distinction, which is related to the size (diameter) of permeability channels in the tubule epithelium, illustrates the primitiveness of the myriapod excretory system, but may also imply that the hindgut of this group offers a particularly suitable site for microbial colonization. The evidence of preliminary surveys (e.g. Figs. 8.3 and 8.8) confirms that an abundant and diverse flora occurs in this location. In both insects and myriapods additional fluid input to the gut takes place via the salivary glands and accompanies ingestion of food. Analyses of salivary fluid have shown that it is generally iso-osmotic with the haemolymph, but unlike the secretion of the Malpighian tubules contains a mucous component and a number of digestive enzymes. The major ionic species are Na^+, K^+ and Cl^- (Day, 1951; Bland & House, 1971).

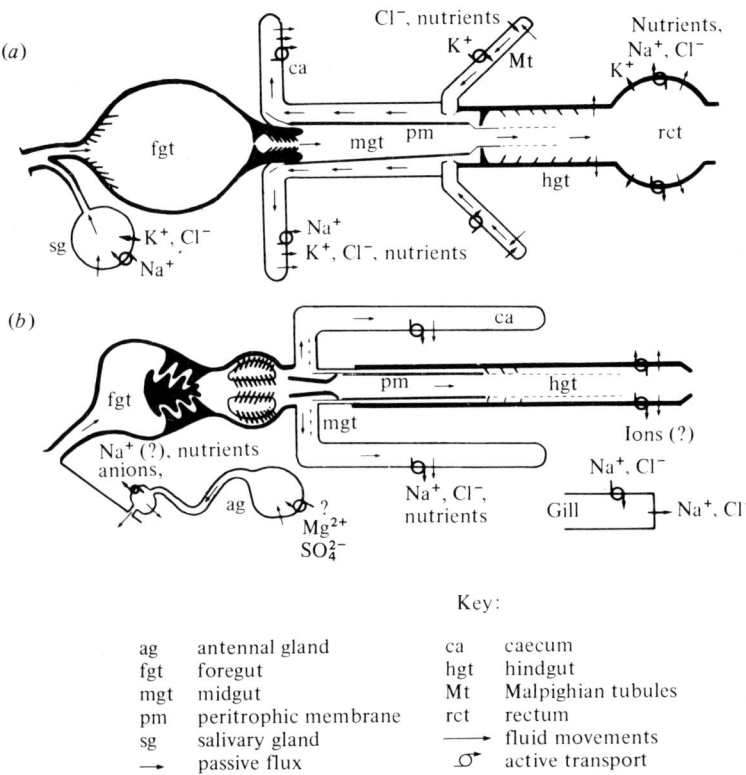

Fig. 8.4. Physiological and structural organization of the gut in insects (*a*) and crustaceans (*b*), showing the major fluid movements and the principal sites of ion and nutrient fluxes. The most important difference is that the insect gut serves as the sole organ of excretion and osmotic regulation in addition to being the digestive tract. As a consequence there are complex internal fluid movements which largely obviate the need for an active efflux of organic nutrients at the absorptive surfaces in the midgut. In crustaceans ionic and osmotic exchanges with the environment can take place at the gill surfaces and via specialized excretory glands not associated anatomically with the alimentary canal. Hence the regulatory role of the gut is reduced and the intestinal environment relatively impoverished for microbial proliferation. Active efflux of organic nutrients occurs in crustaceans.

Since the crustacean gut lacks both salivary glands and Malpighian tubules (Fig. 8.4), it is unlikely that fluid influx can occur at any site other than the midgut caeca. Gastric juices emanating from these diverticula contain Na^+, Ca^{2+}, Mg^{2+}, Cl^-, PO_4^{3-} and SO_4^{2-} in addition to a broad spectrum of digestive enzymes, and are passed forwards into the foregut by antiperistalsis (van Weel, 1970) or added to solid food material passing into the anterior hindgut (Hassall & Jennings, 1975). In malacostracans at least, the fluid is said to be iso-ionic with the serum (Holliday, Mykles, Terwilliger & Dangott, 1980). Although the gills are still considered the principal site of ionic exchanges, evidence is emerging that the posterior midgut, where present, transports some ions and water into the haemolymph, the process being linked to and driven by the active efflux of Na^+ (Ahearn, Maginniss, Song & Tornquist, 1977; Mykles, 1981). The combination of secretion from the caeca and

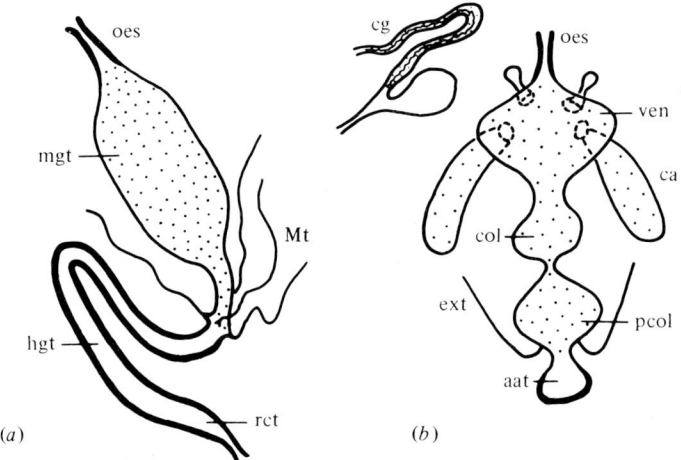

Fig. 8.5. Generalized form of the alimentary canal in millipedes (a) and mites (b) (reconstructed from Hughes, 1959). The cuticle-lined hindgut is more prominent in millipedes; in mites the structure is restricted to the anal atrium. Excretory tubules (ext) or Malpighian tubules (Mt) are present in all millipedes and most mites, excepting the Cryptostigmata. Coxal glands or related structures with an osmoregulatory function are present in the Cryptostigmata, Mesostigmata and Astigmata, but not in other mites. A peritrophic membrane and some form of salivary gland associated with the pre-oral region normally occur in both groups. Relatively little is known of the physiology of the gut in these arthropods. Abbreviations: aat, anal atrium; cg, coxal gland; col, colon; ext, excretory tubule; hgt, hindgut; mgt, midgut; oes, oesophagus; pcol, posterior colon; rct, rectum; ven, ventriculus.

Table 8.1. *pH of the alimentary canal in representative arthropods*

Species	Foregut	Midgut	Hindgut	Citation
(Insecta)				
Periplaneta americana	4.8–6.8	6.1–6.6	6.3–6.8	Greenberg et al. (1970)
Melanoplus bivittatus	5.5	6.7	6.4	Grayson (1951)
Zootermopsis nevadensis	6.8	7.1	7.1	Bignell & Anderson (1980)
(Alkaline gut group)		(highest pH recorded)		
Culex pipiens and other mosquito larvae	—	10.5	7.0	Dadd (1975)
Oryctes nasicornis	8.5	10.4	6.7	Rössler (1961)
Cubitermes severus	6.7	7.5	10.4	Bignell & Anderson (1980)
various Lepidoptera	—	10.3	8.9	Waterhouse (1949)
Tipula abdominalis (larva)	—	11.6	—	Martin et al. (1980)
(Acid gut group)		(lowest pH recorded)		
Lucilia cuprina	7.8	3.3	7.8	Waterhouse (1940)
various Diptera	5.2	2.8	7.3	Greenberg (1968)
(Crustacea)				
Calanus finmarchicus	—	6.0–8.0	—	Mansour-Bek (in van Weel, 1970)
Daphnia magna	—	6.0–6.2	—	As *C. finmarchicus*
Ligia oceanica	—	6.4	—	As *C. finmarchicus*
Astacus fluviatilis	4.7–6.6	—	—	As *C. finmarchicus*
Cancer pagurus	5.8–6.0	—	—	As *C. finmarchicus*
(Acari)				
Phthiracarus sp.	—	5.4–6.6	—	Dinsdale (1974)

uptake in the posterior midgut would therefore secure a local fluid circulation within the mesenteric portion of the gut, but would not provide an irrigation of the hindgut comparable to that which occurs in insects and myriapods. Furthermore, it has been suggested that a gut exudate may be produced in crustaceans which is toxic to bacteria (Boyle & Mitchell, 1978). Some ionic fluxes are thought likely to occur across the hindgut wall, but their direction and significance have yet to be elucidated (Holdich & Ratcliffe, 1970; Mykles, 1979).

Table 8.1 gives pH data for the guts of representative arthropods. In the majority of species investigated values fall within the normal physiological range tolerated by bacteria, the most acidic conditions tending to occur in the foregut and a shift towards the alkaline side

Table 8.2. *Redox potentials (E_h) in the guts of various insects and a millipede, determined with a platinum and calomel electrode combination*

Species	Range of readings in imagos (mV) (8–20 individuals per species)		
	Foregut	Midgut	Hindgut
Locusta migratoria	+117 to +287	+67 to +290	+97 to +232
Periplaneta americana	+251 to +347	+37 to +187	−173 to +157
Cubitermes severus	+107 to +347	—	−223 to +197
Zootermopsis nevadensis	—	—	−298 to −23
Glomeris marginata	—	+267 to +307	+167 to +277

occurring in more posterior regions (House, 1974; van Weel, 1970). Some insects show more extreme conditions which frequently seem to be related to a requirement for the solubilization of recalcitrant dietary materials such as tanned proteins or lignocellulose complexes. A number of insects showing exceptionally alkaline conditions also harbour large populations of intestinal microorganisms (e.g. higher termites, scarabaeid larvae), amongst which some adaptation to the elevated pH might be expected. There is evidence that the viability of non-indigenous ingested organisms is affected by extreme conditions (see below).

Relatively little is known about the oxygen status of arthropod guts, other than that insects engaging in symbiont-mediated, fermentative digestion of wood and related recalcitrant materials have strongly reducing gut conditions, presumably the result of oxygen sequestration by the microorganisms (Bayon & Etievant, 1980; O'Brien & Slaytor, 1982). In other insects electrometric determinations of redox potential indicate that conditions are far from anaerobic, although the occurrence of anaerobic microsites within a predominantly aerobic or microaerophilic gut lumen cannot be excluded (Table 8.2). It must always be borne in mind that arthropods are relatively small animals with higher surface area : volume ratios than occur in practically all vertebrates. In consequence they are more likely to reach equilibrium with the environment unless efficient permeability barriers are established or active internal regulation is carried out. Ritter (1961) proposed that anaerobic conditions were maintained in the hindgut of the wood-feeding cockroach *Cryptocercus punctulatus* by the secretion of the reducing agent glutathione. Unfortunately the work did not establish whether this substance was of animal or microbial origin.

A wide range of nitrogenous compounds are excreted by Malpighian tubules. The most widely encountered is uric acid, but allantoin, urea, ammonia, hypoxanthine, various tryptophan metabolites (kynurenic, 8-hydroxyquinaldic and xanthurenic acids) and proteinaceous materials may also be included in the primary urine reaching the hindgut (Mullins & Cochran, 1973; Cochran, 1975). It is of interest to point out that several quinolones associated with tryptophan metabolism are mutagenic (Kuznezova, 1969) and that ammonia has a well-known toxic effect when present in higher concentrations. Ammonia is the primary excretory product in isopods (Wieser, Schweizer & Hartenstein, 1969).

Internal trituration and transit

Despite the efficiency with which the mouthparts can manipulate, shred or pulverize food materials, a number of insects and crustaceans possess internal trituration mechanisms associated with the foregut. In malacostracans the foregut often has a proventriculus or gastric mill anteriorly and a press/filter apparatus incorporated into the posterior pyloric portion (Fig. 8.4). These crustaceans do not masticate food: mechanical trituration occurs in the gastric mill, accompanied by an enzymatic predigestion. The press and filter allow only the smallest particles to enter the midgut caeca, larger particles being returned to the mill or passed directly into the posterior midgut. Other crustaceans, in which the mill is reduced or absent, either feed on fine material suspended in water or masticate with the mouth parts before swallowing (van Weel, 1970). In many insects the foregut terminates in a sclerotized gizzard which also functions to keep particulate material out of the caeca. The combination of the gizzard and the peritrophic membrane make the caeca of both crustaceans and insects relatively inaccessible to microorganisms; any that are found in this location are therefore likely to be specialized symbionts.

Transit times vary widely according to the nature of the food, ranging from less than one hour for suspension feeders and some herbivores, to up to 48 hours for arthropods feeding on recalcitrant materials. The midgut is usually the least contractile region, showing only weak and relatively infrequent peristaltic movements, but in many cases transit is completed more rapidly than in the foregut or hindgut. One explanation for this is that although the latter regions are relatively motile, peristalsis and antiperistalsis are infrequent here, the majority of movements taking the form of segmentation or compression of the gut tube (Cook & Reinecke, 1973; Bignell, 1981*b*). Such patterns of contraction would

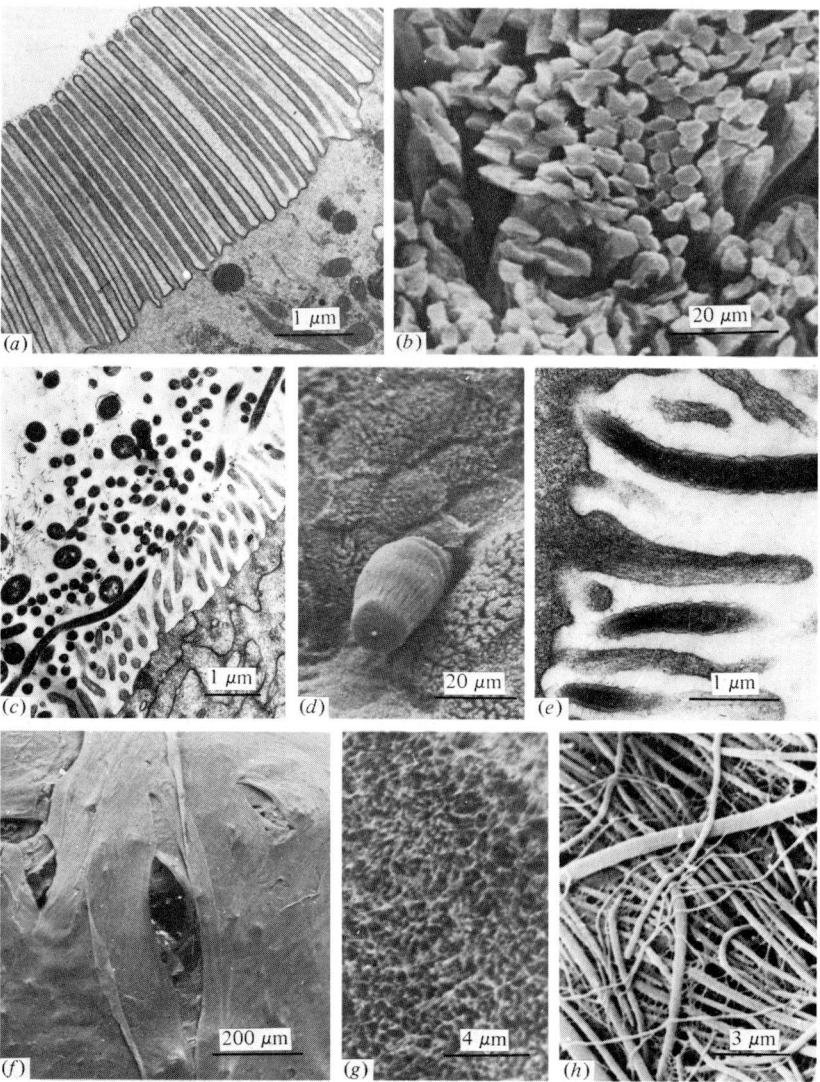

Fig. 8.6. Microorganisms and the mesenteron. (a) Since the mesenteron is the principal region dealing with digestion and absorption, the main structural feature is the elaboration of the epithelial surface into numerous parallel microvilli lacking a cuticular covering; thin section of anterior midgut of *Periplaneta americana*. (b) The length and diameter of midgut microvilli vary considerably, but all epithelial cells are renewed every 24–48 hr and the surface is therefore relatively precarious for microbial colonization; scanning micrograph of the midgut surface of *Glomeris marginata*. (c) Colonization of the ectoperitrophic space does occur in many termites, but the main zone

clearly render the hindgut more suitable for microbial growth by agitating the contents independently of rearward food movement. The combination of motility and internal trituration may account in part for the relative infrequency with which the fungi are able to colonize arthropod guts.

Absorption and the peritrophic membrane

The guts of insects are exceptional amongst higher invertebrates in that absorption does not appear to involve the active transport of organic molecules from the lumen to the haemolymph, either by means of independent carriers or linked to Na^+ efflux in the manner of the vertebrate intestine (Treherne, 1967). While an absorption mechanism dependent only on passive gradients is energetically economical, it is likely to be efficient only when gradients are steep in the aftermath of a meal; during the later stages of digestion these gradients are therefore reinforced by a countercurrent flow of fluid forwards from the Malpighian tubules along the absorptive surfaces of the posterior midgut and caeca (Fig. 8.4; Berridge, 1970; Dow, 1981a). The flow is driven osmotically as a consequence of K^+ transport in the tubule epithelium and a complementary active Na^+ efflux from the lumen of the midgut (Dow, 1981b), sweeping nutrients into the caeca from which passive absorption takes place.

It is unclear at present whether the peritrophic membrane specifically delimits a channel at the periphery of the lumen through which the countercurrent flow takes place, but its functions must include that of preventing the physiologically active zone adjacent to the epithelium

Caption for fig. 8.6. (cont.).
of microbial growth is adjacent to the surface with relatively few cells penetrating between the microvilli; *Cubitermes severus*. (d) Organisms able to attach to the midgut surface include the sporozoan parasite *Gregarina*; posterior caeca of *Schistocerca gregaria*. (e) The most successful mesenteric colonizers are probably spirochaetes, which establish contact along the sides of the microvilli as well as at the base of the intermicrovillar cleft; *Cubitermes severus*. (f) Food in the midgut is enveloped in a peritrophic membrane which isolates the microvilli from direct contact with particulate material and ingested organisms; epithelial face of the membrane in *Schistocerca gregaria*. (g) Chitinous fibrils are laid down initially between the microvilli and subsequently lifted off and consolidated to form the peritrophic membrane; *Schistocerca gregaria*. (h) In rare instances the peritrophic membrane itself serves as an attachment site for filamentous organisms; *Cubitermes severus*.

8 *Arthropod gut as an environment for microorganisms* 220

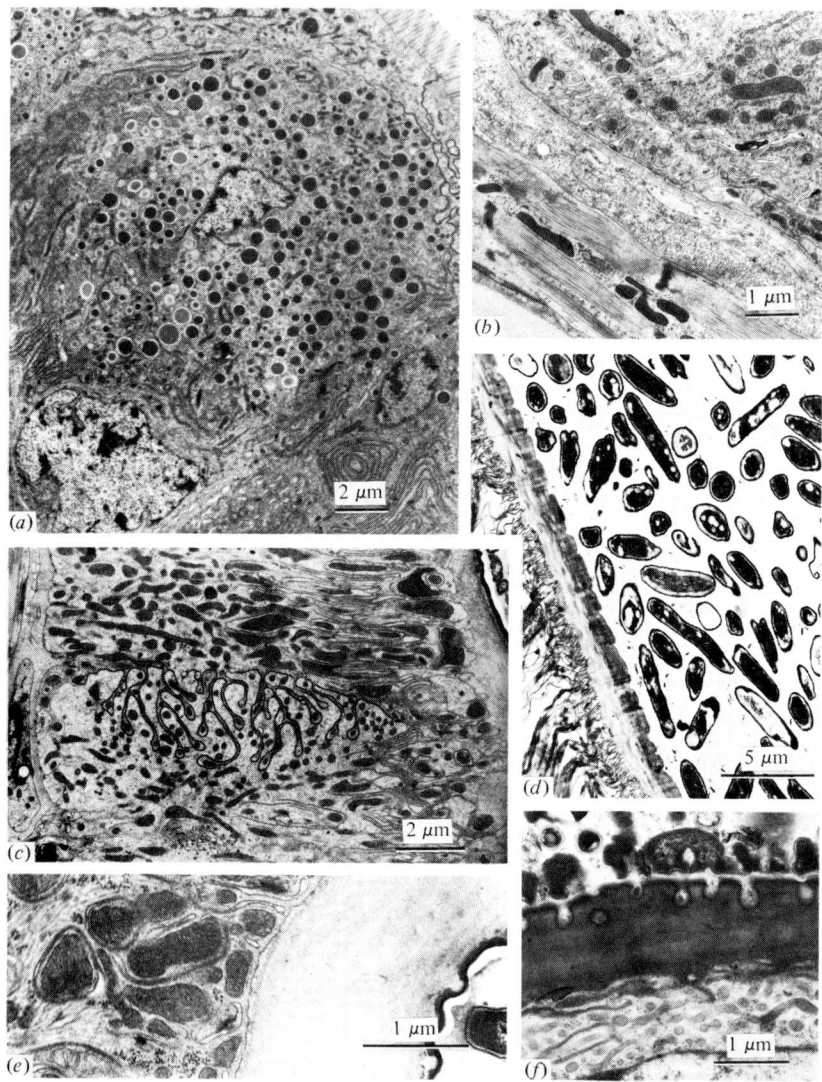

Fig. 8.7. Characterization of epithelial function. (*a*) Secretory regions of the midgut may be recognized by the presence of membrane-bounded vesicles and abundant rough endoplasmic reticulum in the cytoplasm; *Cubitermes severus*. (*b*) Fluid flux in mesenteric tissue may be inferred from the number of extracellular compartments formed within the epithelium by infolding of the basal plasma membrane; *Periplaneta americana*. (*c*) In physiologically active hindgut epithelia, large numbers of mitochondria are usually evident in association with infoldings of the apical plasma membrane which are the sites of intensive local ion transport; *Cubitermes severus* anterior colon. (*d*)

Table 8.3. *Ionic ratios and isolates of actinomycetes in various regions of the gut of the termite* Procubitermes aburiensis

Region	pH	Ratio [K$^+$] : [Na$^+$]	Total Isolates	Novel types Observed	Expected
Crop	6.7	0.6 : 1	21	3	2.8
Mesenteron	6.9	2.0 : 1	28	5	3.6
Mixed segment	8.5	1.7 : 1	9	0	1.2
First proctodaeal segment	9.7	12.9 : 1	6	1	0.8
Third proctodaeal segment	9.1	10.1 : 1	2	0	0.3
Colon	7.5	3.8 : 1	3	0	0.4
Rectum	7.1	3.3 : 1	8	1	1.0
Total			77	10	10.1

Note: The populations were characterized and compared using a 44 test discriminant analysis system, following aerobic isolation on soil extract or starch casein agar media. Profiles were developed from assessments of macroscopic appearance, surface colour and pigmentation, spore formation, sensitivity to UV light, temperature and antibiotics and utilization of starch, sugars and amino acids. Isolates separating from the general population above the 30% level of difference were designated as novel types.

from becoming clogged with particulate material. Fluid and nutrient movements enhance the mesenteric environment for those microorganisms able to circumvent the barrier of the peritrophic membrane; however their strategy of colonization must account for the relatively rapid turnover of epithelial cells and the high enzymatic activity on the caecal microvilli (Day & Powning, 1949; Terra *et al.*, 1979).

Selection and novelty amongst intestinal organisms

Although work on the characterization of arthropod gut floras is in its infancy, some data are available on the effects of extreme conditions on the viability of ingested organisms. For example, in the

Caption for fig. 8.7. (*cont.*).
Permeability of cuticles can be assessed by examination of the electron-dense outer layer of the epicuticle: discontinuity or thinning suggests fluid flux, and associated adherent organisms will usually be present; posterior hindgut of *Oniscus asellus*. (*e*) epicuticular thinning associated with dome-like protrusions into the gut lumen; colon of *Periplaneta americana*. (*f*) epicuticular thinning associated with deep pitting; posterior colon of *Procubitermes aburiensis*.

Table 8.4. *Comparisons of the numbers of novel actinomycetes from various sources associated with the termite* Procubitermes aburiensis

Sources	Total Isolates	Ratio a : b	Novel types		χ^2	Probability (P)
			Observed	Expected		
a. Soil substratum b. Termite gut	57	35 : 22	2 : 20	13.3 : 8.7	24.1	<0.001
a. Mound material b. Termite gut	57	35 : 20	1 : 20	12.9 : 8.1	28.4	<0.001
a. Soil substratum b. Mound material	40	20 : 20	7 : 6	6.5 : 6.5	0.08	n.s.
a. *Procubitermes* gut b *Cubitermes* gut	49	30 : 19	4 : 0	2.4 : 1.5	2.5	n.s.

Note: In each case the populations were analysed in pairs and the ratio of novel types was compared with the ratio of population sizes in a χ^2 test. It is unlikely that any of these isolated organisms form a significant part of the symbiotic gut flora.

soil-feeding termite *Procubitermes aburiensis* a sharp rise in pH, from 6.9 to 9.7, occurs between the midgut and the first proctodaeal segment as a result of an active influx of K^+ across the gut wall (Table 8.3). The number of actinomycetes isolatable from various parts of the gut falls markedly after the midgut, suggesting that the elevated pH affects the viability of these (ingested) organisms. Discriminant analysis of the populations showed that the number of novel isolates did not depart very far from expectation for any particular gut region, so that the adverse effects of the alkaline conditions would seem to be general rather than selective. However, when the general population isolated from within the gut was compared with populations obtained by the same isolation methods from parent soil and termite mound material, significant novelty was discovered amongst the gut organisms (Table 8.4). In particular the intestinal populations showed a more variable response to carbohydrate source and amino acid utilization than did the soil population, and there was some difference in the percentage of various morphological types between the two populations. Interestingly, mound populations did not differ significantly from those of the parent soil, nor was the population isolated from the gut of *Procubitermes aburiensis* distinguishable from that of a related sympatric termite, *Cubitermes severus*. Infrequently branched prokaryotic filaments of unknown affiliation are prominent amongst the *in situ* gut flora of termites and millipedes (Fig. 8.8).

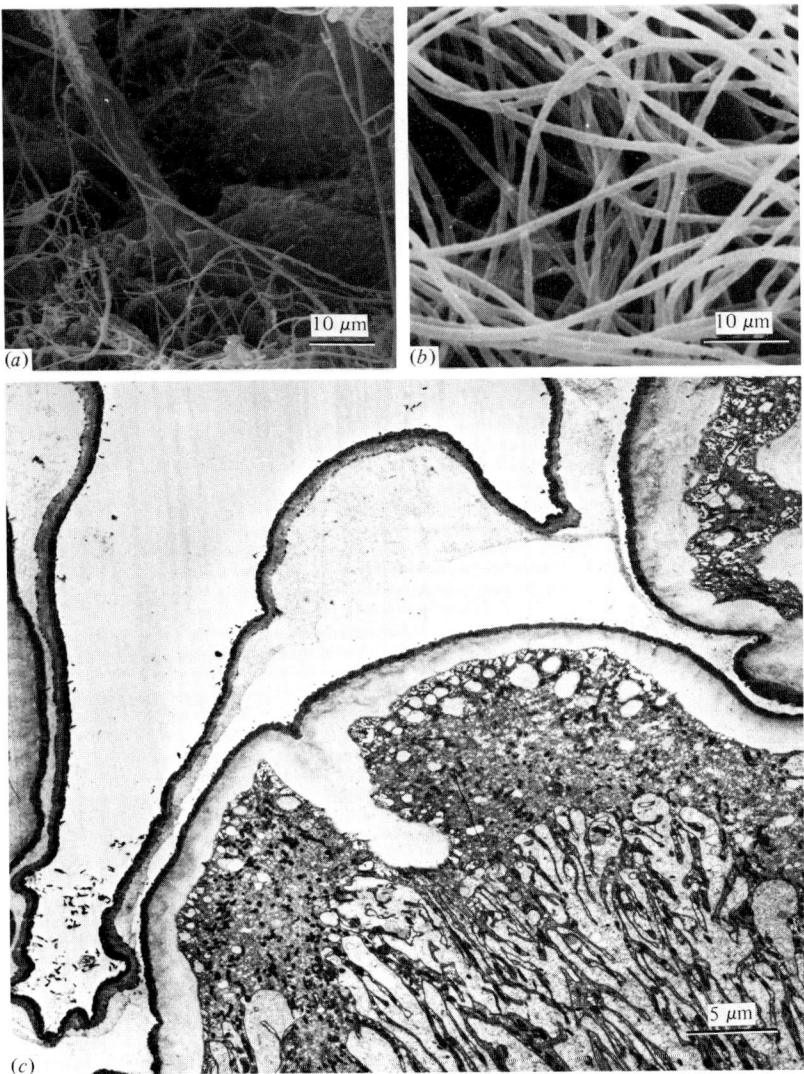

Fig. 8.8. Very thin, infrequently branching filaments occur widely in the guts of insects and millipedes. (a) Hindgut of *Procubitermes aburiensis*. (b) hindgut of *Tachypodiulus* sp. The filament is inherently adapted to the intestinal environment since stability requires only terminal attachment while most of the organism is free to penetrate the food bolus. (c) Fragmentation of filaments may enable the organisms to survive the ecdysial process in which the entire gut cuticle is shed, after partial digestion, and voided with the faeces, leaving a new cuticular surface available for colonization; moulting in the hindgut of *Oniscus asellus*.

Conclusions

While much work remains to be done on the nature and dynamics of arthropod gut floras it appears that key factors enhancing the suitability of this habitat for microbial colonization include the stability and relative permanence of the cuticular intima and a variety of stimuli arising from host physiological processes. The latter include the secretion of readily metabolizable nutrients and excretory products, mitigation of unstirred layers by fluid efflux, buffering to favourable pH and redox conditions and agitation of the lumen contents by contractions of the gut wall.

Acknowledgments. Characterizations of actinomycetes summarized in Tables 8.3 and 8.4 were carried out by Dr Roger Crosse of Glaxo Group Research Ltd, using isolates prepared by the author and Dr J. M. Anderson, and micrographs of millipede and isopod guts were prepared by Hakon Oskarsson in the Department of Biological Sciences, University of Exeter. Financial support by the Natural Environment Research Council to J.M.A. is acknowledged for both sources of material.

References

Ahearn, G. A., Maginniss, L. A., Song, Y. K. & Tornquist, A. (1977). Intestinal water and ion transport in freshwater malacostracan prawns (Crustacea). In *Water Relations in Membrane Transport in Plants and Animals*, ed. A. M. Jungreis, T. K. Hodges, A. Kleinzeller & S. G. Schultz, pp. 129–42. New York: Academic Press.

Bayon, C. (1971). La cuticule proctodéale de la larve d'*Oryctes nasicornis* (Coleoptères, Scarabeides). Étude au microscope électronique à balayage. *Journal de Microscopie*, **11**, 353–70.

Bayon, C. & Etievant, P. (1980). Methanic fermentation in the digestive tract of a xylophagous insect: *Oryctes nasicornis* L. larva (Coleoptera, Scarabaeidae). *Experientia*, **36**, 154–5.

Berridge, M. J. (1970). A structural analysis of intestinal absorption. *Symposium of the Royal Entomological Society of London*, **5**, 135–51.

Bignell, D. E. (1981*a*). Comments on functions of the peritrophic membrane in haematophagous insects. *Parasitology*, **82**, 95–7.

Bignell, D. E. (1981*b*). Nutrition and digestion. In *The American Cockroach*, ed. W. J. Bell & K. G. Adiyodi, pp. 57–86. London: Chapman & Hall.

Bignell, D. E. & Anderson, J. M. (1980). Determination of pH and oxygen status in the guts of lower and higher termites. *Journal of Insect Physiology*, **26**, 183–8.

Bignell, D. E., Oskarsson, H. & Anderson, J. M. (1980). Colonization of the epithelial face of the peritrophic membrane and the ectoperitrophic space by actinomycetes in a soil-feeding termite. *Journal of Invertebrate Pathology*, **36**, 426–8.

Bland, K. P. & House, C. R. (1971). Function of the salivary glands of the cockroach *Nauphoeta cinerea*. *Journal of Insect Physiology*, **17**, 2069–84.
Boyle, P. J. & Mitchell, R. (1978). Absence of micro-organisms in crustacean digestive tracts. *Science*, **200**, 1157–9.
Bracke, J. W., Cruden, D. L. & Markovetz, A. J. (1979). Intestinal microbial flora of the American cockroach *Periplaneta americana*, L. *Applied and Environmental Microbiology*, **38**, 945–55.
Cochran, D. G. (1975). Excretion in insects. In *Insect Biochemistry and Function*, ed. D. J. Candy & B. A. Kilby, pp. 177–282. London: Chapman & Hall.
Cook, B. J. & Reinecke, J. P. (1973). Visceral muscles and myogenic activity in the hindgut of the cockroach *Leucophaea maderae*. *Journal of Comparative Physiology*, **84**, 95–118.
Dadd, R. H. (1970). Arthropod digestion. In *Chemical Zoology*, vol. 5, ed. M. Florkin & B. Scheer, pp. 35–95. New York: Academic Press.
Dadd, R. H. (1975). Alkalinity within the midgut of mosquito larvae with alkaline-active digestive enzymes. *Journal of Insect Physiology*, **21**, 1847–53.
Day, M. F. (1951). The mechanism of secretion by the salivary gland of the cockroach *Periplaneta americana*. *Australian Journal of Scientific Research (B)*, **4**, 136–43.
Day, M. F. & Powning, R. F. (1949). A study of the processes of digestion in certain insects. *Australian Journal of Scientific Research (B)*, **2**, 175–215.
Dinsdale, D. (1974). The digestive activity of a phthiracarid mite mesenteron. *Journal of Insect Physiology*, **20**, 2247–60.
Dow, J. A. T. (1981*a*). Countercurrent flows, water movements and nutrient absorption in the locust midgut. *Journal of Insect Physiology*, **27**, 579–85.
Dow, J. A. T. (1981*b*). Ion and water transport in locust alimentary canal: evidence from *in vivo* electrochemical gradients. *Journal of Experimental Biology*, **93**, 167–79.
Grayson, J. M. D. (1951). Acidity–alkalinity in the alimentary canal of twenty insect species. *Virginia Journal of Science (N.S.)*, **2**, 46–59.
Greenberg, B. (1968). Micro-potentiometric pH determinations of muscoid maggot digestive tracts. *Annals of the Entomological Society of America*, **61**, 365–8.
Greenberg, B., Kowalski, J. & Karpus, J. (1970). Micro-potentiometric pH determinations of the gut of *Periplaneta americana* fed three different diets. *Journal of Economic Entomology*, **63**, 1795–7.
Hassall, M. & Jennings, J. B. (1975). Adaptive features of gut structure and digestive physiology in the terrestrial isopod *Philoscia muscorum* (Scopoli) 1763. *Biological Bulletin*, **149**, 348–64.
Holdich, D. M. & Ratcliffe, N. A. (1970). A light and electron microscope study of the hindgut of the herbivorous isopod *Dynamene bidentata* (Crustacea: Peracarida). *Zeitschrift für Zellforschung und mikroscopische Anatomie*, **111**, 209–27.
Holliday, C. W., Mykles, D. L., Terwilliger, R. C. & Dangott, L. J. (1980). Fluid secretion by the midgut caeca of the crab *Cancer magister*. *Comparative Biochemistry and Physiology*, **67A**, 259–63.
House, H. L. (1974). Digestion. In *The Physiology of Insecta*, 2nd edn, vol. 5, ed. M. Rockstein, pp. 63–117. New York: Academic Press.
Hughes, T. E. (1959). *Mites or the Acari*. University of London : Athlone Press.
Kuznezova, L. E. (1969). Mutagenic effect of 3-hydroxykynurenine and 3-hydroxyanthranilic acid. *Nature (London)*, **222**, 484–5.
Le Berre, R. (1967). Les membranes péritrophique chez les arthropodes. Leur rôle dans la digestion et leur intervention dans l'evolution d'organismes parasitaires. *Cahier ORSTOM (Series d'Entomologie Medicine et Parasitologie)*, **5**, 147–204.

Lockey, K. H. (1980). Insect cuticular hydrocarbons. *Comparative Biochemistry and Physiology*, **65B**, 457–62.
Maddrell, S. H. P. (1977). Insect Malpighian tubules. In *Transport of Ions and Water in Animals*, ed. B. L. Gupta, R. B. Moreton, J. L. Oschman & B. J. Wall, pp. 541–69. London: Academic Press.
Maddrell, S. H. P. (1981). The functional design of the insect excretory system. *Journal of Experimental Biology*, **90**, 1–15.
Maddrell, S. H. P. & Gardiner, B. O. C. (1980). The permeability of the cuticular lining of the insect alimentary canal. *Journal of experimental Biology*, **85**, 227–37.
Martin, M. M., Martin, J. S., Kukor, J. J. & Merritt, R. W. (1980). The digestion of protein and carbohydrate by the stream detritivore *Tipula abdominalis* (Diptera, Tipulidae). *Oecologia*, **46**, 360–4.
Mullins, D. E. & Cochran, D. G. (1973). Nitrogenous excretory materials from the American cockroach. *Journal of Insect Physiology*, **19**, 1007–18.
Mykles, D. L. (1979). Ultrastructure of alimentary epithelia of lobsters, *Homarus americanus* and *H. gammarus*, and crab, *Cancer magister*. *Zoomorphologie*, **92**, 201–15.
Mykles, D. L. (1981). Ionic requirements of transepithelial potential difference and net water flux in the perfused midgut of the American lobster *Homarus americanus*. *Comparative Biochemistry and Physiology*, **69A**, 317–20.
Neville, A. C. (1975). *Biology of the Arthropod Cuticle*. Berlin: Springer-Verlag.
Noirot, C. & Noirot-Timothée, C. (1969). La cuticle proctodéale des Insects. *Zeitschrift für Zellforschung und mikroskopische Anatomie*, **101**, 477–509.
Noirot, C. & Noirot-Timothée, C. (1976). Fine structure of the rectum in cockroaches (Dictyoptera): general organization and intercellular junctions. *Tissue and Cell*, **8**, 345–68.
O'Brien, R. W. & Slaytor, M. (1982). Role of micro-organisms in the metabolism of termites. *Australian Journal of Biological Science*, **35**, 239–62.
Ofek, I. & Beachey, E. H. (1980). General concepts and principles of bacterial adherence in animals and man. In *Bacterial Adherence*, ed. E. H. Beachey, pp. 1–29. London: Chapman & Hall.
Richards, A. G. (1978). The chemistry of insect cuticle. In *Biochemistry of Insects*, ed. M. Rockstein, pp. 205–32. New York: Academic Press.
Richards, A. G. & Richards, P. A. (1977). The peritrophic membranes of insects. *Annual Review of Entomology*, **22**, 219–40.
Ritter, H. (1961). Glutathione-controlled anaerobiosis in *Cryptocercus*, and its detection by polarography. *Biological Bulletin*, **121**, 330–46.
Rössler, M. F. (1961). Ernahrungsphysiologishe Untersuchungen an Scarabaeiden larven *Oryctes nasicornis* L. und *Melolantha melolantha* L. *Journal of Insect Physiology*, **6**, 62–80.
Terra, W. R. & Ferreira, C. (1981). The physiological role of the peritrophic membrane and trehalase: digestive enzymes in the midgut and excreta of starved larvae of *Rhynchosciara*. *Journal of Insect Physiology*, **27**, 325–31.
Terra, W. R., Ferreira, C. & de Bianchi, A. G. (1979). Distribution of digestive enzymes among the endo- and ectoperitrophic spaces and midgut cells of *Rhynchosciara* and its physiological significance. *Journal of Insect Physiology*, **25**, 487–94.
Treherne, J. E. (1967). Gut absorption. *Annual Review of Entomology*, **12**, 43–58.
van Weel, P. B. (1970). Digestion in Crustacea. In *Chemical Zoology*, vol. 5, ed. M. Florkin & B. Scheer, pp. 97–115. New York: Academic Press.
Wall, B. J. (1977). Fluid transport in the cockroach rectum. In *Transport of Ions and Water in Animals*, ed. B. L. Gupta, R. B. Moreton, J. L. Oschman & B. J. Wall, pp. 599–612. London: Academic Press.

References

Waterhouse, D. F. (1940). Studies on the physiology and toxicology of blowflies. 5. The hydrogen-ion concentration in the alimentary canal. *Pamphlet of the Council of Scientific and Industrial Research of Australia*, **102,** 7–27.

Waterhouse, D. F. (1949). The hydrogen-ion concentration in the alimentary canal of larval and adult Lepidoptera. *Australian Journal of Scientific Research (B)*, **2,** 428–37.

Wieser, W., Schweizer, G. & Hartenstein, R. (1969). Patterns in the release of gaseous ammonia by terrestrial isopods. *Oecologia*, **3,** 390–400.

9
Physiological aspects of destructive pathogenesis in insects by fungi: a speculative review

A. K. CHARNLEY

School of Biological Sciences, University of Bath, Claverton Down, Bath BA2 7AY, Avon, UK

Key words: entomopathogen; parasitic fungi; histopathology; extracellular enzymes

Introduction

Aspects of the general biology of entomopathogenic fungi have been reported many times (e.g. MacLeod, 1963; Madelin, 1963, 1966; McEwen, 1963; Müller-Kögler, 1965; Roberts & Yendol, 1971; Ferron, 1978; Roberts & Humber, 1981). However, no general reviews concerning the physiological mechanisms underlying pathogenicity have appeared since that of Madelin (1963), and the present article seeks to fill the gap. This will be done by following the changing interrelationships which occur from initial contact between fungus and insect to the death of the latter. Where appropriate, reference will be made to pathogenesis by fungi of other arthropods, including Crustacea and Acarina. Useful analogies can be drawn between fungal pathogenesis in insects and in plants. Consequently comparisons will be made with the more extensive knowledge of phytopathogenesis to help identify areas for future research. An important element in such comparisons concerns interaction between the fungus and the exposed surfaces of insects and plants, and a diagram illustrating the principal components of insect and plant cuticles is provided for reference in Fig. 9.1.

Pre-infection development
Attachment of the spore to the cuticle

A spore must retain contact with the surface of an insect's integument long enough for germination and subsequent invasion to occur. This is particularly critical in situations such as occur on the cuticular lining of the foregut and hindgut of the mosquito, *Culex*

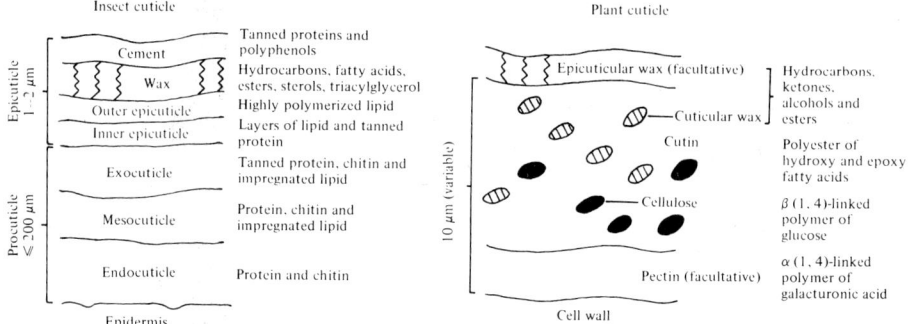

Fig. 9.1. The structure of insect and plant cuticles. Notes:
1. Proteins are the major components of insect cuticles and many different molecules are present in the same cuticle. 2. Tanning or sclerotization is the cross-linking of the proteins by aromatic compounds such as N-acetyldopamine; it hardens insect cuticle. 3. In soft-bodied insects (e.g. lepidopteran larvae) the exocuticle is thin and indistinguishable from the epicuticle. The arthrodial membrane at joints and between segments has no exocuticle. 4. Chitin exists in the form of a chitin–protein complex. 5. Chitin and cellulose are both $\beta(1,4)$-linked hexose polymers. Chitin is made up of N-acetylglucosamine units and cellulose of glucose units. Comparisons have been made between the enzyme species necessary to cause hydrolysis of the two polymers (see Stirling et al., 1979.) (Left-hand side of figure, see Neville, 1975, Locke, 1974; right-hand side, after Ende & Linskens, 1974.)

fatigans, where establishment of *Culicinomyces clavosporus* occurs against a background of almost continuous peristaltic and antiperistaltic movement (Sweeney, 1975; Jones, 1960). Although the adhesive properties of the mucilaginous coat of *Entomophthora* spp. and the slime drop of *Verticillium* spp. conidia are well-known (Webster, 1980), for most entomopathogenic fungi the physical and chemical characteristics of spore and cuticular surfaces responsible for attachment remain uncertain (Roberts & Humber, 1981). However, the importance of this step is clear from the fact that very few conidia of a hypovirulent mutant of *Metarhizium anisopliae* attached to the perispiracular valves of the mosquito, *Culex pipiens*, compared with those of the virulent wild type (Al-Aidroos & Roberts, 1978).

Host specific attachment occurs with zoospores of *Coelomomyces psorophorae* (Zeobold, Whisler, Shemanchuk & Travland, 1979), which adhere promptly to the cuticle of susceptible mosquitoes, but not to that of resistant species. Vesicles in the zoospores disappeared during attachment indicating host specific secretion of an adhesive substance

(Travland, 1979a). The distribution of cysts on the exuvia of mosquito larvae were different from those of the intact integument indicating that the zoospores responded to a site cue mediated by differences in cuticular texture (Travland, 1979b). Nyhlen & Unestam (1980) similarly detected preferential aggregation of cysts of *Aphanomyces astaci* on crayfish cuticle, in this case on or in the vicinity of wounds, perhaps due to the leakage of attractant chemicals.

Spore germination

While some evidence indicates that certain isolates of entomopathogenic fungi can germinate in water without nutrients (e.g. Clerk, 1969), most suggests that they cannot (e.g. Gabriel, 1959; Getzin, 1961; Selhime & Muma, 1966; Clark, Kellen, Fukuda & Lindegren, 1968). The different results may be due to deficiencies in experimental technique. Thus although Lefebvre (1931) reported up to 90% germination of *Beauveria bassiana* in distilled water, this may have been due to failure to wash the conidia prior to transfer from culture. More recent results indicate that washed conidia only germinate in a medium containing an energy source (e.g. glucose), subsequent growth being additionally dependent on supplementation with a nitrogen source. Extensive growth occurred on *N*-acetylglucosamine or glucosamine, indicating that these compounds could be used as sole sources of carbon, nitrogen, and energy (Smith & Grula, 1981). A similar result has been obtained with *Metarhizium anisopliae* (St Leger, Charnley and Cooper, unpublished). Of the compounds most likely to be found on the surface of insect cuticle, some long chain fatty acids, lanolin (wax) and hydrocarbons from light oil could be used by *B. bassiana* as energy sources. However, no correlation was found between pathogenicity of mutant strains for the corn earworm, *Heliothis zea*, and the ability to utilize lanolin as sole carbon source (Smith & Grula, 1981).

The lipolytic ability of entomopathogenic fungi may give them some advantage over other fungi on the cuticle. Schaerffenberg (1964) found that, by contrast with airborne contaminants, germination and growth of *M. anisopliae* and *Beauveria* spp. was favoured by the addition of lipid to non-sterile media.

There is little information on the availability of nutrients on the surface of insect cuticle. Notini & Mathlein (1944) found that conidia of *M. anisopliae* germinated slowly on regions of the cuticle of the goat moth, *Cossus cossus*, that had been treated with lipid solvents, but rapidly on intact cuticle, while ether-soluble compounds from the surface of the cuticle of desert locust, *Schistocerca gregaria*, stimulated germination of *M. anisopliae* (Veen, 1968). Nolla (1929) has recorded that homogenates of aphids stimulated germination of spores of *Acrostalagmus aphidum*, the degree of stimulation depending on the species of aphid.

Successful germination not only requires assimilation of utilizable nutrients but tolerance of any toxic compounds present. Indirect evidence for the presence of inhibitory substances comes from studies where removal of epicuticular lipids mechanically or by chemical means enhances pathogenicity (Sussman, 1951, 1952a; Koidsumi, 1957; Evlakhova & Shekhurina, 1963), or permits germination or invasion of otherwise innocuous fungi; (Nyhlen & Unestam, 1975; Smith, Pekrul & Grula, 1981). However, wiping the integument of the yellow flour beetle, *Tenebrio molitor*, with ether, or abrading it with fine carborundum paper, did not aid penetration of isolated sclerites by *B. bassiana*, *M. anisopliae* or *Aspergillus flavus* (Robinson, 1966).

Koidsumi (1957) extracted lipids which were fungicidal or fungitoxic (depending on concentration) from the cuticles of the Japanese silkworm, *Bombyx mori*, and the stem borer, *Chilo simplex*. The most active were medium chain saturated fatty acids, which were presumed to be caprylic or capric acids. Lipids extracted from strains of *B. mori* resistant to *A. flavus* had greater antifungal activity than those from susceptible strains (Koidsumi & Wada, 1955). Evlakhova & Shekhurina (1963) also found that ether extracts of the shield bug, *Eurygaster integriceps*, depressed spore germination and mycelial growth of *B. bassiana*. Caprylic acid is also the major free fatty acid in extracts from the surface of *H. zea* and the armworm, *Spodoptera frugiterda*, where it has a fungistatic action, inhibiting germination of *B. bassiana* conidia for varying lengths of time, related to its concentration and the type of growth medium (Smith & Grula, 1982).

However, until the *in vivo* concentrations of these substances have been determined, their significance cannot be certain, since they may be too low to have any effect (Latgé & Vey, 1974). In the latter case alternative explanations must be sought for the effect of epicuticular disruption on fungal pathogenesis. The epicuticle is considered to be the principal barrier to water loss (Beament, 1967) and chronic dehydration occurs if it is abraded (Ebling, 1976). This could favour the fungus either by weakening the insect (Steinhaus, 1956), increasing humidity around the spore, favouring hyphal penetration, or releasing water soluble nutrients.

Schabel (1976a, 1978) detected a significant germination of *M. anisopliae* on surface sterilized as opposed to non-sterilized individuals of the beetle, *Hylobius pales*, and suggested this was due to antibiosis. Treatments designed to remove epicuticular lipids could adversely affect the surface microflora, thus reducing antibiosis and enhancing fungal pathogenicity. However, Walstad, Anderson & Stambaugh (1970) had earlier reported that both *B. bassiana* and *M. anisopliae* germinated readily on the unsterilized body of *H. pales*. Schabel's results may have been due to the disruptive effect of the sterilant (aqueous zephiran chloride) on water soluble antifungal compounds in the cuticle. Smith *et al.* (1981) found that integuments of *H. zea* were more readily colonized by non-pathogenic fungi after washing with water.

Spores are clearly at risk during the period of exposure on the cuticle prior to germination, and it is not surprising that hypervirulence in certain mutants of *M. anisopliae* pathogenic for *Culex pipiens* correlated with speed of germination (Al-Aidroos & Roberts, 1978; Al-Aidroos & Seifert, 1980). Recently we have found that presoaking conidia of *M. anisopliae* in distilled water for 40 hours prior to providing a suitable nutrient source halved the time to germination; an observation of potential practical value (Fig. 9.2).

Pre-penetration growth

Entomopathogens vary in the extent of surface growth prior to penetration. *Entomophthora aphidis* invades the cuticle of the pea aphid, *Aphis pisum*, close to the conidium (Brobyn & Wilding, 1977), while considerable growth of *M. anisopliae* can occur over *H. pales* and wireworm larvae before penetration (McCauley, Zacharuk & Tinline, 1968; Schabel, 1978). Surface growth is often most extensive on hard cuticle. Whatever the cause, this behaviour may either allow location of thinner or softer areas of cuticle for penetration (Lefebvre, 1934) or enhance invasiveness (inoculum potential) via synergism with other hyphae (Schabel, 1978). The latter appears to occur on pupae of the silkworm, *Bombyx mori*, where hyphae of *B. bassiana* aggregate into clumps before penetrating *en masse*, a behaviour not observed on the thinner larval cuticle (Takahashi, 1958). Hyphae from *B. bassiana* penetrated the cuticle of non-cephalic regions of *Heliothis zea* soon after germination, but not on the head where germination was followed by errant hyphal extension (Pekrul & Grula, 1979).

Directional hyphal growth following surface features, such as occurs with leaf- and root-infecting plant pathogens (Wynn & Staples, 1981), has not been recognized, although it is an interesting possibility in relation to species-specific microsculpture (Hinton, 1970).

The nature of the stimulus or stimuli causing germ tube orientation towards the cuticle also remains to be established, although errant growth of certain mutants of *B. bassiana* on non-cephalic regions of *H. zea* indicates that some form of chemical recognition may be a prerequisite (Pekrul & Grula, 1979). Evidence is convincing for many phytopathogens that the entire infection structure (appressorium, infection peg, vesicle and infection hypha) is elicited by a single stimulus.

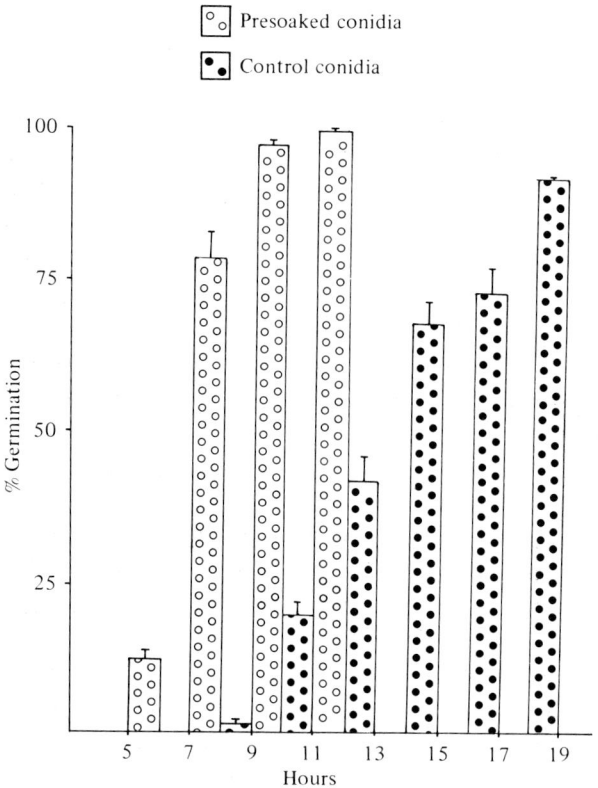

Fig. 9.2. The effect of presoaking in distilled water on the germination time of *Metarhizium anisopliae* (ME1). Conidia ($2 \times 10^7 \text{ml}^{-1}$) were presoaked in distilled water for 40 hr at 20 °C, then incubated in nutrient broth for a given period of time before the percentage germination of 10×200 conidia was determined. (Dillon & Charnley, unpublished.)

Emmett & Parbery (1975) have reviewed most of the factors influencing appressorial formation in phytopathogenic fungi, except concerning emergence of the infection peg which is controlled by the localization of a weak point in the appressoria floor in contact with the cuticle (Wynn & Staples, 1981). By analogy the following may be of relevance to entomopathogens: surface hardness or softness (thigmotrophic response), cuticular architecture, contact with surface waxes or exudates, and exhaustion of endogenous reserves. Also fungistatic compounds on the cuticle may impair germ tube elongation and thus enhance appressorial formation.

Appressorium-like structures, that is swellings at hyphal apices formed on the cessation of germ tube elongation, are produced by *Paecilomyces farinosus* on *Tenebrio molitor* (Robinson, 1966; Madelin, Robinson & Williams, 1967), *M. anisopliae* on *Hylobius pales* (Schabel, 1976a) and on wireworm larvae (Zacharuk, 1970), *Aspergillus parasiticus* on the bagworm moth, *Thyridopteryx ephemeraeformis* (Berisford & Tsao, 1975), and *Entomophthora fresenii* on *Aphis fabae* (Brobyn & Wilding, 1977). Such structures are absent from *Entomophthora apiculata* on the cabbage looper, *Trichoplusia ni* (Lambiase & Yendol, 1977), *Aphanomyces astaci* on the crayfish, *Astacus astacus* (Nyhlen & Unestam, 1975), and *Entomophthora aphidis* on *A. pisum* (Brobyn & Wilding, 1977). One isolate of *B. bassiana* produces appressoria on the colorado beetle *Leptinotarsa decemlineata* (Vey & Fargues, 1977), but these are not formed by another isolate on *Heliothis zea* (Pekrul & Grula, 1979). Germ tubes of the same isolate of *M. anisopliae* produce appressoria on the surface of wireworms but penetrate spiracles directly (McCauley *et al.*, 1968). In the light of such variation, it may be useful to adopt a broader definition of the term appressorium: a germ tube or hyphal tip without regard to morphology, which has the ability to initiate host penetration (Emmett & Parbery, 1975).

Principal roles of the appressorium include attachment and production of infection pegs, but it may also aid penetration by liberating enzymes (Zacharuk, 1970). Firm adhesion of germ tubes and appressoria of phytopathogens to a suitable surface has often been attributed to a mucilaginous secretion (Marks, Berbee & Riker, 1965; Paus & Raa, 1973), which in *Colletotrichum graminicola* may be a hemicellulose (Lapp & Skoropad, 1978). Appressoria of entomopathogenic fungi may also produce adhesive mucilage (Madelin *et al.*, 1967; Zacharuk, 1970; Schabel, 1978).

Penetration

Site of penetration

The arthrodial membrane at joints and between segments is often the favoured site of penetration by entomopathogenic fungi (MacLeod, 1963; David, 1967; Schabel, 1978). However, there are reports of invasion via segmental cuticle (Yendol & Paschke, 1965; Veen, 1966; McCauley *et al.*, 1968; Mohamed, Sikorowski & Bell, 1978), sense organs (McCauley *et al.*, 1968; Mohamed *et al.*, 1978) and tracheae (Lepesme, 1938; Hedland & Pass, 1968). The question of *per os* infection will be considered in a later section. There is little evidence for penetration via natural channels such as pore canals (David, 1967).

Although highly tanned cuticle can be penetrated (Takahashi, 1958), the thickness and composition does influence pathogenicity. Penetration of flies, mosquitoes and aphids by *Entomophthora* spp. occurs readily through any part of the cuticle, but of larger insects only through arthrodial membranes (MacLeod, 1963), which are thin, non-sclerotized, and afford a protected site of localized high humidity where the spore is less likely to be dislodged (David, 1967). Since the thickness of a cuticle increases in successive instars (Richards, 1951), earlier instars are more susceptible to fungi than later ones (Schaerffenberg, 1957; Getzin, 1961).

Penetration of the cuticle is generally considered to involve enzymic and mechanical components (Ferron, 1978; Roberts & Humber, 1981), the evidence for which is reviewed in the following sections.

Histology

Enzymic involvement in penetration may occur even prior to germination. Sannasi & Oliver (1971) noted that the epicuticle of the mite *Dinothrombium giganteum* around a conidium of *Aspergillus flavus* became granular and pale. Wallengren & Johansson (1929) made similar observations with *M. anisopliae*, parasitic on the European cornborer, *Ostrinia nubilalis* (= *Pyrausta nubilalis*). During the course of an ultrastructural study Nyhlen & Unestam (1975) found that the outer layers of crayfish epicuticle were 'corroded' and folded back at the point of attachment of an *Aphanomyces astaci* spore (Fig. 9.3*a*). Subsequent penetration of the inner epicuticle by the penetration peg appeared to be a combination of enzymic degradation and mechanical pressure (Fig. 9.3*b*). The wax layer also disappeared under the appressorium of *M. anisopliae* on wireworm larval cuticle. An irregular walled cavity formed in the epicuticle beneath the appressorium, but there was no evidence of mechanical displacement of the surrounding epicuticle (Zacharuk, 1970) (Figs. 9.3c, d). Further evidence for the enzymic degradation of the epicuticle comes from a scanning electronmicroscope study which revealed a clear circular hole around the germ tube of *B. bassiana* at the point of entry into larval *Heliothis zea* (Pekrul & Grula, 1979) (Fig. 9.3e).

Implicit in the above observations is the ability of the fungus concerned to secrete lipases and/or wax-degrading enzymes; indirect evidence for this comes from the work of Takahashi (1958) who suspended a 'large amount' of *B. bassiana* conidia in physiological saline for 4 days. After filtration, normal larvae of *B. mori* were immersed in the filtrate for 24 hours at 28 °C. When they were transferred to ammoniacal silver hydroxide the surface of integument blackened more quickly than did that of the controls, suggesting that the silver solution penetrated through to layers with silver reducing capability because the surface wax had been removed by enzymes in the filtrate. More convincing evidence for the *in vivo* digestion of lipid comes from the positive histochemical test for lipase activity at the point of entry of *Entomophthora coronata* into the wax moth, *Galleria mellonella* (Gabriel, 1968a).

Evidence for the enzymic degradation of the procuticle is not consistent from one study to another. This is presumably a reflection of variability in both insect cuticular structure and fungal synthetic capabilities. McCauley *et al.* (1968) found changes in the staining reactions of elaterid larval cuticle around the penetration pegs of *M. anisopliae*. The clearing zones were mostly semicircular and associated with the appressorium, suggesting diffusion of enzymes from there. Similar clearing zones have been found around the penetration pegs of *M. anisopliae* infecting other insect cuticles (Wallengren & Johansson, 1929; Notini & Mathlein, 1944; Robinson, 1966), but in these cases the zones are of uniform diameter along the length of the peg suggesting that enzymes are secreted solely from the tip of the infection thread. In contrast Vey & Fargues (1977) found no evidence for histolytic action during the penetration of the outer lamellae of the procuticle of *Leptinotarsa decemlineata* by *B. bassiana*.

Although McCauley *et al.* (1968) and Zacharuk (1970) concluded that, apart from the initial stages, penetration of the procuticle was largely mechanical, there is histochemical evidence for degradation of the inner procuticle for other fungus–insect combinations. The endocuticle of queens of the termite, *Odontotermes obesus*, is periodic

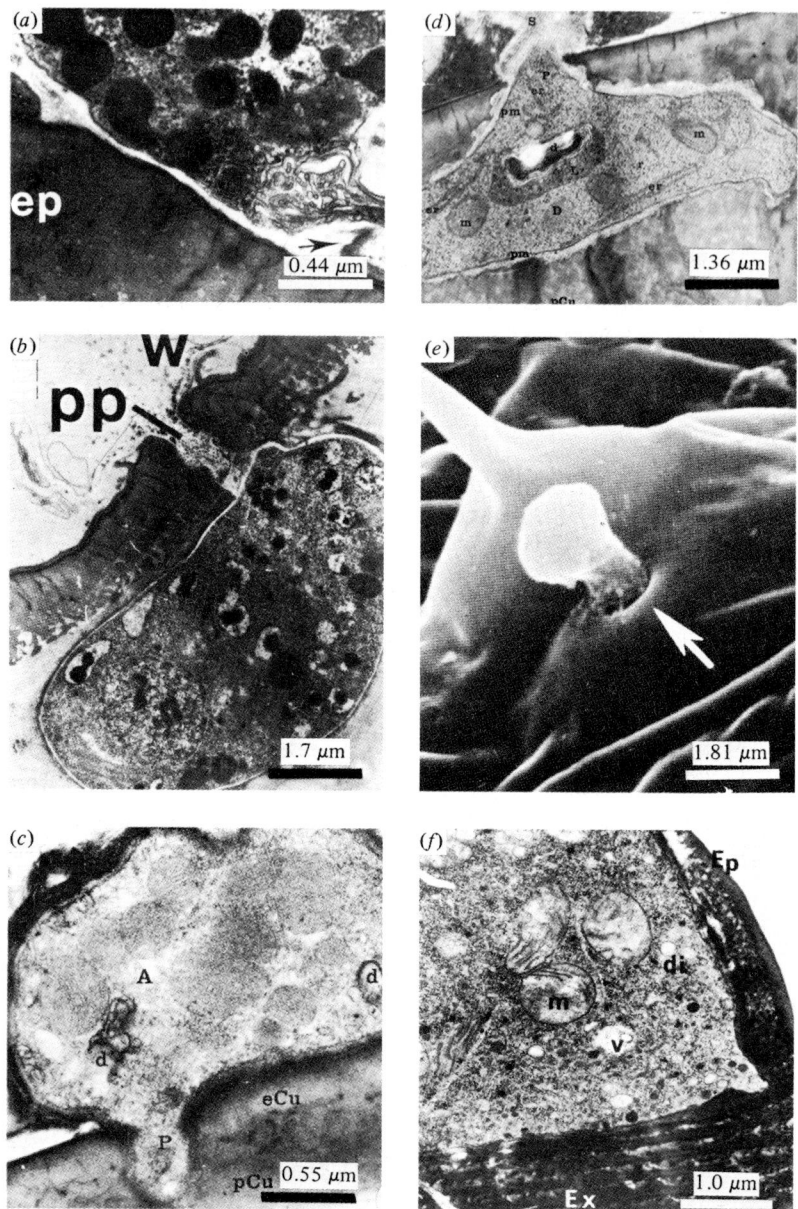

Fig. 9.3. (a) Encysted zoospore of *Aphanomyces astaci* on crayfish (*Astacus astacus*) cuticle. The lipid layer of the epicuticle has been lysed and folded back (arrow) at the point of contact. ep, epicuticle. (From Nyhlen & Unestam, 1975.) (b) Passage of the penetration peg of *A. astaci* through the epicuticle of the crayfish by a combination of

acid-Schiff (PAS) negative, but becomes PAS positive after infection with *Aspergillus flavus* (Sannasi, 1969). Similar changes have been reported by Benz (1963) in the mite, *Malacosoma alpicola*, infected by *Spicaria* sp., and by Sannasi & Oliver (1971) in the mite, *Dinothrombium giganteum*, attacked by *A. flavus*. Benz (1963) suggested that this might be comparable to the PAS positive reaction of endocuticle when digested by moulting fluid which contains proteinase and chitinase activity (Wigglesworth, 1957). However, proteases may free amino groups in chitin to give a positive PAS reaction, so that their action alone could explain the effect of fungal penetration on the PAS reaction of arthropod cuticle (Delachambre, 1969).

Gabriel (1968a) suggested a further interpretation of the PAS reaction, namely that removal of chitin by the pathogen would produce PAS negative reaction from the invaded portion of the integument. His own study of the penetration of *Galleria mellonella* larvae by *E. coronata* showed little difference in the PAS reactivity of diseased and control cuticles, which suggested that if chitinolytic enzymes are being secreted by the fungus, this occurs in a limited way. The results of a mercury/bromophenol blue test for protein also indicated restricted hydrolysis of protein around the invading hyphae.

Endocuticles of *Bombyx mori* and the mite *Dinothrombium giganteum* infected with *B. bassiana* and *A. flavus* respectively, become acidophilic and stain with fuchsin in contrast to the basophilic aniline blue reaction of intact endocuticle (Takahashi, 1958; Sannasi & Oliver,

Caption for fig. 9.3. (*cont.*).
enzymatic degradation and mechanical pressure. pp, penetration peg; w, electron transparent wall. (From Nyhlen & Unestam, 1975.) (*c*) Penetration peg of *Metarhizium anisopliae* has penetrated the cuticle of *Limonius californicus* histolytically. Initial pressure by the appressorium caused slight indentation of the outer epicuticle. A, appressorium; d, dictyosome; eCu, epicuticle; P, penetration peg; pCu, procuticle. (From Zacharuk, 1970.) (*d*) Penetration peg and plate of *Metarhizium anisopliae* in the procuticle of *Hypolithus bicolor*. Note the histolysis of the surrounding epicuticle and procuticle. d, dictyosome; D, penetration plate; er, endoplasmic reticulum; L, lipoid body; m, mitochondrion; P, penetration peg; pCu, procuticle; pm, plasmalemma; r, ribosome; S, septum. (From Zacharuk, 1970.) (*e*) Scanning electronmicrograph showing a circular hole around the germ tube of *Beauveria bassiana* on the cuticle of *Heliothis zea* indicating enzymatic hydrolysis (arrow). (From Pekrul & Grula, 1979.) (*f*) Hypha of *Entomophthora apiculata* cleaving apart lamellae of the exocuticle of *Trichoplusia ni*. di, dictyosome; Ep, epicuticle; Ex, exocuticle; m, mitochondrion; v, vacuole. (From Lambiase & Yendol, 1977.)

1971). The change in staining reaction may be a consequence of enzymolysis, implicating fungal enzymes in the penetration process. However, the switch from aniline blue to fuchsin staining must reflect a decrease in the magnitude of the intermolecular spaces, associated with the relative molecular sizes of the two stains, which is the opposite to that expected if enzymic degradation is taking place (Neville, 1975).

Overall, it is apparent that the histochemical studies to date have contributed little to our understanding of enzymic processes during penetration, due to the poorly defined chemistry of the staining reactions employed. A more profitable approach would be the use of the more precise immunohistochemical techniques which are now available (see below).

The penetration peg is often reported to be thin and constricted within the outer layers of the procuticle, with the growing tip swelling in the inner layers (Robinson, 1966; Brobyn & Wilding, 1977). This has been interpreted as the result of the hypha being under pressure during mechanical penetration of the outer region of the cuticle, followed by expansion in the cavity created within the softer inner layers by enzymic action (Robinson, 1966).

Indentation or displacement of cuticular lamellae is another clear indication of a mechanical component to the penetration process (Wallengren & Johansson, 1929; Lefebvre, 1934; Sannasi, 1969; Zacharuk, 1970; Nyhlen & Unestam, 1975). Although not universal (e.g. Takahashi, 1958; McCauley et al., 1968), it is a consistent feature that may accompany signs of histolysis. The primacy of mechanical disruption in the penetration of *Trichoplusia ni* by *E. apiculata* is clear from the way the growing hyphae wedge procuticular layers apart (Lambiase & Yendol, 1977) (Fig. 9.3*f*). However, in the latter case, fungal enzymes may be responsible for the marked destruction of the epithelium in the area of initial penetration. Such epidermal lysis has frequently been noted in insects infected by entomopathogenic fungi (Vey & Fargues, 1977) but may alternatively be ascribed to the action of low molecular weight mycotoxins (Vey, Quiot & Vago, 1973; see also below).

Enzyme studies in vitro

The three main constituents of insect cuticle, apart from tanning agents, are lipids, protein and chitin (see Fig. 9.1). Extracellular enzymes capable of digesting these substrates are produced on solid media by *M. anisopliae*, *Cordyceps militaris*, *B. bassiana* and *Aspergillus*

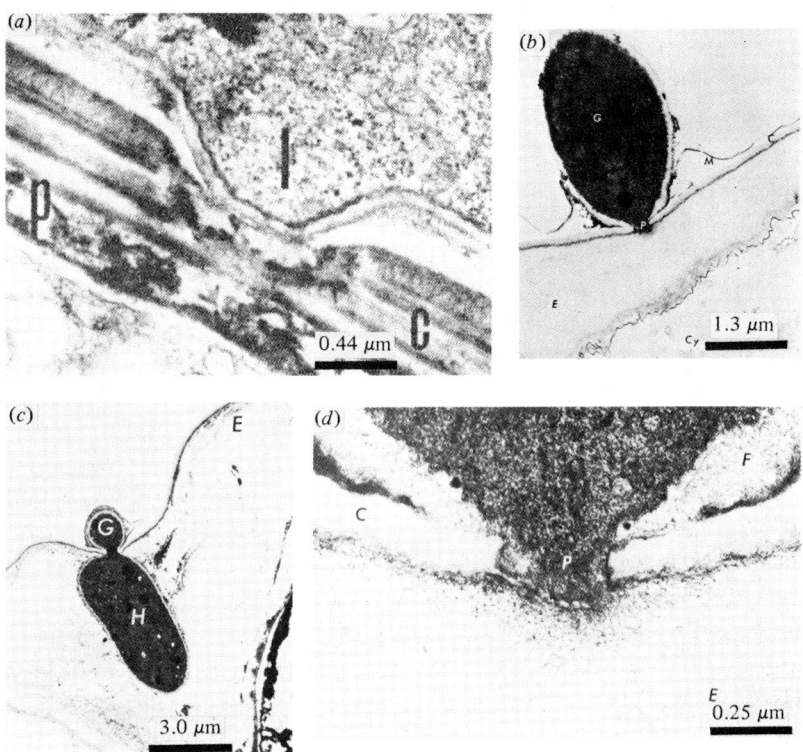

Fig. 9.4. (a) Enzymatic digestion of barley (*Hordeum vulgare*) epidermal cell wall by an infection peg (I) of *Erysiphe graminis*. No signs of stress are seen in the microfibrillar portion of the host. c, cell wall; p, papilla. (From Edwards & Allen, 1970.) (b) Penetration of *Botrytis cinerea* into the broad bean (*Vicia faba*). The infection peg (P) is almost through the cuticle (C). Epithelium (E) is unaffected apart from the area in front of the peg. A large amount of mucilage is present (M). Cy, cytoplasm. (From McKeen, 1974.) (c) The infection peg of *Botrytis cinerea* has enlarged after passage through the cuticle of *Vicia faba*. Epidermal cell (E) is swollen and has pushed out the cuticle. G, germ tube; H, infection hypha. (From McKeen, 1974.) (d) Enlargement of part of 9.4b. The infection peg (P) has a plasma membrane but no fungal wall (F). C, cuticle; E, epidermis. (From McKeen, 1974.)

flavus (Huber, 1958), several *Entomophthora* spp. (Gabriel, 1968b), and *Nomuraea rileyi* (Mohamed et al., 1978). Entomopathogenic fungi also release enzymes in liquid culture (Leopold & Samsinakova, 1970; Samsinakova, Misikova & Leopold, 1971; Latgé, 1974). However, the relevance of these enzymes *in vivo* is not clear: constituents of the culture medium can markedly affect the extracellular protein pattern

(Grula et al., 1978), while the timing of enzyme production relative to the growth stage of the fungus may debar it from influencing pathogenicity. The position could be clarified by examining the effect of purified fungal enzymes on insect cuticle, but this has not been done, the only studies having used semi-pure commercial enzyme preparations. Thus Samsinakova et al. (1971) used mechanically cleared, water-washed integument of larval *Galleria mellonella*, and showed that a mixture of three commercial enzymes of mixed origin brought about the complete breakdown of lipids, chitin and protein, resulting in cuticle decolourization and disintegration. If the enzymes were applied sequentially (each allowed to act for 24 hours) the order lipase, protease, chitinase produced the most efficient degradation of cuticle. Chitinase was only effective if applied after protease, indicating that the chitin framework was protected by a covering of protein. Smith et al. (1981) boiled whole *Heliothis zea* larvae in 1% sodium dodecyl sulphate for 2–4 hours producing transparent 'ghosts', in which all internal tissues had been digested. Chemical analyses of such ghosts revealed a composition of proteins and chitins with an absence of reducing sugars and glucans. Complete disintegration of the ghosts was achieved with a sequence of protease followed by chitinase. Pretreatment of the ghost with a wax (lanolein), but not a crude-oil preparation, protected them from enzyme attack.

A particularly useful 'model system' is the degradation of insect cuticle by endogenous enzymes at the moult, when only a proportion of the old cuticle is shed, the rest being digested and recycled. The *modus operandi* of this process in larvae of *Manduca sexta* has been studied by Bade and her co-workers (Bade, 1974; Bade & Shoukimas, 1974; Bade & Stinson, 1978). Multiple steps are involved when cuticle passes from the premoult stage, when it lacks endogenous chitinase activity, to the stage when tightly bound chitinase is degrading chitin. A plausible sequence is that active moulting fluid first penetrates the outer endocuticle, 'trypsin-like' protease then unmasks cuticular chitin, chitinase and chitobiase degrade chitin to N-acetylglucosamine, and finally chitinase is degraded by a neutral protease (Bade & Stinson, 1978). Smith et al. (1981) found that cathepsin, chymotrypsin, pepsin, lysing enzyme or trypsin participated with equal facility in the degradation of *Heliothis zea* ghosts.

The interaction between protease and chitinase defined by these studies is not unexpected when related to the structure of insect cuticle, where chitin exists as a protein-chitin complex (Neville, 1975). The

Penetration 243

synthesis and/or release of protease during insect moulting may precede the elaboration of the bulk of the chitinase in moulting fluid (Phillips & Loughton, 1976; Bade & Stinson, 1978). A similar sequence occurred in culture fluid when *M. anisopliae* was grown on locust cuticle presumably reflecting the physical availability of the cuticular polymers (see above), resulting in induction as successive substrates became available during progressive degradation (Fig. 9.5).

A parallel situation occurs with plant cell walls, where polysaccharides are also not amenable to the action of hemicellulases and cellulases of *Colletotrichum lindemuthianum* and *Trichoderma viride* until polygalacturonide of the cell wall has been degraded by endo-pectic enzymes

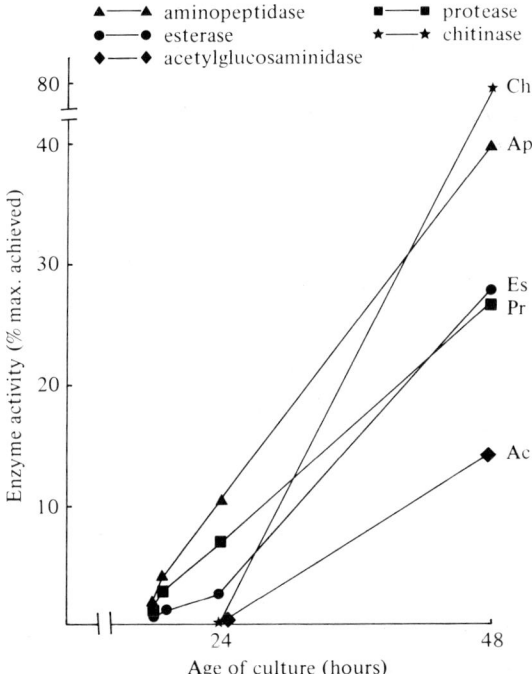

Fig. 9.5. Formation of extracellular enzymes by *Metarhizium anisopliae* (ME1) grown at 27 °C in a medium containing 1% unground locust cuticle. Cuticle was prepared by the method of Andersen (1980); a basal salts solution was also provided. Esterase, protease and aminopeptidase activities appeared at approximately the same time (18 hr). However, chitinase and acetylglucosaminidase appeared later (24 hr). The same sequence of production occurred whether or not the medium was buffered with a non-biodegradable buffer, 2-(-N-morpholino) ethanesulphonic acid. (From St Leger, Charnley & Cooper, unpublished.)

(Karr & Albersheim, 1970). This contrasts with reports that single polysaccharidases of *Verticillium albo-atrum* degraded most of the glycosidic linkages in cell walls from tomato stems without the modifying action of other enzymes (Cooper, Rankin & Wood, 1978). However, the order of polysaccharidase synthesis in cultures containing cell walls may also occur in infected tissues (Cooper, 1976).

Chitin and many of the cuticular proteins are insoluble (Neville, 1975) and as such could not induce synthesis of enzymes until after slight degradation by low levels of basal enzyme. Once sufficient inducer has been released by basal enzyme to initiate synthesis, the process should be autocatalytic. A similar model has been proposed to account for the production of cell wall degrading enzymes (Cooper, 1976) and for cutinase synthesis during plant cuticle penetration by fungal pathogens (Kolattukudy, 1980). Basal synthesis could be quite important in pathogenesis, as implied by the production of arabinases by plant pathogens but not by saprophytic fungi in the absence of inducers (Fuchs, Jobson & Wouts, 1965).

Nevertheless, little is known of the mechanisms regulating the synthesis of cuticle-degrading enzymes of entomopathogenic fungi. The few studies that have been attempted contribute little to our understanding, as the experiments have involved long-term trials in batch cultures under unfavourable conditions, in which effects of catabolite repression (CR), growth rate, autolysis, changes in pH and medium composition have been ignored or insufficiently considered. Thus fungi have been grown on high levels of energy sources (e.g. glucose 2.5%, starch 2.5% and corn steep 2% (Samsinakova *et al.*, 1971)), far in excess of that required to effect CR of synthesis of typical catabolic enzymes (*c.* 0.01%; Cooper, 1976). Under such conditions attempts have been made to correlate the minute amounts of enzyme obtained with virulence (Samsinakova & Misikova, 1973). It is axiomatic that the enzymic potential is not realized under these conditions.

Recently St Leger, Charnley and Cooper (unpublished) have followed the production of extracellular chitinase and protease by *M. anisopliae* (strain ME1) more rigorously. In batch cultures high chitinase activity is only achieved when chitin is supplied as substrate. Chitinase yield is also influenced by the accessibility of the chitin (inversely related to particle size and concentration). In a medium containing chitin the absence of an inorganic nitrogen source does not reduce the production of chitinase, indicating that chitin acts simultaneously as a carbon and nitrogen source. Growth on <0.5% *N*-

acetylglucosamine resulted in slightly greater chitinase activity than with non-chitinous substrates. Chitinase production was repressed with >0.5% N-acetylglucosamine or when additional carbon sources such as carbohydrates, olive oil and proteins were added to a medium containing chitin. The above results are consistent with chitinase being an inducible enzyme, in which case limited chitinase production in non-inducing media (e.g. 1% sucrose) may be due to basal synthesis.

It is unlikely that chitin itself is the inducer as it is too large a molecule to enter the fungal cells. Although the monomer N-acetylglucosamine (or oligomers) is a likely candidate, batch shaking cultures containing 0.2–1% produced only small amounts of chitinase activity (see above). In order to reduce the possibility of CR, potential inducers were supplied in restricted cultures from diffusion capsules (Pirt, 1971). These are simply filled with carbohydrate solutions and added to shake cultures, into which sugars diffuse at linear rates through a semipermeable membrane; diffusion rates are determined by the internal carbohydrate concentration and the number of membranes in the capsules. This enables the manipulation of growth rates in culture (Cooper & Wood, 1975).

The amount of chitinase obtained after 7 days with N-acetylglucosamine as inducer is never as great as in batch cultures with chitin as sole carbon and nitrogen source because a restricted culture produces starvation conditions. However, N-acetylglucosamine and glucosamine were far more potent inducers than other sugars tested, e.g. glucose (see Table 9.1).

The pH activity spectrum of the protease complex of *M. anisopliae* (ME1) produced in liquid culture exhibits two optima, pH 5.5 and pH 8.0 (see Fig. 9.6). These may represent two distinct proteases, the relative production of each depending on the substrate provided. Similar extracellular acid and alkaline proteases have been found in the culture fluids of *B. bassiana* (Leopold, Samsinakova & Misikova, 1973) and of another strain of *M. anisopliae* (Kucera, 1980).

The alkaline protease was produced in the presence of all nitrogen sources tested including $NaNO_3$. However, the inorganic source did not allow the production of the acidic enzyme (Fig. 9.6). Organic non-protein sources such as bactopeptone and casamino acids caused slight production, but significant activity was only achieved when protein was present in the medium. With the 'whole locust' homogenate the acidic protease predominated over the alkaline (cf. Kucera, 1980). The two species of protease exhibited similar patterns of catabolite repression

Table 9.1. *Induction of synthesis of chitin degrading enzymes of* Metarhizium anisopliae *(ME1) by a restricted supply of sugars*

	Enzyme activities[a]			
Inducer	Chitinase	Chitosanase	Chitobiase	Acetylglucosaminidase
N-acetyl-glucosamine	100	86	100	100
Glucosamine	66	100	85	93
Glucose	8	9	44	48

[a]Enzyme activities are expressed as a % of the maximum specific activities attained in culture after 7 days exposure to different sugars.

Sugars were fed from diffusion capsules at linear rates of 2 mg 100 ml^{-1} hr^{-1} to shake cultures of *M. anisopliae* inoculated with mycelium from 3 day cultures grown on sucrose/basal salts medium. The amount of different sugars in the medium did not rise above 0.07 mg ml^{-1} indicating that they were being utilized by the fungi at approximately the rate of diffusion. Consequently growth was limited by the rate of supply of nutrient sugars and was considerably less than in media containing 'unrestricted' amounts of a carbon source.

(From St Leger, Charnley & Cooper, unpublished.)

Table 9.2 *Various carbon and nitrogen sources which affect the growth and protease production of* Metarhizium anisopliae *(ME1)*

1.[a] Compounds which increased the rate of growth but repressed the production of protease:		
	Casamino acids	1%
	L-alanine	0.05–0.1 M
	L-glycine	0.01 M
	L-tyrosine	0.1 M
	Glycyl-glycine	0.1 M
	L-cysteine	0.1 M
	Glucose	0.055 M
	N-acetylglucosamine	0.055 M
2. Compounds which had only slight effect on the growth but repressed protease production:		
	Urea	0.1 M
	NH$_4$Cl	0.05 M
3. Compounds which had little effect on either growth or protease production:		
	Lactose	0.055 M
	Xylose	0.055 M

[a]Activity expressed with respect to that in 1% casein + basal salts medium.

Penetration

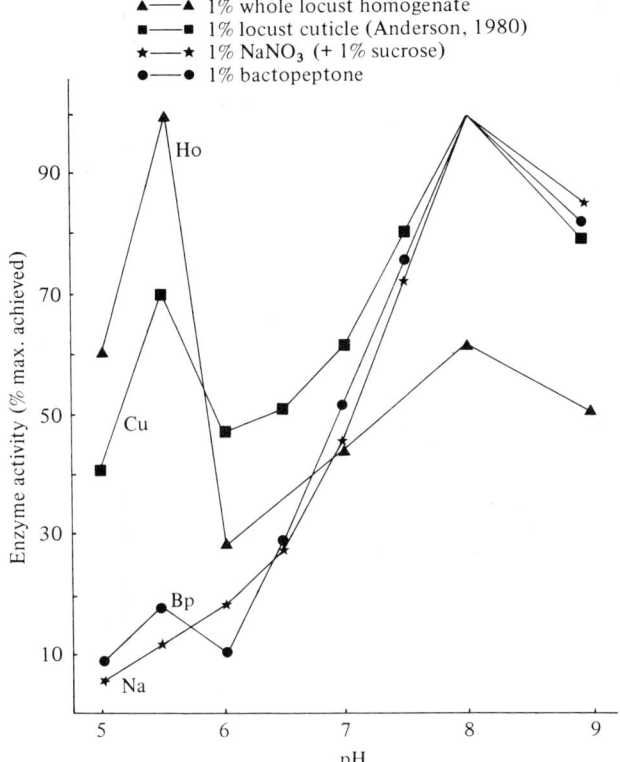

Fig. 9.6. pH optima of the proteolytic complex of *Metarhizium anisopliae* (ME1) produced on different nitrogen media. A basal salts solution was also supplied in each case. (From St Leger, Charnley & Cooper, unpublished.)

(Table 9.2), while CR of the acidic protease occurred in the presence of potential inducers.

The dermatophytic fungi *Microsporum canis* (O'Sullivan & Mathison, 1971) and *Trichophyton rubrum* (Meevootisom & Niederprucm, 1979) also produce extracellular protease with protein as sole source of carbon and nitrogen; repression occurring upon the addition of more readily available sources of C and N such as amino acids. In *Neurospora crassa* protease induction by protein is balanced by nutrilite repression (Cohen & Drucker, 1977), while extracellular proteolytic enzymes of *Aspergillus nidulans* are regulated by de-repression, release only occurring when the medium is deficient in carbon, nitrogen or sulphur (Cohen, 1973). Although urea and NH_4^+ repressed protease production by *M. anisopliae* at 0.05–0.1 M, only a 30% reduction occurred.

The possibility that virulence may be correlated (at least in part) with cuticle-degrading enzyme activity has stimulated several studies, with conflicting results. Pavlyushin (1978) found low lipolytic and proteolytic activity in strains of *B. bassiana* with low virulence towards *Galleria mellonella*, while high virulent strains had high enzyme activity. On the other hand Yanagita (1980) found no relationship between activity of lipase and cellulase in strains of *Aspergillus flavus-oryzae* and pathogenicity to *Bombyx mori* larvae, but chitinolytic activity did parallel pathogenicity. Samsinakova & Misikova (1973) also found that isolates of *B. bassiana*, *M. anisopliae* and *Aspergillus parasiticus* with highest infectivity towards *G. mellonella* larvae produced the highest amount of chitinase. Nine strains of *Beauveria brogniartii* (= *Beauveria tenella*) which gave greater than 32% mortality of *Melolontha melolontha* possessed lipolytic activity, while seven strains with less than 32% mortality did not show this activity (Paris & Segretain, 1975). Although Bajan *et al.* (1979) found that the pathogenicity of 21 out of 36 strains of *B. bassiana* against *G. mellonella* was concordant with the sequence of chitinase production, that of the remaining 15 was not.

Unfortunately comparisons between isolates for pathogenicity and production of enzymes may only reveal the great variability within a species for numerous factors, many of which may influence but be unrelated to cuticle-degrading enzyme synthesis. Induction of mutants with a common genetic background is an alternative approach which is beginning to be exploited. Paris & Ferron (1979) have found a link between virulence and lipase in mutants of *B. brogniartii*. However, mutants deficient in other respects were also avirulent, suggesting that, as one might expect, pathogenicity was a function of many attributes. Similarly Pekrul & Grula (1979) found that low entomopathogenicity of *B. bassiana* mutants against *H. zea* larvae was not simply a consequence of the lack of a suitable enzyme 'cocktail'. All mutants possessed varying levels of the three major enzyme activities regardless of their pathogenicity and some mutants contained very high levels of certain enzyme activities yet were poor pathogens.

The speed of induction of cuticle-degrading enzymes could be important in pathogenesis, enabling cuticle penetration before host defence or repair mechanisms became effective. In view of this it would be of interest to compare the pathogenicity of a range of mutants of a parasitic species differing with respect to regulation of one or more cuticle-degrading enzymes. In addition to the comparison of mutants productive or deficient in enzyme synthesis, attention could be paid to inducibility, repressibility and basal synthesis, as Cooper (1976) has suggested for cell wall degrading enzymes in phytopathogens.

Albersheim, Jones & English (1969) proposed that because polysaccharidase induction by cell wall sugars and their action on polysaccharides was highly specific, cell wall polysaccharides might be involved in specific resistance to fungal infection in plants. However, there is a considerable body of evidence against such a role in specific interactions (Cooper, 1976). It seems equally unlikely that strain specificity of entomopathogenic fungi is a function of a singular response to nuances of cuticular composition. Insect cuticle, like plant cell walls, may lack sufficient chemical variation to account for differences in susceptibility to strains of a given fungus. Chitin is composed of the one monomer, N-acetylglucosamine (with a proportion of glucosamine residues) (Neville, 1975). The protein fraction consists of many different molecules (Vincent, 1980) but what is known about their regulation (see above) argues against specificity of control. Cuticular lipids, however, are perhaps a sufficiently heterogeneous group to contribute to the specificity of the host–parasite relationship (Jackson & Blomquist, 1976).

Cuticular penetration – an overview

Histological studies demonstrate involvement of mechanical and enzymic processes during penetration of cuticle by entomopathogenic fungi (see above). Available evidence suggests that a similar dual mechanism is employed by phytopathogens (McKeen, Smith & Bhattacharya, 1969; Edwards & Allen, 1970; McKeen, 1974; Aist, 1976) (Fig. 9.4*a, b, c, d*). However, the histochemical evidence for the *in vivo* production of cuticle-degrading enzymes is better for phytopathogens than for entomopathogens. Lipase of *Bremia lactucae* produced during penetration of lettuce cotyledon has been localized cytochemically in the part of the appressorial wall in contact with the host cuticle (Duddridge & Sargent, 1978). A cutinase from the extracellular fluid of *Fusarium solani* grown on cutin has been purified and characterized (Purdy & Kolattukudy, 1975*a, b*). Subsequent ultrastructural studies with ferritin-conjugated antibody prepared against cutinase showed that *F. solani* secreted a cutinase during the infection of its host (Shaykh, Soliday & Kolattukudy, 1977). Furthermore infection was drastically reduced if the conidial suspension was mixed with antiserum prepared against purified cutinase prior to inoculation of the intact surface. The antiserum, however, had no effect if the stem surface was wounded (Maiti & Kolattukudy, 1979; Köller, Allan & Kolattukudy, 1982). Clearly such an approach could be used profitably with entomopathogens once cuticle-degrading enzymes have been purified.

The timing of extracellular enzyme production with respect to the fungal growth cycle is particularly important if *in vitro* studies are to be related to what occurs *in vitro*. Söderhäll & Unestam (1975) found that protease release by *Aphanomyces astaci* (a parasite of crayfish) was mostly from old autolysing hyphae and suggested that *in vivo* it was responsible for subsequent weakening of the cuticle already invaded by the mycelium. Correlation between declining dry weight of fungus and increase in extracellular protease by *Cordyceps militaris* has also been interpreted as progressive release of enzyme during mycelial lysis rather than secretion (Latgé, 1974). In contrast, using alkaline phosphatase activity as an indicator of cell lysis, protease production by *M. anisopliae* occurs well in advance of autolysis (St Leger, Charnley and Cooper, unpublished).

It is difficult to equate the apparently limited area of enzyme degradation in insect cuticle by parasitic fungi (see Fig. 9.3d) with the release of extracellular enzymes. The comparable problem in phytopathology has been considered by Cooper (1982). Activity could be reduced by binding and the effects of inhibitors in the cuticle, while movement could be restricted by molecular sieving (large enzymes and small intermolecular spaces). The alternative, that cuticle-degrading enzymes are primarily wall bound, does not appear to have been extensively considered, but at least in *C. militaris* there is little endocellular chitinase or proteinase activity (Latgé, 1974).

Few attempts appear to have been made to assay extracellular enzyme production by germinating conidia, and available techniques may not be sensitive enough to detect the small amounts of enzyme produced. This is unfortunate because it is axiomatic that the cuticle is penetrated by young rather than mature hyphae. Söderhäll, Svensson & Unestam (1978) found protease in ungerminated and germinated spores of *Aphanomyces astaci*, whereas no chitinase was demonstrable until the germ tube hyphal complex began to branch *in vitro* after 12–18 hours, suggesting that the delay corresponded to the time taken for a germ tube to penetrate the epicuticle. Unfortunately they used a medium containing 0.1% peptone to germinate spores, and our studies have shown that chitinase secretion by *M. anisopliae* is repressed and acid protease slightly enhanced by bactopeptone (St Leger, Charnley and Cooper, unpublished). The extended incubation period required to detect enzyme activity (protease after 2 weeks and chitinase after 12 weeks) also casts doubt on the significance of their results.

To date the majority of histochemical, ultrastructural and *in vitro* enzyme studies on entomopathogenic fungi have been performed on non-sclerotized cuticle, usually in endopterygote larvae, because of the problems of fixing exocuticle. However, as has been mentioned, there are reports of fungal penetration of sclerotized cuticle. Lipke & Geoghegan (1971) have shown a high degree of resistance to enzymolysis in the sclerotized pronotal cuticle of larvae of the American cockroach, *Periplaneta americana*. Trypsin, pepsin, collagenase, papain, panprotease, and pronase hydrolysed less than 3% of the available linkages in the cuticular protein, while a variety of carbohydrases had little or no effect. They concluded that generalized protection by hydrogen bond formation occurred during the dehydration that is a feature of sclerotization. In the light of this, if enzymic degradation does play a part in the penetration of sclerotized cuticle then the entomopathogenic fungus or its enzymes must have special properties. Although purely speculative, a reduction of the pH of the cuticle by a fungal metabolite could loosen the bonds between protein molecules (equivalent to the rehydration of the cuticle) and favour the action of hydrolytic enzymes, perhaps in a similar way to the proposed mechanism of cuticle plasticization (Reynolds, 1975). In support of this hypothesis, oxalic acid is produced *in vitro* by *B. bassiana* (Cordon & Schwartz, 1962) and by *M. anisopliae* (Lihnell, 1944), and crystals of it have been found on the surface of *Bombyx mori* killed by *B. bassiana* (Kodaira, 1961). However, not all sclerotized cuticle may be as resistant to digestion as that of *P. americana*. Chitin in the exuvia of the crayfish *Orconectes sanborni* is readily hydrolysed by chitinase (Stevenson, 1969).

It has been argued elsewhere that strain specificity is unlikely to be a consequence of enzyme induction in response to changes in cuticular composition. However, some comparisons of the pathogenicity of topically applied and injected spores suggest that cuticle *per se* may play a part. Ferron & Diomandé (1969) found that the Scarabid beetle, *Orycetes monoceros*, was susceptible to topically applied spores only from strains of *M. anisopliae* isolated from *Orycetes* spp. Additional isolates were only pathogenic by injection. It remains to be seen whether the cuticle provides a specific chemical barrier or is simply a general physical barrier which, by delaying and perhaps staggering the arrival of penetrant hyphae, optimizes the efficiency of the haemocytic defences.

Infection via the gut

Although entomopathogenic fungi normally invade through the surface integument, infection via the gut has been recorded (Ferron, 1978). Indeed Sussman (1952a) considered that the hindgut was the main site of entry of *Aspergillus flavus* into *Bombyx mori*, while Masera (1957) could only infect *B. mori* with a fed dose of *M. anisopliae*. The gut also appears to be the sole target for the needle-shaped ascospores of *Monosporella unicuspidata*, parasitic on the biting midge *Dasyhelea obscura* (Keilin, 1920), and for *Culicinomyces clavosporus* pathogenic for *Culex fatigans* (Sweeney, 1975). Further examples of gut penetration include the invasion of the termite, *Reticulitermes flavipes*, by *B. bassiana* (Bao & Yendol, 1971) and *Entomophthora coronata* (Yendol & Paschke, 1965). However, failure to induce mycosis *per os* is common (Dieuzeide, 1925; Glaser, 1926; Lepesme, 1938).

Ingestion of conidia may cause death by toxicosis rather than by mycosis (Madelin, 1963). Death of *Apis mellifera* occurred prior to extensive growth of *A. flavus* mycelia in the lumen or penetration of the gut wall (Burnside, 1930; Toumanoff, 1931). More recently Crisan (1971) found a correlation between the percentage of disrupted conidia of *M. anisopliae* in the gut of *Culex pipiens* and the proportion of larvae exhibiting symptoms of intoxication, suggesting that digestion of conidia within the gut released an insecticidal toxin.

Few studies on gut penetration have excluded integumentary infection by surface sterilization, without which histological evidence is unconvincing (Schabel, 1976b). Recent studies where this precaution was taken support an oral route for penetration of *M. anisopliae* into *Hylobius pales* (Schabel, 1976b) and *Schistocerca gregaria* (Veen, 1966), but do not provide evidence for infection via the intestine. However, conidia of *B. bassiana* were observed to germinate in the lumen and penetrate the midgut of surface sterilized fire ants, *Solenopsis richteri* (Broome, Sikorowski & Norment, 1976). Gabriel (1959) induced mycoses in *Galleria mellonella*, *B. mori*, and *Tenebrio molitor*, by injecting conidia of *M. anisopliae* into the foregut, but in many cases bacterial infection also occurred due to accidental puncture of the gut.

Madelin (1963) has considered those elements likely to affect the outcome of *per os* infection: spores must retain viability in the presence of digestive enzymes, germinate in the conditions prevailing in the gut, remain in undisturbed contact with part of the gut wall for sufficient time to allow germination and penetration, and be capable of penetrating the gut wall. Although there is little information on any of these factors, the low confirmed incidence of gut penetration suggests that these prerequisites for successful invasion are not often met.

The foregut (including buccal cavity) and hindgut constitute the principal sites of gut penetration. Both are of ectodermal origin and have a cuticular intima, which may aid attachment. On the other hand, though the midgut is usually lined with a chitinous peritrophic membrane, this is continuously delaminated and attached spores may have insufficient time to penetrate before they are excreted.

In many insects the gut environment appears to be unfavourable for fungal germination. Conidia of *M. anisopliae* fail to germinate in the gut of the beetles *Orycetes rhinoceros* and *Hylobius pales* (Schabel, 1976*b*). Although passage through the gut of *H. pales* was not detrimental (Schabel, 1976*b*), conidia of *M. anisopliae* and *B. bassiana* were inhibited from germinating by the intestinal contents of several insect species *in vitro* (Gabriel, 1959). Such observations have been attributed to either a high gut pH (Gabriel, 1959) or the nutritional unsuitability of the gut (Lepesme, 1938). However, neither of these factors could account for low germination and loss of viability of *M. anisopliae* (ME1) conidia during passage through the gut of *Schistocerca gregaria* (Dillon & Charnley, 1982) (see Table 9.3). The pH of the locust alimentary canal proved to be optimal for germination (pH 5.5–7.0), all regions were aerobic, and boiled foregut fluid was efficacious for germination. However, untreated gut fluid was markedly inhibitory (see Table 9.3). The effect was not due to the diet as homogenization with feed-material (wheat seedling) did not affect germination. A direct action of or competition with the gut flora also tended to be discounted because filtered fluid (Millipore, 0.22 μm) was still inhibitory. In the light of the above, it would seem most likely that the action of digestive enzymes was the cause of fungitoxicity. However, locusts do not appear to have fungal wall-degrading enzymes (Morgan, 1976). It is interesting to note that the detrimental effect of human gastric juice on conidia of *Nomuraea rileyi in vitro* was attributable to low pH rather than to the effect of enzymes (Ignoffo & Garcia, 1978).

The results are best explained in terms of a heat-labile toxin. Recent experiments have shown that gut fluid from axenic insects does not inhibit germination (Table 9.3). Since germ-free locusts are physiologically comparable to conventional insects (Hunt and Charnley, unpublished), the gut bacterial flora may be responsible for the production of the hypothetical factor.

Table 9.3. *The effect of passage through the gut of the desert locust* Schistocerca gregaria *on the viability of conidia of* Metarhizium anisopliae *(ME1)*

Treatment	% Germination				
	Experimental	n	Control	n	
Conidia passed through the gut[a]	26.6	14	78.3	11	
Conidia incubated with homogenized wheat seedling[a]	81.3	11	78.6	11	NS
Conidia incubated in foregut fluid	13.3	27	78.8	27	
Conidia incubated in filtered foregut fluid (Millipore 0.22 μm)[b]	9.1	12	88	12	
Conidia incubated in heated foregut fluid (100 °C for 5 min)[b]	72	24	60	23	NS
Unheated experimental control[b]	3.1	16			
Conidia incubated in foregut fluid[b]	1.6	25	71.6	25	
Conidia incubated in foregut fluid from axenic animals[b]	97.8	18	92.8	18	NS

Incubation was in the presence of Sabouraud's dextrose agar[a] or nutrient broth[b] (1 : 4 dilution, fluid : broth) for 20 hours. NS, not significantly different from the control. All other values were significant at $P < 0.01$. The inhibition is possibly fungitoxic rather than fungistatic as the effect was not reduced up to 3 days. (From Dillon & Charnley, unpublished.)

Invasion of the haemocoel and the cause of death

The invasion of the insect cannot be considered successful until the fungus reaches the epidermis, because the inoculum may be lost with the exuvia if ecdysis intervenes (Zacharuk, 1973b; Vey & Fargues, 1977). *Aphanomyces astaci* is unusual because it remains largely within the cuticle of *Astacus astacus* (Nyhlen & Unestam, 1975).

Passage of a penetration tube may occur without apparent effect on the epidermis (Travland, 1979a) or cause widespread histolysis (Zacharuk, 1971; Vey & Fargues, 1977). Initial growth of the fungus may be localized around the epidermis at the point of entry (Hurpin & Vago, 1958) but finally the fungus proceeds to invade the rest of the insect. On entering the haemocoel penetrant hyphae of certain entomopathogenic fungi continue filamentous development, as with *Entomophthora coronata* in the termite, *Reticulitermes flavipes* (Yendol & Paschke, 1965), and *Aspergillus flavus* in the silkmoth *Hyalophora cecropia* (= *Platysamia cecropia*) (Sussman, 1952a). However, most entomopathogenic fungi sooner or later produce a variety of structures collectively referred to as hyphal bodies (Prasertphon & Tanada, 1968) which circulate in the haemolymph before germinating into mycelium. These are generally ovoid (true blastospores) or truncated hyphal fragments produced by budding or separation from the initial penetrant hyphae, which often proliferate further by division (Martin, 1969; Sweeney, 1975, Brobyn & Wilding, 1977; Mohamed et al., 1978). A further specialized case is the production of wall-less protoplasts, e.g. by *E. egressa* in the haemolymph of *Choristoneura fumiferana* (Tyrrell, 1977).

Degradation of host tissue may range from limited (Brobyn & Wilding, 1977) to extensive (Mohamed et al., 1978) prior to death, but often fungal growth is confined to the haemolymph during this phase (Sweeney, 1975; Prasertphon & Tanada, 1968). However, *E. coronata* and *E. apiculata* are found in the fatbody, cuticle and other tissues of *G. mellonella*, probably because the hyphal bodies are too large to circulate in the haemolymph (Prasertphon & Tanada, 1968).

Electron microscopy studies may reveal greater penetration of host tissues than is apparent from light microscopy (Zacharuk, 1970; Zacharuk, 1971; Zacharuk, 1973a). Indeed Zacharuk (1973a) observed ultrastructural changes in wireworm tissue in advance of penetrating hyphae of *M. anisopliae*. Similar disruption of cellular organelles occurred *in vitro* when haemocytes of *Orycetes rhinoceros* were exposed to culture filtrates of *M. anisopliae*; this suggests the action of a mycotoxin (Vey & Quiot, 1975).

Several low molecular weight toxins have been identified from cultures of entomopathogenic fungi, particularly the Deuteromycetes (Roberts, 1980). However, most of these toxins have not been isolated from parasitized insects and their relevance to mycosis remains to be established. An exception are the destruxins, produced by *M. anisopliae*. Suzuki, Kawakami & Tamura (1971) isolated 0.0004 µmoles larva^{-1} of destruxins A and B from *B. mori* the day before the animals died. This was approximately 25% of the injected pure dose of destruxin required to cause paralysis and death of larvae (Kodaira, 1961). The behaviour elicited by these chemicals was comparable to that observed in *B. mori* prior to death from *M. anisopliae* (Roberts, 1966), thus establishing destruxins as the probable cause of death of silkworms infected with *M. anisopliae* (Roberts, 1980).

Insects vary in their susceptibility to destruxins just as strains of *M. anisopliae* vary in the level of production. However, the quantity of destruxin produced *in vitro* correlated directly with virulence for mosquito larvae of several mutants of *M. anisopliae* (Al-Aidroos & Roberts, 1978).

Premature death of *Reticulitermes flavipes* from infection with *E. coronata* (Prasertphon, 1967) was subsequently correlated with the *in vitro* production of toxic proteases (Prasertphon & Tanada, 1969). Recently *B. bassiana* and *M. anisopliae* have also been shown to produce proteases *in vitro* which are lethal when injected into *G. mellonella* (Kucera & Samsinakova, 1968; Kucera, 1980). Whether or not such proteases are toxic *in vivo*, they may be more relevant as agents of cuticle hydrolysis during penetration or digestion of protein following host death (Roberts, 1980).

It is unlikely that death from mycosis is the result of a single lethal lesion. Indeed it is difficult to define the point of death of an insect, since heart and brain are not essential to life and lungs are absent. Host physiology must be chronically disrupted by the growth of the hyphae through tissues and the blockage of the haemocoel by hyphal bodies (Madelin, 1963), resulting in stress reactions possibly including auto-intoxication (Sternberg, 1963). In those cases where fungi overcome their hosts after limited growth, toxins may play a significant part in host death, perhaps particularly by causing neuromuscular dysfunction (Bao & Yendol, 1971; Bell, 1974; Roberts, 1980). Roberts (1980) found that an injected dose of destruxins caused immediate tetanus in larvae of *G. mellonella*, while Evlakhova & Rakitin (1968) observed reduced activity in the thoracic nerve cord of the locust, *Locusta migratoria manilensis*, infected with *Aspergillus flavus*. Symptoms such as sluggishness, decreased irritability, inability to recover from an inverted position, partial or general paralysis may all occur in the later stages of fungal infection (Madelin, 1963), and are not dissimilar from those produced by synthetic neurotoxic insecticides (Wilkinson, 1976). Toxins may elicit further pathological effects by causing water loss from cells in general and through Malpighian tubules and midgut in particular (Zacharuk, 1973*a*). Chronic spiracular water loss has also been recorded in a parasitized insect (Sussman, 1952*b*). Water loss through permanently open spiracles, chronic diuresis and a 'leaky integument' is a characteristic feature of death from nerve-acting insecticides (Maddrell & Reynolds, 1972; Gerolt, 1976).

Roberts (1980) has suggested that, in general, the lower entomopathogenic fungi (e.g. *Coelomomyces*, *Lagenidium* and *Entomophthora*) overcome susceptible hosts primarily by 'starvation' – that is they use up soluble host reserves – rather than via toxins. This is supported by the work of Domnas, Giebel & McInnis, (1974) on *Lagenidium giganteum*-infected *Culex pipiens quinquefasciatus*, where the level of free amino acids, sugars and proteins and the activity of a number of host enzymes were significantly reduced compared with controls. Unfortunately these authors did not report any effect of infection on feeding behaviour: loss of appetite is an early symptom of many mycoses and presumably results in a general starvation of the insect (Bell, 1974). This certainly occurs in *Heliothis zea* infected with *B. bassiana*, where the effects of infection on haemolymph proteins, amino acid and carbohydrate levels are indistinguishable from those of starvation (Cheung & Grula, 1982). However, Gardner, Sutton & Noblet (1979) suggested that changes in the electrophoretic pattern of haemolymph proteins of several Lepidoptera spp. infected with *B. bassiana* were not entirely due to cessation of feeding.

Host response to invasion
Cuticle

Dark patches in the cuticle surrounding a penetrant hypha are characteristic early symptoms of infection (Wallengren & Johansson, 1929; Nirula, Radha & Menon, 1955; Takahashi, 1958; Gabriel, 1968*a*; Aoki & Yanase, 1970; Domnas *et al.*, 1974; Brobyn & Wilding, 1977; Schabel, 1976*a*) but are not universal (McCauley *et al.*, 1968). In those studies where histochemical tests have been performed, the pigment has been identified as melanin (Gabriel, 1968*b*; Gotz & Vey, 1974). The phenomenon is not confined to the surrounding cuticle since the hyphal walls may become heavily melanized (Nyhlen & Unestam, 1980) or enveloped by a melanotic capsule (Gotz & Vey, 1974).

Given the known susceptibility of microorganisms to phenols it has generally been assumed that the melanization reactions have antifungal effects. However, melanic reactions form part of an arthropod's non-specific response to wounding (Lai-Fook, 1966), and there is little need to question the source of the melanic substrate polyphenoloxidase or to hypothesize that the anomalous pigmentation is due to endocrine dysfunction in the host brought about by the fungus (Sannasi & Oliver, 1971). Also there appear to be few instances where melanization prevents invasion. However, extreme blackening of the cuticle of the mosquito, *Culiseta inornata*, under appressoria of *Coelomomyces psorophorae* correlated both with a failure to penetrate and disintegration of the young hypha. Conversely when penetration did occur only a mild reaction was seen (Travland, 1979*b*). Furthermore Nyhlen & Unestam (1980) found a positive correlation between the degree of melanization in the cuticle of crayfish species and resistance to *Aphanomyces astaci*.

In soft crayfish cuticle, phenoloxidase is located in the outer layers, close to the epicuticle, but the phenolic substrate for melanization of the parasite is apparently translocated from the epidermis (Unestam & Ajaxon, 1976). The enzyme will attach to many foreign bodies including fungal walls (Söderhall, Hall, Unestam & Nyhlen, 1979) and the activation of phenoloxidase is elicited by β-1,3-glucans present in the fungal wall of *Aphanomyces astaci* (Unestam & Söderhäll, 1977; Söderhäll & Unestam, 1979). A polysaccharide from *Colletotrichum lindemuthianum* causes browning and phytoalexin production when applied to cut surfaces of bean. It also consists predominantly of 3- and 4-linked glucosyl residues, possibly β-linked (Anderson-Prouty & Albersheim, 1975).

Heavily melanized hyphae of *A. astaci* in crayfish cuticle are still viable, though mycelial growth is almost completely inhibited (Nyhlen & Unestam, 1980). *In vitro* tests confirm that several quinones and a naphthalene derivative inhibit growth of *A. astaci* (Söderhall & Ajaxon, 1982). Although quinones are known to inhibit proteases and chitinases (Kuo & Alexander, 1967; Bull, 1970), *A. astaci* protease is unaffected (Söderhäll & Ajaxon, 1982).

Haemolymph

Three types of defence reaction to fungal invasion have been observed within insect haemolymph: phagocytosis, encapsulation – i.e. capsule, 'giant' cell and nodule formation (Salt, 1970; Whitcomb, Shapiro & Granados, 1974) – and humoral encapsulation (Gotz & Vey, 1974).

Haemocytes have been observed to accumulate around the point of entry of a fungus through the epidermis (Vey & Fargues, 1977); they may also be attracted from a distance by the presence of hyphae in the haemolymph (Vey & Quiot, 1976). Kawakami (1965) clearly observed phagocytosis of *Paecilomyces* (= *Isaria*) *fumosorosea* during early stages of hyphal body formation in the haemolymph of *Bombyx mori*. However, Glaser (1926) reported that with *M. anisopliae* in *B. mori*, the fungus was too large for ingestion by phagocytes. This was not a problem in the cockchafer, *Melolontha melolontha*, where *M. anisopliae* was actively phagocytized (Hurpin & Vago, 1958). Conidia of *Aspergillus flavus* were also phagocytized when injected into *Hyalophora cecropia*, but the reaction was more intense when the non-pathogen *A. niger* was introduced (Sussman, 1952a).

The accumulation of haemocytes around a fungus to form a capsule is often accompanied by the disintegration of the innermost cells of the capsule envelope and the deposition of melanin, (Whitcomb et al., 1974). Haemocytic encapsulation has been described for *Torula nigra* in the Ichneumonid fly, *Exeristes comstokii* (Bucher & Bracken, 1966), for *Mucor hiemalis* growing through wounds in *Galleria mellonella* (Vey, 1968), for conidia of *Aspergillus niger* injected into *G. mellonella* (Vey & Vago, 1969), and for *B. bassiana* in *Galleria mellonella* (Vey & Vago, 1971).

An alternative form of encapsulation, humoral encapsulation, occurs in some dipterous larvae without direct participation of blood cells (Gotz & Vey, 1974). This involves deposition of a melanin–protein complex on the surface of the fungus; it is effective against injected conidia of *M. hiemalis* and *A. niger* into *Chironomus* larvae, but not against large inocula of *B. bassiana* (Gotz & Vey, 1974).

The extensive haemocytic response to infection by Deuteromycete fungi (*M. anisopliae, B. bassiana, A. flavus*) is not always equalled with virulent entomopathogens of the Zygomycotina and Mastigomycotina. Although Klein & Coppel (1973) observed encapsulation of *Entomophthora tenthredinis* by agglomerated blood cells in the sawfly, *Diprion similis*, Brobyn & Wilding (1977) found no agglomeration of *E. aphidis* in *Aphis pisum*. Similarly whereas crayfish haemocytes readily encapsulate hyphae of *Aphanomyces astaci in vitro* (Unestam & Nylund, 1972), another member of the Mastigomycotina, *Coelomomyces punctatus*, is not attacked by haemocytes of the mosquito, *Anopheles quadrimaculatus* (Powell, 1976).

The ability of the host to respond may depend, in part, on whether or not protoplasts form the parasitic stage of the fungus (Roberts & Humber, 1981). Haemocytes of the eastern hemlock looper, *Lambdina fiscellaria fiscellaria*, adhered strongly to hyphae and hyphal bodies of *Entomophthora egressa in vitro* but not to protoplasts either *in vivo* or *in vitro* (Dunphy & Nolan, 1980). Protoplasts are part of the natural life cycle of *E. egressa* in the spruce budworm, *Choristoneura fumiferana*, (Tyrrell, 1977; Chapter 11 this volume) and of *E. grylli* in grasshoppers (MacLeod, Tyrell & Welton, 1980). These structures are also produced during the early stages of mycosis in *A. quadrimaculatus* by *C. punctatus* (Powell, 1976). Dunphy & Nolan (1980) have suggested that proposed strategies for parasitoid evasion of host haemocyte response (Vinson, 1977) could be applied to the protoplasts of *E. eggressa*. These include the acquisition of host haemolymph molecules by the surface of the protoplast, the innate possession of chemically unreactive surface, or the evolution of molecular mimicry.

Although many Deuteromycotina initiate a widespread active response by host haemocytes, this may not prevent rapid infection (Hurpin & Vago, 1958), and strain specificity may affect the outcome. Fargues, Roberts & Vey (1976) found that an injected inoculum of ten conidia of species-specific strains of *M. anisopliae* killed larvae of the beetles *Orycetes rhinoceros* or *Cetonia auratia*. However the minimum lethal dose was 10^4 conidia per larva in non-species specific combinations, even though haemocytes of *O. rhinoceros* encapsulated the conidia of each strain equally after 24 hours. However, after 4 days, whilst the non-adapted strain was still encapsulated, the adapted strain had grown out from the enveloping haemocytes, possibly aided by an anti-haemocytic toxin, the existence of which has been shown *in vitro* by Vey & Quiot (1975).

Acknowledgments. I would like to thank Dr L. Jacobs for critical reading of the manuscript, Dr R. M. Cooper and Dr S. E. Reynolds for helpful comments, SERC and Tate & Lyle Ltd for financial support, and Miss S. Leaman for patient decoding of the manuscript.

References

Aist, J. R. (1976). Cytology of penetration and infection. In *Encyclopedia of Plant Physiology*, vol. 4, ed. R. Heitefuss & P. H. Williams, pp. 198–221. New York: Springer-Verlag.

Al-Aidroos, K. & Roberts, D. W. (1978). Mutants of *Metarrhizium anisopliae* with increased virulence towards mosquito larvae. *Canadian Journal of Genetics and Cytology*, **20**, 211–19.

Al-Aidroos, K. & Seifert, A. M. (1980). Polysaccharide and protein degradation, germination, and virulence against mosquitoes in the entomopathogenic fungus, *Metarrhizium anisopliae*. *Journal of Invertebrate Pathology*, **36**, 29–34.

Albersheim, P., Jones, T. M. & English, P. D. (1969). Biochemistry of the cell wall in relation to infective processes. *Annual Review of Phytopathology*, **7**, 171–94.

Andersen, S. O. (1980). Cuticular sclerotization. In *Cuticle Techniques in Arthropods*, ed. T. A. Miller, pp. 185–217. New York: Springer-Verlag.

Anderson-Prouty, A. & Albersheim, P. (1975). Host–pathogen interactions. VIII. Isolation of a pathogen-synthesized fraction rich in glucan that elicits a defense response in the pathogen's host. *Plant Physiology*, **56**, 286–91.

Aoki, J. & Yanase, K. (1970). Phenol oxidase activity in the integument of the silkworm *Bombyx mori* infected with *Beauveria bassiana* and *Spicaria fumoso-rosea*. *Journal of Invertebrate Pathology*, **16**, 459–64.

Bade, M. L. (1974). Localization of moulting chitinase in insect cuticle. *Biochimica et biophysica acta*, **372**, 474–7.

References

Bade, M. L. & Shoukimas, J. J. (1974). Neutral metal chelator-sensitive protease in insect moulting fluid. *Journal of Insect Physiology*, **20**, 281–90.

Bade, M. L. & Stinson, A. (1978). Digestion of cuticle chitin during the moult of *Manduca sexta* (Lepidoptera: Sphingidae). *Insect Biochemistry*, **9**, 221–31.

Bajan, C., Kalalova, S., Kmitowa, K., Samsinakova, A. & Wojciechowska, M. (1979). The relationship between infectious activities of entomophagous fungi and their production of enzymes. *Bulletin de l'Académie polonaise des sciences. Classe II. Série des Sciences biologiques*, **27**, 963–8.

Bao, L-L. & Yendol, W. G. (1971). Infection of the eastern subterranean termite, *Reticulitermes flavipes*, with the fungus *Beauveria bassiana*. *Entomophaga*, **16**, 343–52.

Beament, J. W. L. (1967). Lipid layers and membrane models. In *Insects and Physiology*, ed. J. W. L. Beament & J. E. Treherne, pp. 303–14. London: Oliver & Boyd.

Bell, J. V. (1974). Mycoses. In *Insect Diseases*. vol. 1, ed. G. E. Cantwell, pp. 185–236. New York: Marcel Dekker.

Benz, G. (1963). Physiopathology and histochemistry. In *Insect Pathology: An Advanced Treatise*, vol. 1, ed. E. A. Steinhaus, pp. 299–338. New York: Academic Press.

Berisford, Y. C. & Tsao, C. H. (1975). Appressoria formation by *Aspergillus parasiticus* on bagworm cuticle. *Annals of the Entomological Society of America*, **68**, 1111–12.

Brobyn, P. J. & Wilding, N. (1977). Invasive and developmental processes of *Entomophthora* species infecting aphids. *Transactions of the British Mycological Society*, **69**, 349–66.

Broome, J. R., Sikorowski, P. P. & Norment, B. R. (1976). A mechanism of pathogenicity of *Beauveria bassiana* on larvae of the imported fire ant *Solenopsis richteri*. *Journal of Invertebrate Pathology*, **28**, 87–91.

Bucher, G. E. & Bracken, G. K. (1966). Fungus disease of adult parasitic insects caused by *Torula nigra* (Marpmann). *Journal of Invertebrate Pathology*, **8**, 193–204.

Bull, A. T. (1970). Inhibition of polysaccharases by melanin: enzyme inhibition in relation to mycolysis. *Archives of Biochemistry and Biophysics*, **137**, 345–56.

Burnside, C. E. (1930). Fungus diseases of the honey bee. *U.S. Department of Agriculture Technical Bulletin*, **149**, 1–43.

Cheung, P. Y. K. & Grula, E. A. (1982). *In vivo* events associated with entomopathology of *Beauveria bassiana* for the corn earworm *Heliothis zea*. *Journal of Invertebrate Pathology*, **39**, 303–13.

Clark, T. B., Kellen, W. R., Fukuda, T. & Lindegren, J. E. (1968). Field and laboratory studies on the pathogenicity of the fungus *Beauveria bassiana* to three genera of mosquitoes. *Journal of Invertebrate Pathology*, **11**, 1–7.

Clerk, G. C. (1969). Influence of soil extracts on the germination of conidia of the fungi *Beauveria bassiana* and *Paecilomyces farinosus*. *Journal of Invertebrate Pathology*, **13**, 120–4.

Cohen, B. L. (1973). Regulation of intracellular and extracellular, neutral and alkaline proteases in *Aspergillus nidulans*. *Journal of General Microbiology*, **79**, 311–20.

Cohen, B. L. & Drucker, H. (1977). Regulation of exocellular protease in *Neurospora crassa*: induction and repression under conditions of nitrogen starvation. *Archives of Biochemistry and Biophysics*, **182**, 601–13.

Cooper, R. M. (1976). Regulation of synthesis of cell wall-degrading enzymes of plant pathogens. In *Cell Wall Biochemistry Related to Specificity in Host–Plant Pathogen Interactions*, ed. B. Solheim & J. Raa, pp. 163–211. Tromsø, Norway: Universitetsforlaget.

Cooper, R. M. (1982). The mechanisms and significance of enzymic degradation of host cell walls. In *Biochemical Plant Pathology*, ed. J. A. Callow. John Wiley, New York (In press).

Cooper, R. M., Rankin, B. & Wood, R. K. S. (1978). Cell wall-degrading enzymes of vascular wilt fungi. II. Properties and modes of action of polysaccharidases of *Verticillium albo-atrum* and *Fusarium oxysporum* f. sp. *lycopersici*. *Physiological Plant Pathology*, **13**, 101–34.

Cooper, R. M. & Wood, R. K. S. (1975). Regulation of synthesis of cell wall-degrading enzymes by *Verticillium albo-atrum* and *Fusarium oxysporum* f. sp. *lycopersici*. *Physiological Plant Pathology*, **5**, 135–6.

Cordon, T. C. & Schwartz, J. H. (1962). The fungus *Beauveria tenella*. *Science*, **138**, 1265–6.

Crisan, L. U. (1971). Mechanism responsible for release of toxin by *Metarrhizium* spores in mosquito larvae. *Journal of Invertebrate Pathology*, **17**, 260–4.

David, W. A. L. (1967). The physiology of the insect integument in relation to the invasion of pathogens. In *Insects and Physiology*, ed. J. W. L. Beament & J. E. Treherne, pp. 17–35. London: Oliver & Boyd.

Delachambre, J. (1969). La réaction de la chitine à l'acide-periodique-schiff. *Histochemie*, **20**, 58–67.

Dieuzeide, R. (1925). Les champignons entomophytes du genre *Beauveria* Vuillemin. Contribution a l'étude de *Beauveria effusa* Vuill. parasite du Doryphore. *Annales des Epiphyties*, **11**, 185–219.

Dillon, R. J. & Charnley, A. K. (1982). The locust gut as an environment for germination of spores of the entomogenous fungus *Metarrhizium anisopliae*. *III International Colloquium on Invertebrate Pathology*, 106 (Abstr.).

Domnas, A., Giebel, P. E. & McInnis, T. M. (1974). Biochemistry of mosquito infection: preliminary studies of biochemical change in *Culex pipiens quinquefasciatus* following infection with *Lagenidium giganteum*. *Journal of Invertebrate Pathology*, **24**, 293–304.

Duddridge, J. A. & Sargent, J. A. (1978). A cytochemical study of lipolytic activity in *Bremia lactucae* Regel. during germination of the conidium and penetration of the host. *Physiological Plant Pathology*, **12**, 289–96.

Dunphy, G. B. & Nolan, R. A. (1980). Response of Eastern Hemlock Looper hemocytes to selected stages of *Entomophthora egressa* and other foreign particles. *Journal of Invertebrate Pathology*, **36**, 71–84.

Ebling, W. (1976). Insect integument: a vulnerable organ system. In *The Insect Integument*, ed. H. R. Hepburn, pp. 383–400, Amsterdam: Elsevier.

Edwards, H. H. & Allen, P. J. (1970). A fine-structure study of the primary infection process during infection of barley by *Erysiphe graminis* f. sp. *hordei*. *Phytopathology*, **60**, 1504–9.

Emmett, R. W. & Parbery, D. G. (1975). Appressoria. *Annual Review of Phytopathology*, **13**, 147–67.

Ende, van den G. & Linskens, H. F. (1974). Cutinolytic enzymes in relation to pathogenesis. *Annual Review of Phytopathology*, **12**, 247–58.

Evlakhova, A. A. & Rakitin, A. A. (1968). Effects of experimental mycosis on electrical activity of nerve chain of *Locusta migratoria* Manilensis Mey. *Dokladȳ Akademii nauk SSSR*, **178**, 81–4.

Evlakhova, A. A. & Shekhurina, T. A. (1963). Antifungal action of the cuticle of *Eurygaster integriceps*. *Dokladȳ Akademii nauk SSSR*, **148**, 977–8.

Fargues, J., Roberts, P. -H. & Vey, A. (1976). Role of the integument and of the cellular defense of host coleoptera in the specificity of the entomopathogenic strains of *Metarrhizium anisopliae* fungi imperfecti. *Comptes rendus hebdomadaire des séances de l'Académie des sciences*, **282**, 2223–6.

Ferron, P. (1978). Biological control of insect pests by entomogenous fungi. *Annual Review of Entomology*, **23**, 409–42.
Ferron, P. & Diomandé, T. (1969). Sur la spécificité à l'égard des insectes de *Metarrhizium anisopliae* (Metsch.) Sorokin (Fungi Imperfecti) en fonction de l'origine des souches de ce champignon. *Comptes rendus hebdomadaire des séances de l'Académie des sciences*, **268**, 331–2.
Fuchs, A., Jobson, J. A., & Wouts, W. M. (1965). Arabanases in phytopathogenic fungi. *Nature, London*, **206**, 714–15.
Gabriel, B. P. (1959). Fungus infections of insects via the alimentary tract. *Journal of Insect Pathology*, **1**, 319–30.
Gabriel, B. P. (1968a). Histochemical study of the insect cuticle infected by the fungus *Entomophthora coronata*. *Journal of Invertebrate Pathology*, **11**, 82–9.
Gabriel, B. P. (1968b). Enzymatic activities of some entomopathogenic fungi. *Journal of Invertebrate Pathology*, **11**, 70–81.
Gardner, W. A., Sutton, R. M. & Noblet, R. (1979). Effects of infection by *Beauveria bassiana* on haemolymph proteins of noctuid larvae. *Annals of the Entomological Society of America*, **72**, 224–8.
Gerolt, P. (1976). The mode of action of insecticides: accelerated water loss and reduced respiration in insecticide-treated *Musca domestica* L. *Pesticide Science*, **7**, 604–20.
Getzin, L. W. (1961). *Spircaria rileyi* (Farlow) Charles, an entomogenous fungus of *Trichoplusia ni* (Hubner). *Journal of Insect Pathology*, **3**, 2–10.
Glaser, R. W. (1926). The green muscardine disease in silkworms and its control. *Annals of the Entomological Society of America*, **19**, 180–92.
Gotz, P. & Vey, A. (1974). Humoral encapsulation in Diptera (Insecta): defense reactions of *Chironomus* larvae against fungi. *Parasitology*, **68**, 193–205.
Grula, E. A., Burton, R. L., Smith, R., Mapes, T. L., Cheung, P. Y. K., Pekrul, S., Champlin, F. R., Grula, M. & Abegaz, B. (1978). Biochemical basis for the pathogenicity of *Beauveria bassiana*. In *Proceedings of the first joint USA/USSR Conference on the Production, Selection and Standardization of Entomopathogenic Fungi*, ed. C. M. Ignoffo, pp. 192–216. Washington DC: American Society for Microbiology.
Hedland, R. C. & Pass, B. C. (1968). Infection of the alfalfa weevil, *Hypera postica*, by the fungus *Beauveria bassiana*. *Journal of Invertebrate Pathology*, **11**, 25–34.
Hinton, H. E. (1970). Some little known surface structures. In *Insect Ultrastructure*, Symposium of the Royal Entomological Society vol. 5, ed. A. C. Neville, pp. 41–59. Oxford: Blackwell.
Huber, J. (1958). Untersuchungen zur Physiologie insektentötender Pilze. *Archiv für Mikrobiologie*, **29**, 257–76.
Hurpin, B. & Vago, C. (1958). Les maladies du Hanneton commun (*Melolontha melolontha* L.). *Entomophaga*, **3**, 285–330.
Ignoffo, C. M. & Garcia, C. (1978). *In vitro* inactivation of conidia of the entomopathogenic fungus *Nomuraea rileyi* by human gastric juice. *Environmental Entomology*, **7**, 217–18.
Jackson, L. L. & Blomquist, G. J. (1976). Insect waxes. In *Chemistry and Biochemistry of Natural Waxes*, ed. P. E. Kolattukudy, pp. 201–33. Amsterdam: Elsevier.
Jones, J. C. (1960). The anatomy and rhythmical activities of the alimentary canal of *Anopheles* larvae. *Annals of the Entomological Society of America*, **53**, 459–74.
Karr, A. L. & Albersheim, P. (1970). Polysaccharide-degrading enzymes are unable to attack plant cell walls without prior action by a 'wall-modifying enzyme'. *Plant Physiology*, **46**, 69–80.

Kawakami, K. (1965). Phagocytosis in muscardine diseased larvae of the silkworm *Bombyx mori* L. *Journal of Invertebrate Pathology*, **7**, 203–8.

Keilin, D. (1920). On a new saccharomycete *Monosporella unicuspidata* Gen. N. Nom, n. sp., parasitic in the body cavity of a dipterous larva (*Dasyhelea obscura* Winnertz). *Parasitology*, **12**, 83–91.

Klein, M. G. & Coppel, H. C. (1973). *Entomophthora tenthredinis*, a fungal pathogen of the introduced pine sawfly in north-western Wisconsin. *Annals of the Entomological Society of America*, **66**, 1178–80.

Kodaira, Y. (1961). Biochemical studies on the muscardine fungi in the silkworms, *Bombyx mori*. *Journal of the Faculty of Textile Science and Technology of Shinshu University*, **29**, 1–68.

Koidsumi, K. (1957). Antifungal action of cuticular lipids in insects. *Journal of Insect Physiology*, **1**, 40–51.

Koidsumi, K. & Wada, Y. (1955). Studies on the antimicrobial function of insect lipids. IV. Racial difference in the antifungal activity in the silkworm integument. *Japanese Journal of Applied Zoology*, **20**, 184–90.

Kolattukudy, P. E. (1980). Biopolyester membranes of plants: cutin and suberin. *Science*, **208**, 990–1000.

Köller, W., Allan, C. R. & Kolattukudy, P. E. (1982). Role of cutinase and cell wall degrading enzymes in infection of *Pisum sativum* by *Fusarium solani* f. sp. *pisi*. *Physiological Plant Pathology*, **20**, 47–60.

Kucera, M. (1980). Proteases from the fungus *M. anisopliae* toxic for *Galleria mellonella* larvae. *Journal of Invertebrate Pathology*, **35**, 304–10.

Kucera, M. & Samsinakova, A. (1968). Toxins of the entomophagous fungus *Beauveria bassiana*. *Journal of Invertebrate Pathology*, **12**, 316–20.

Kuo, M. J. & Alexander, M. (1967). Inhibition of the lysis of fungi by melanins. *Journal of Bacteriology*, **94**, 624–9.

Lai-Fook, J. (1966). The repair of wounds in the integument of insects. *Journal of Insect Physiology*, **12**, 195–226.

Lambiase, J. T. & Yendol, W. G. (1977). The fine structure of *Entomophthora apiculata* and its penetration of *Trichoplusia ni*. *Canadian Journal of Microbiology*, **23**, 452–64.

Lapp, M. S. & Skoropad, W. P. (1978). Nature of adhesive material of *Colletotrichum graminicola* appressoria. *Transactions of the British Mycological Society*, **70**, 221–3.

Latgé, J. P. (1974). Proteolytic and chitinolytic activities of *Cordyceps militaris*. *Entomophaga*, **19**, 41–53.

Latgé, J. P. & Vey, A. (1974). Study on the pathogenesis of a mycosis due to *Cordyceps militaris* on two Lepidoptera. *Annales de la Société entomologique de France*, **10**, 149–59.

Lefebvre, C. L. (1931). Preliminary observations on two species of *Beauveria* attacking the corn borer *Pyrausta nubilalis*. *Phytopathology*, **21**, 1115–28.

Lefebvre, C. L. (1934). Penetration and development of the fungus, *Beauveria bassiana*, in the tissues of the corn borer. *Annals of Botany*, **48**, 441–52.

Leopold, J. & Samsinakova, A. (1970). Quantitative estimation of chitinase and several other enzymes in the fungus *Beauveria bassiana*. *Journal of Invertebrate Pathology*, **15**, 34–42.

Leopold, H., Samsinakova, A. & Misikova, S. (1973). Enzymic character of toxic substances released by the entomophagous fungus *Beauveria bassiana* and their stimulation of their production. *Zentralblatt für Bakteriologie, Parasitenkunde, Infectionskrankheiten und Hygiene (Abteilung II)*, **128**, 31–41.

Lepesme, P. (1938). Recherches sur une aspergillose des Acridiens. *Bulletin de la Société d'historie Naturelle de l'Afrique du Nord*, **29**, 372–81.
Lihnell, D. (1944). Grömykos förorsakad av *Metarrhizium anisopliae* (Metsch.) Sorok. II. Fysiologiska undersökningar över grönmykosens svamp. *Statens Växtskyddsanstalt Meddelande*, **43**, 59–90.
Lipke, H. & Geoghegan, T. (1971). Enzymolysis of sclerotized cuticle from *Periplaneta americana* and *Sarcophaga bullata*. *Journal of Insect Physiology*, **17**, 415–25.
Locke, M. (1974). The structure and formation of the integument of insects. In *The Physiology of Insecta*, 2nd edn, vol. VI, ed. M. Rockstein, pp. 123–213. New York: Academic Press.
McCauley, V. J. E., Zacharuk, R. Y. & Tinline, R. D. (1968). Histopathology of green muscardine in larvae of four species of Elateridae (Coleoptera). *Journal of Invertebrate Pathology*, **12**, 444–59.
McEwen, F. L. (1963). *Cordyceps* infections. In *Insect Pathology: An Advanced Treatise*, vol. 2, ed. E. A. Steinhaus, pp. 273–90. New York: Academic Press.
McKeen, W. E. (1974). Mode of penetration of epidermal cell walls of *Vicia faba* by *Botrytis cinerea*. *Phytopathology*, **64**, 461–7.
McKeen, W. E., Smith R. & Bhattacharya, P. K. (1969). Alterations of the host wall surrounding the infection peg of powdery mildew fungi. *Canadian Journal of Botany*, **47**, 701–6.
MacLeod, D. M. (1963). Entomophthorales infections. In *Insect Pathology: An Advanced Treatise*, vol. 2, ed. E. A. Steinhaus, pp. 189–231. New York: Academic Press.
MacLeod, D. M., Tyrrell, D. & Welton, M. A. (1980). Isolation and growth of the grasshopper pathogen, *Entomophthora grylli*. *Journal of Invertebrate Pathology*, **36**, 85–9.
Maddrell, S. H. P. & Reynolds, S. E. (1972). Release of hormones in insects after poisoning with insecticides. *Nature*, **236**, 404–6.
Madelin, M. F. (1963). Diseases caused by hyphomycetous fungi. In *Insect Pathology: An Advanced Treatise*, vol. 2, ed. E. A. Steinhaus, pp. 233–71. New York: Academic Press.
Madelin, M. F. (1966). Fungal parasites of insects. *Annual Review of Entomology*, **11**, 423–48.
Madelin, M. F., Robinson, R. K. & Williams, R. J. (1967). Appressorium-like structures in insect-parasitizing Deuteromycetes. *Journal of Invertebrate Pathology*, **9**, 404–12.
Maiti, I. B. & Kolattukudy, P. E. (1979). Prevention of fungal infection of plants by specific inhibition of cutinase. *Science*, **205**, 507–8.
Marks, G. C., Berbee, J. G. & Riker, A. J. (1965). Direct penetration of leaves of *Populus tremuloides* by *Colletotrichum gloeosporioides*. *Phytopathology*, **55**, 408–12.
Martin, W. W. (1969). A morphological and cytological study of development in *Coelomomyces punctatus* parasitic in *Anopheles quadrimaculatus*. *Journal of Elisha Mitchell Scientific Society*, **85**, 59–72.
Masera, E. (1957). *Metarrhizium anisopliae* (Metchnikoff) Sorokin, parassita del baco da seta. *Annali della sperimentazione agraria*, **11**, 281–95.
Meevootisom, V. & Niederpruem, D. J. (1979). Control of exocellular proteases in dermatophytes and especially *Trichophyton rubrum*. *Sabouraudia*, **17**, 91–106.
Mohamed, A. K. A., Sikorowski, P. P. & Bell, J. V. (1978). Histopathology of *Nomuraea rileyi* in larvae of *Heliothis zea* and *in vitro* enzymatic activity. *Journal of Invertebrate Pathology*, **31**, 345–52.
Morgan, M. R. J. (1976). A qualitative survey of the carbohydrases of the alimentary tract

of the migratory locust, *Locusta migratoria migratorioides*. *Journal of Insect Physiology*, **21**, 1045–53.
Müller-Kögler, E. (1965). *Pilzkrankheiten bei Insekten*. Berlin: Parey
Neville, A. C. (1975). *Biology of the Arthropod Cuticle*. Berlin: Springer-Verlag.
Nirula, K. K., Radha, K. & Menon, K. P. U. (1955). The green muscardine disease of *Orycetes rhinoceros* L. *Indian Coconut Journal*, **9**, 3–10.
Nolla, J. A. B. (1929). *Acrostalagmus aphidum* Oud., and aphid control. *Journal of the Department of Agriculture Porto Rico*, **13**, 59–72.
Notini, G. & Mathlein, R. (1944). Grönmykos fororsakad av *Metarrhizium anisopliae* (Metsch.) Sorok. I. Grönmykosen som biologiskt insektbekampningsmedel. *Statens Växtskyddsanstalt Meddelande*, **43**, 1–58.
Nyhlen, L. & Unestam, T. (1975). Ultrastructure of the penetration of the crayfish integument by the fungal parasite *Aphanomyces astaci*, Oomycetes. *Journal of Invertebrate Pathology*, **26**, 353–66.
Nyhlen, L. & Unestam, T. (1980). Wound reactions and *Aphanomyces astaci* growth in crayfish cuticle. *Journal of Invertebrate Pathology*, **36**, 187–97.
O'Sullivan, J. & Mathison, G. E. (1971). The localization and secretion of a proteolytic enzyme complex by the dermatophytic fungus *Microsporum canis*. *Journal of General Microbiology*, **68**, 319–26.
Paris, S. & Ferron P. (1979). Study of the virulence of some mutants of *Beauveria brogniartii* (= *Beauveria tenella*). *Journal of Invertebrate Pathology*, **34**, 71–7.
Paris, S. & Segretain, G. (1975). Physiological characters of *Beauveria tennella* in relation to the virulence of strains of the fungus to the larvae of the common cockchafer, *Melolontha melolontha*. *Entomophaga*, **20**, 135–8.
Paus, F. & Raa, J. (1973). An electron microscope study of infection and disease development in cucumber hypocotyls inoculated with *Cladosporium cucumerinum*. *Physiological Plant Pathology*, **3**, 461–4.
Pavlyushin, V. A. (1978). Virulence mechanisms of the entomopathogenic fungus *Beauveria bassiana*. In *Proceedings of the 1st joint USA/USSR Conference on the Production, Selection and Standardization of Entomopathogenic Fungi*, ed. C. M. Ignoffo, pp. 153–72. Washington DC: American Society for Microbiology.
Pekrul, S. & Grula, E. A. (1979). Mode of infection of the corn earworm (*Heliothis zea*) by *Beauveria bassiana* as revealed by scanning electron microscopy. *Journal of Invertebrate Pathology*, **34**, 238–47.
Phillips, D. R. & Loughton, B. G. (1976). Cuticle protein in *Locusta migratoria*. *Comparative Biochemistry and Physiology*, **55B**, 129–35.
Pirt, S. J. (1971). The diffusion capsule, a novel device for the addition of a solute at a constant rate to a liquid medium. *Biochemical Journal*, **121**, 293–7.
Powell, M. J. (1976). Ultrastructural changes in the cell surface of *Coelomomyces punctatus* infecting mosquito larvae. *Canadian Journal of Botany*, **54**, 1419–37.
Prasertphon, S. (1967). Mycotoxin production by species of *Entomophthora*. *Journal of Invertebrate Pathology*, **9**, 281–2.
Prasertphon, S. & Tanada, Y. (1968). The formation and circulation, in *Galleria*, of hyphal bodies of entomophthoraceous fungi. *Journal of Invertebrate Pathology*, **11**, 260–80.
Prasertphon, S. & Tanada, Y. (1969). Mycotoxins of entomophthoraceous fungi. *Hilgardia*, **39**, 581–600.
Purdy, R. E. & Kolattukudy, P. E. (1975a). Hydrolysis of plant cuticle by plant pathogens. Purification, amino acid composition and molecular weight of two isoenzymes of cutinase and a nonspecific esterase from *Fusarium solani*. *Biochemistry*, **14**, 2824–31.

Purdy, R. E. & Kolattukudy, P. E. (1975b). Hydrolysis of plant cuticle by plant pathogens. Properties of cutinase I, cutinase II, and a nonspecific esterase isolated from *Fusarium solani pisi*. *Biochemistry*, **14**, 2832–40.

Reynolds, S. E. (1975). The mechanism of plasticization of the abdominal cuticle in *Rhodnius*. *Journal of Experimental Biology*, **62**, 81–98.

Richards, A. G. (1951). *The Integument of Arthropods*. Minnesota University Press.

Roberts, D. W. (1966). Toxins from the entomogenous fungus *Metarrhizium anisopliae*. II. Symptoms and detection in moribund hosts. *Journal of Invertebrate Pathology*, **8**, 222–7.

Roberts, D. W. (1980). Toxins of entomopathogenic fungi. In *Microbial Control of Insects, Mites and Plant Diseases*, vol. 2, ed. H. D. Burges, pp. 441–63. New York: Academic Press.

Roberts, D. W. & Humber, R. A. (1981). Entomogenous fungi. In *Biology of Conidial Fungi*, vol. 2, ed. G. T. Cole & B. Kendrick, pp. 201–36. New York: Academic Press.

Roberts, D. W. & Yendol, W. G. (1971). Use of fungi for microbial control of insects. In *Microbial Control of Insects and Mites*, ed. H. D. Burges & N. W. Hussey, pp. 125–49. New York: Academic Press.

Robinson, R. K. (1966). Studies on penetration of insect integument by fungi. *Pest Articles and News Summaries*, **12**, 131–42.

Salt, G. (1970). The Cellular Defence Reactions of Insects. *Cambridge Monographs of Experimental Biology*, **16**. Cambridge University Press.

Samsinakova, A. & Misikova, S. (1973). Enzyme activities in certain entomophagous representatives of Deuteromycetes (Moniliales) in relationship to their virulence. Česká mykologie, **27**, 55–60.

Samsinakova, A., Misikova, S. & Leopold, J. (1971). Action of enzymatic systems of *Beauveria bassiana* on the cuticle of the greater wax moth larvae (*Galleria mellonella*). *Journal of Invertebrate Pathology*, **18**, 322–30.

Sannasi, A. (1969). Studies of an insect mycosis. I. Histopathology of the integument of the infected queen of the mound-building termite *Odontotermes obesus*. *Journal of Invertebrate Pathology*, **13**, 4–10.

Sannasi, A. & Oliver, J. H. (1971). Integument of the velvet-mite, *Dinothrombium giganteum*, and histopathological changes caused by the fungus *Aspergillus flavus*. *Journal of Invertebrate Pathology*, **17**, 354–65.

Schabel, H. G. (1976a). Green muscardine disease of *Hylobius pales* (Herbst) (Coleoptera: Curculionidae). *Zeitschrift für angewandte Entomologie*, **81**, 413–21.

Schabel, H. G. (1976b). Oral infection of *Hylobius pales* by *Metarrhizium anisopliae*. *Journal of Invertebrate Pathology*, **27**, 377–83.

Schabel, H. G. (1978). Percutaneous infection of *Hylobius pales* by *Metarrhizium anisopliae*. *Journal of Invertebrate Pathology*, **31**, 180–7.

Schaerffenberg, B. (1957). *Beauveria bassiana* (Vuill.) Link als parasit des kartoffelkäfers (*Leptinotarsa decemlineata* Say). *Anzeiger für Schaedlingskunde*, **30**, 69–74.

Schaerffenberg, B. (1964). Biological and environmental conditions for the development of mycoses caused by *Beauveria* and *Metarrhizium*. *Journal of Insect Pathology*, **6**, 8–20.

Selhime, A. G. & Muma, M. H. (1966). Biology of *Entomophthora floridina* attacking *Eutetranychus banksi*. *Florida Entomologist*, **49**, 161–8.

Shaykh, M., Soliday, C. & Kolattukudy, P. E. (1977). Proof for the production of cutinase by *Fusarium solani* f. *pisi* during penetration into its host, *Pisum sativum*. *Plant Physiology*, **60**, 170–2.

Smith, R. J. & Grula, E. A. (1981). Nutritional requirements for conidial germination and hyphal growth of *Beauveria bassiana*. *Journal of Invertebrate Pathology*, **37**, 222–30.

Smith, R. J. & Grula, E. A. (1982). Toxic components on the larval surface of the corn earworm (*Heliothis zea*) and their effects on germination and growth of *Beauveria bassiana*. *Journal of Invertebrate Pathology*, **39**, 15–22.

Smith, R. J., Pekrul, S. & Grula, E. A. (1981). Requirement for sequential enzymatic activities for penetration of the integument of the corn earworm (*Heliothis zea*). *Journal of Invertebrate Pathology*, **38**, 335–44.

Söderhäll, K. & Ajaxon, R. (1982). Effect of quinones and melanin on mycelial growth of *Aphanomyces* spp. and extracellular protease of *Apharomyces astaci*, a parasite on crayfish. *Journal of Invertebrate Pathology*, **39**, 105–9.

Söderhäll, K., Hall, L., Unestam, T. & Nyhlen, L. (1979). Attachment of phenoloxidase to fungal cell walls in arthropod immunity. *Journal of Invertebrate Pathology*, **34**, 285–94.

Söderhäll, K., Svensson, E. & Unestam, T. (1978). Chitinase and protease activities in germinating zoospore cysts of a parasitic fungus, *Aphanomyces astaci*, Oomycetes. *Mycopathologia*, **64**, 9–11.

Söderhäll, K., & Unestam, T. (1975). Properties of extracellular enzymes from *Aphanomyces astaci* and their relevance in the penetration process of crayfish cuticle. *Physiologia Plantarum*, **35**, 140–6.

Söderhäll, K. & Unestam, T. (1979). Activation of serum prophenoloxidase in arthropod immunity. The specificity of cell wall glucan activation and activation by purified fungal glycoproteins of crayfish phenoloxidase. *Canadian Journal of Microbiology*, **25**, 406–14.

Steinhaus, E. A. (1956). Stress as a factor in insect disease. *Proceedings of the 10th International Congress on Entomology*, pp. 725–30.

Sternberg, J. (1963). Autointoxication and some stress phenomena. *Annual Review of Entomology*, **8**, 19–38.

Stevenson, J. (1969). Sclerotin in the crayfish cuticle. *Comparative Biochemistry and Physiology*, **30**, 503–8.

Stirling, J. L., Cook, G. A. & Pope, A. M. S. (1979). Chitin and its degradation. In *Fungal walls and hyphal growth*. British Mycological Society Symposium 2, ed. J. H. Burnett & A. P. J. Trinci, pp. 169–88. Cambridge University Press.

Sussman, A. S. (1951). Studies of an insect mycosis. I. Etiology of the disease. *Mycologia*, **43**, 338–50.

Sussman, A. S. (1952a). Studies of an insect mycosis. III. Histopathology of an aspergillosis of *Platysamia cecropia*. *Annals of the Entomological Society of America*, **45**, 233–45.

Sussman, A. S. (1952b). Studies of an insect mycosis. IV. The physiology of the host–parasite relationship of *Platysamia cecropia* and *Aspergillus flavus*. *Mycologia*, **44**, 493–505.

Suzuki, A., Kawakami, K. & Tamura, S. (1971). Detection of destruxins in silkworm larvae infected with *Metarrhizium anisopliae*. *Agricultural and Biological Chemistry*, **35**, 1641–43.

Sweeney, A. W. (1975). The mode of infection of the insect pathogenic fungus *Culicinomyces* in larvae of the mosquito *Culex fatigans*. *Australian Journal of Zoology*, **23**, 49–57.

Takahashi, Y. (1958). Studies on the cuticle of the silkworm, *Bombyx mori* L. XI. Penetration of hyphae of the fungus, *Beauveria bassiana* (Bals.) Vuill., through the larval and pupal cuticles. *Annotationes zoologicae japonenses*, **31**, 13–21.

Toumanoff, C. (1931). Actions des champignons entomophytes sur les abeilles. *Annales de parasitologie humaine et comparée*, **9**, 462–82.

Travland, L. B. (1979a). Structures of the motile cells of *Coelomomyces psorophorae* and function of the zygote in encystment on host. *Canadian Journal of Botany*, **57**, 1021–35.
Travland, L. B. (1979b). Initiation of infection of mosquito larvae (*Culiseta inornata*) by *Coelomomyces psorophorae*. *Journal of Invertebrate Pathology*, **33**, 95–105.
Tyrrell, D. (1977). Occurrence of protoplasts in the natural life cycle of *Entomophthora egressa*. *Experimental Mycology*, **1**, 259–63.
Unestam, T. & Ajaxon, R. (1976). Phenol oxidase in soft cuticle and blood of crayfish compared with that in other arthropods and activation of the phenol oxidase by fungal and other cell walls. *Journal of Invertebrate Pathology*, **27**, 287–95.
Unestam, T. & Nylund, J. E. (1972). Blood reactions *in vitro* in crayfish against a fungal parasite, *Aphanomyces astaci*. *Journal of Invertebrate Pathology*, **19**, 94–106.
Unestam, T. & Söderhäll, K. (1977). Soluble fragments from fungal cell walls elicit defence reactions in crayfish. *Nature, London*, **267**, 45–6.
Veen, K. H. (1966). Oral infection of second instar nymphs of *Schistocerca gregaria* by *Metarrhizium anisopliae*. *Journal of Invertebrate Pathology*, **3**, 254–6.
Veen, K. H. (1968). Recherches sur la maladie, due à *Metarrhizium anisopliae* chez criquet pelerin. *Mededelingen van de Landbouwhoogeschool te Wageningen*, **68**, 1–77.
Vey, A. (1968). Réactions de defense cellulaire dans les infections de blessures à *Mucor hiemalis* Wehmer. *Annales des Epiphyties*, **19**, 695–702.
Vey, A. & Fargues, J. (1977). Histological and ultrastructural studies of *Beauveria bassiana* infection in *Leptinotarsa decemlineata* larvae during ecdysis. *Journal of Invertebrate Pathology*, **30**, 207–15.
Vey, A. & Quiot, J. M. (1975). In vitro effect of *Metarrhizium anisopliae* toxins on the haemocytic reaction of *Oryctes rhinoceros*, Coleoptera. *Comptes rendus hebdomadaire des séances de l'Académie des sciences*, **280**, 931–5.
Vey, A. & Quiot, J. M. (1976). Action toxique du champignon *Mucor hiemalis* sur les cellules d'insectes en culture *in vitro*. *Entomophaga*, **21**, 275–9.
Vey, A., Quiot, J. M. & Vago, C. (1973). Mise en évidence et étude de l'action d'une mycotoxine, la beauvericine, sur des cellules d'insectes cultiveés *in vitro*. *Comptes rendus hebdomadaire des séances de l'Académie des sciences*, **276**, 2489–92.
Vey, A. & Vago, C. (1969). Recherches sur la guérison dans les infections cryptogamiques d'insectes. Infection à *Aspergillus niger* V. Teigh. chez *Galleria mellonella* L. *Annales de Zoologie – Ecologie Animal*, **1**, 121–6.
Vey, A. & Vago, C. (1971). Réaction anticryptogamique de type granulome chez les insectes. *Annales de l'Institut Pasteur*, **121**, 527–32.
Vincent, J. F. V. (1980). Insect cuticle: a paradigm for natural composites. In *The Mechanical Properties of Biological Materials*. XXXIV Sociсty for Experimental Biology Symposium, ed. J. F. V. Vincent & J. D. Currey, pp. 183–211. Cambridge University Press.
Vinson, S. B. (1977). Insect host responses against parasitoids and the parasitoid's resistance with emphasis on the Lepidoptera–Hymenoptera association. In *Comparative Pathobiology: Invertebrate Immune Responses*, ed. C. A. Bulla & T. C. Cheng, vol. 3, pp. 103–25. New York: Plenum Press.
Wallengren, H. & Johansson. R. (1929). On the infection of *Pyrausta nubilalis* Hb. by *Metarrhizium anisopliae* (Metsch.). *International Corn Borer Investigations, Scientific Reports*, **2**, 131–45.
Walstad, J. D., Anderson, R. F. & Stambaugh, W. J. (1970). Effects of environmental conditions on two species of muscardine fungi (*Beauveria bassiana* and *Metarrhizium anisopliae*). *Journal of Invertebrate Pathology*, **16**, 221–6.

Webster, J. (1980). *Introduction to Fungi*. 2nd Edition. Cambridge University Press.
Whitcomb, R. F., Shapiro, M. & Granados, R. R. (1974). Insect defense mechanisms against micro-organisms and parasitoids. In *Physiology of Insects* vol. V, ed. M. Rockstein, pp. 447–536. New York: Academic Press.
Wigglesworth, V. B. (1957). The physiology of insect cuticle. *Annual Review of Entomology*, **4**, 1–16.
Wilkinson, C. F. (ed.) (1976). *Insecticides, Biochemistry and Physiology*. New York: Plenum Press.
Wynn, W. K. & Staples, R. C. (1981). Tropisms of fungi in host recognition. In *Plant Disease Control*, ed. R. C. Staples, pp. 45–71. New York: John Wiley.
Yanagita, T. (1980). The formaldehyde resistance of *Aspergillus* fungi attacking silkworm larvae. 4. relationship between pathogenicity to silkworm larvae and chitinase activity of *Aspergillus flavus oryzae*. *Journal of Sericulture Science of Japan*, **49**, 440–5.
Yendol, W. G. & Paschke, J. D. (1965). Pathology of an *Entomophthora* infection in the Eastern subterranean termite *Reticulitermes flavipes*. *Journal of Invertebrate Pathology*, **7**, 414–22.
Zacharuk, R. Y. (1970). Fine structure of the fungus *Metarrhizium anisopliae* infecting three species of larval Elateridae. III. Penetration of the host integument. *Journal of Invertebrate Pathology*, **15**, 372–96.
Zacharuk, R. Y. (1971). Fine structure of the fungus *Metarrhizium anisopliae* infecting three species of larval elateridae Coleoptera. IV. Development within the host. *Canadian Journal of Microbiology*, **17**, 525–9.
Zacharuk, R. Y. (1973a). Electron-microscope studies of the histopathology of fungal infections by *Metarrhizium anisopliae*. *Miscellaneous Publications of the Entomological Society of America*, **9**, 112–19.
Zacharuk, R. Y. (1973b). Penetration of the cuticular layers of elaterid larvae (Coleoptera) by the fungus, *Metarrhizium anisopliae*, and notes on a bacterial invasion. *Journal of Invertebrate Pathology*, **21**, 101–6.
Zeobold, S. L., Whisler, H. C., Shemanchuk, J. A. & Travland, J. B. (1979). Host specificity and penetration in the mosquito pathogen *Coelomomyces psorophorae*. *Canadian Journal of Botany*, **57**, 2766–70.

10
The transmission of Dutch elm disease: a study of the processes involved

JOAN F. WEBBER AND C. M. BRASIER

Forest Research Station, Alice Holt Lodge, Wrecclesham, Farham, Surrey GU10 4LH, UK

Key words: Dutch elm disease; *Scolytus*; *Ceratocystis*; beetle–fungus interaction

Introduction

Dutch elm disease first appeared in north-west Europe in the early part of this century (Guyot, 1921), and as a result of the movement of diseased timber quickly spread to neighbouring parts of Europe and to North America (Clinton & McCormick, 1936). By 1927 it was established that the disease was caused by a fungus which invaded the vascular system (Schwarz, 1922; Wollenweber, 1928), but an element of controversy continued to surround the question of disease transmission. Schwarz (1922) considered that the disease was spread by rain and entered through leaf stomata, while others believed it was wind dispersed and invaded through wounds in the bark of otherwise healthy elms (Westerdijk & Buisman, 1929; Smucker, 1935).

The first report to suggest that the disease might be insect transmitted was that of Marchal (1927), who realized that the bark beetles so often found in dead elms might spread the fungus to healthy trees. Many other invertebrates found on elm, including bark weevils and mites, were also proposed as possible vectors (Jacot, 1934, 1936; Collins, Buchanan, Witten & Hoffman, 1936), but the critical evidence that bark beetles were indeed the vectors of the disease came as a result of a series of elegant experiments by C. Buisman and J. J. Fransen in the 1930s involving the caging of beetles on young elms (Fransen, 1931*a*; Fransen & Buisman, 1935). Since then, however, very few of the details of the transmission process have been elucidated.

In recent years with the appearance of new forms of the pathogen, further massive epidemics of the disease have spread across much of Europe, south-west Asia and North America (Brasier, 1979, 1983a). In Britain alone, some 20 million elms died in the decade 1970–1980, showing all too clearly how devastating this combination of beetle vector and pathogen can be on a susceptible host population. Inevitably this has accelerated the search for new control methods and highlighted the current poor state of knowledge of disease transmission. With the future of the elm now under serious threat in many areas (Brasier, 1983a), the need to improve our understanding of the transmission process and its role in maintaining the balance of host and pathogen has become a matter of urgency.

More practical considerations aside, further studies on the transmission of the disease are needed to develop a better conceptual model of this pathogen–vector system, to promote a deeper understanding of its evolution and dynamics, and to provide a basis for comparison with similar pathogen–vector relationships. With this in mind, we have attempted in the present investigation to examine the transmission of Dutch elm disease as a series of ecological processes.

Distribution of elm, pathogen, and vectors

Some 30 species of elm (*Ulmus*) are found throughout the northern hemisphere. Broadly speaking the elms of North America are the most susceptible and those of Eastern Asia the most resistant to Dutch elm disease (Heybroek, 1981). The causal pathogen of Dutch elm disease is the Ascomycete fungus *Ceratocystis* (*Ophiostoma*) *ulmi* which is present throughout the epidemic areas already mentioned, and also occurs at least as far east as Tashkent and Kashmir. Whether *C. ulmi* occurs in Eastern Asia is not known, but this area is often considered to be its likely centre of origin (Heybroek, 1966).

Research over the past decade has shown that *C. ulmi* exists in nature not as a continuum of variation within one population, but as three genetically isolated sub-populations or sub-groups: the more weakly pathogenic non-aggressive strain, responsible for the first epidemics of the disease in Europe and North America; and the Eurasian (EAN) and North American (NAN) races of the aggressive strain, responsible for the current epidemic. Each sub-group has its own characteristic properties and range of variation. The aggressive and non-aggressive strains may be sub-species (Gibbs & Brasier, 1973; Brasier, 1982a, b; 1983a).

The species of bark beetle responsible for spreading the disease varies to some extent from region to region, but all belong to the family

Distribution of elm, pathogen and vectors

160 mm

Fig. 10.1. The larger European elm bark beetle *Scolytus scolytus*. This species is the principal vector of Dutch elm disease in Europe. (Copyright Forestry Commission.)

Scolytidae. In Europe there are two principal vectors, *Scolytus scolytus* (Fig. 10.1) and *S. multistriatus*, the larger and smaller European elm bark beetles respectively. A third species of intermediate size, *S. laevis*, is a vector in parts of northern Europe (Lekander, Bejer-Peterson, Kangas & Bakke, 1977) and has also recently been found in Britain (Atkins, O'Callaghan & Kirby, 1981). Three small bark beetles, *S. pygmaeus*, *S. kirschi* and *S. ensifer*, may also be involved to some extent in central and southern Europe and in south-west Asia (Lanier & Peacock, 1981). In North America the disease is transmitted principally by *S. multistriatus*, which was introduced into the country around 1910 (Chapman, 1910), and to a lesser extent by the native American elm bark beetle *Hylurgopinus rufipes* (Lanier & Peacock, 1981).

The disease cycle and the transmission event

Transmission takes place when beetles emerge from their breeding material in spring and early summer and fly to the tops of healthy elms to feed on the young, sappy bark in twig crotches (Fig. 10.2a). Some of the wounds made during crotch feeding become

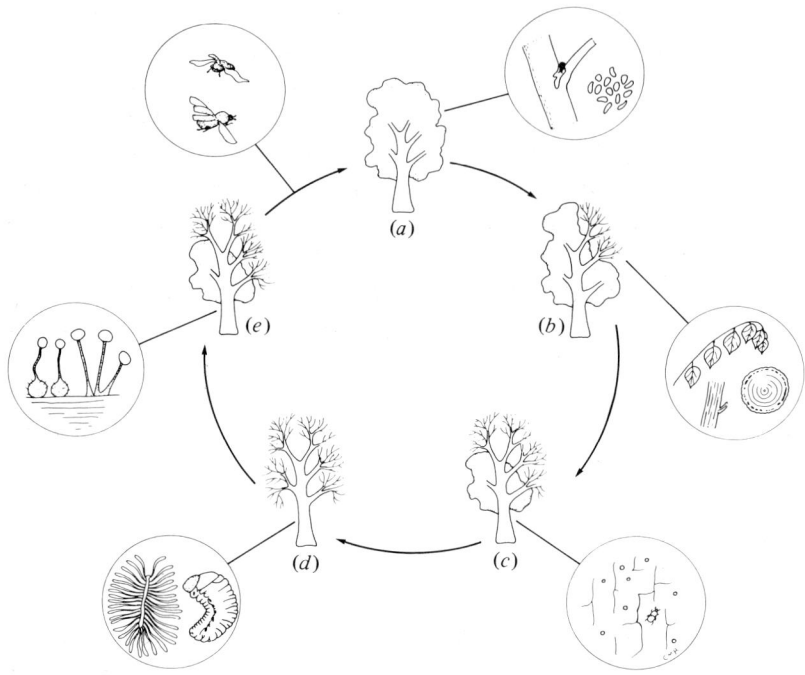

Fig. 10.2. Life cycle of fungus and beetle in Dutch elm disease. (a) Adult beetles emerge in spring and summer from the bark of dead and dying elm, carrying spores of *Ceratocystis ulmi*, and feed in the twig crotches of healthy elms. (b) As a result of beetle feeding the pathogen may be introduced into healthy elms, and infected twigs wilt and show characteristic streaks or spots. (c) Trees weakened by disease become breeding sites for beetles. (d) Beetle larvae cut galleries in the bark. (e) *C. ulmi* fruits in the breeding galleries. (Redrawn from Peace, 1962.)

contaminated with spores of *C. ulmi* carried by the beetles and this occasionally results in the pathogen gaining entry into the host. Such infection leads to the development and spread of the fungus within the xylem vessels of the tree, and a severe wilt usually ensues (Fig. 10.2b).

Once a tree is dead or dying from the disease, its inner bark affords ideal breeding material for the beetle vectors, and female beetles enter the bark to excavate a breeding gallery and lay their eggs (Fig. 10.2c). As the new generation of beetle larvae develops during the autumn and winter the bark also becomes thoroughly colonized by *C. ulmi*, which sporulates heavily during most of the period of larval development (Fig. 10.2d, e). The fungus forms three spore stages in and around the beetle breeding galleries in diseased elm bark (Fig. 10.3): vegetative conidia, produced on the mycelium (Fig. 10.3a, b); synnemiospores, produced in a mucilaginous droplet at the end of a stalk or synnemata (Fig. 10.3c, d); and ascospores, also embedded in a sticky matrix, produced from a long-necked perithecium (the sexual stage) shown in Fig. 10.3e, f. Whereas asexual conidia or synnematospores produced by a single mycelium will be of the same genotype, each ascospore from a single perithecium is likely to be genetically unique.

In the following spring and summer, the larvae pupate and a new generation of young beetles emerges from the bark and undertakes crotch feeding (Fig. 10.2e). A proportion of the beetles carry an inoculum of *C. ulmi* spores, enabling the disease cycle to be continued. In warmer climates, more than one generation of beetles may be produced in a single season.

Gaps in our knowledge of the transmission process

Despite knowledge of the basic Dutch elm disease cycle, too many important questions have remained unanswered, or even unasked, about the various links in the transmission process. It is not known, for example, whether the fungus colonizing the xylem of a tree which is later used as beetle breeding material contributes to the spore inoculum present on the new generation of beetles emerging from that tree. Nor is it sufficiently understood how or when individual beetles obtain their inoculum of *C. ulmi*; or what is the number and type of spores usually carried by emerging beetles and the length of time such spores can retain their potential to cause infection. These and a number of similar questions will now be considered in relation to recent experimental evidence.

Origin of inoculum

Pathogenic and saprophytic phases of C. ulmi

During the annual cycle of Dutch elm disease, apart from the time that *C. ulmi* is present on beetles during their period of flight and twig crotch feeding, the pathogen exists either in a pathogenic phase, involving invasion and spread in the xylem of the host tree, or in a saprophytic phase during which the fungus colonizes the beetle breeding galleries present in the inner bark (phloem) of dying or recently dead trees (Lea, 1977; Gibbs & Smith, 1978). A fundamental issue is whether the spore inoculum carried away from dead trees by newly emerging beetles is derived from the saprophytic or pathogenic phase, or both. To

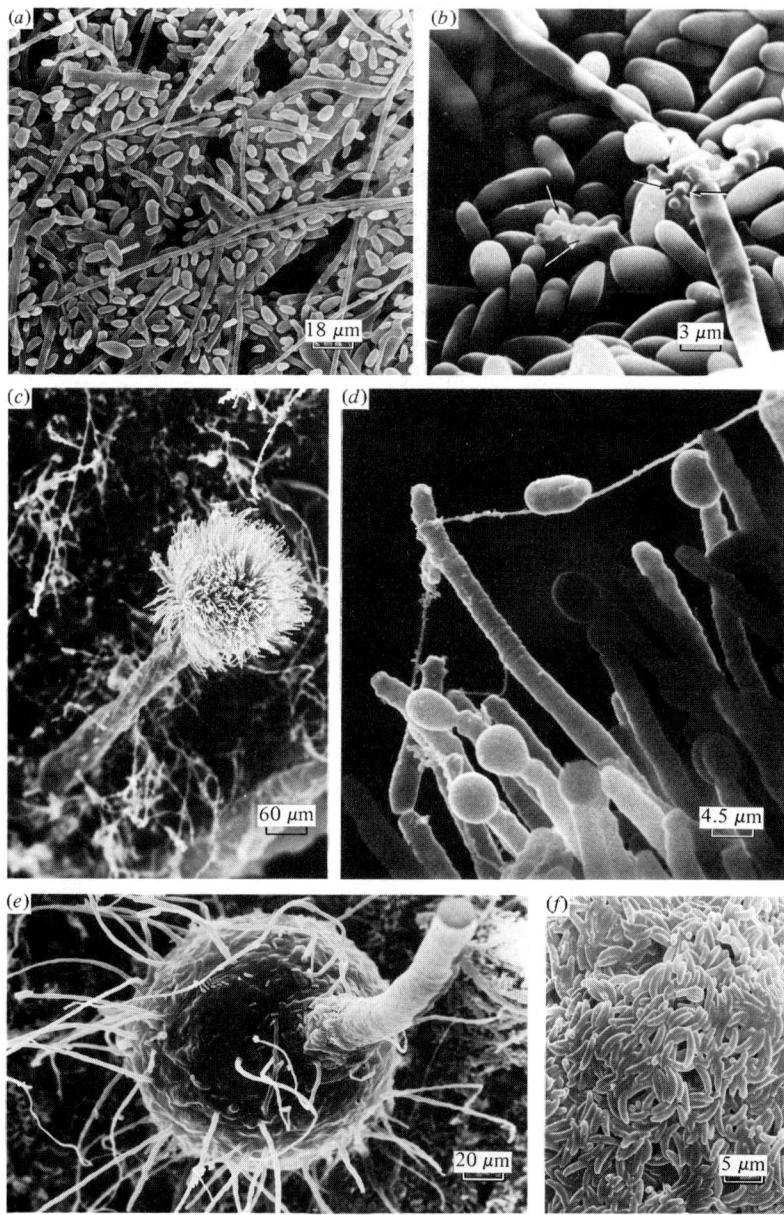

Fig. 10.3. Spore stages produced by *C. ulmi*. (*a*) Vegetative or mycelial conidia (*Sporothrix* form). (*b*) Mycelial conidia and conidiophores with denticles (arrowed) indicating the sympodial development of the spores. (*c*) Synnema. (*d*) Head of a synnema showing synnematospores without the mucilagenous covering, after preparation of the material for the SEM. (*e*) Flask-shaped perithecium. (*f*) Crescent shaped ascospores, extruded from the neck of the perithecium.

examine this problem, the fungal component of the disease cycle will now be considered more closely.

Newly emerged beetles carrying spores of the fungus which first fly to and feed in the tops of healthy trees (Fig. 10.3a) afterwards fly to dead or dying trees to breed, presumably still carrying spores of *C. ulmi*. The colonization of the bark around the beetle breeding galleries is therefore thought to originate mainly from spores brought into the bark by beetles entering to breed (Clinton & McCormick, 1936; Fransen, 1939; Brasier & Gibbs, 1975), and not from the outward spread of 'pathogenic' *C. ulmi* from the xylem. In addition, it is rarely appreciated that twig crotch feeding is not always obligatory to the vectors (e.g. Choudury, 1979). Some beetles, whether as a result of a genetically different behavioural pattern or a response to environment, emerge from the bark of one tree and shortly afterwards enter the bark of the same or another suitable brood tree and excavate breeding galleries without crotch feeding. Spores of *C. ulmi* carried by such beetles must therefore leave the bark of one tree and enter the bark of another without ever being involved in infection of a healthy tree. It could be said, therefore, that *C. ulmi* can exist via a recurrent bark to bark cycle in the saprophytic phase (Fig. 10.4a) without ever being involved in the pathogenic phase of the disease.

trees occurs as a result of beetles crotch feeding *en route* (Fig. 10.4b). This amounts to an expansion of the fungus from the saprophytic to the pathogenic phase. But it has often been supposed that once an individual of *C. ulmi* enters this phase it remains 'locked up' within the xylem, even after the death of the tree, and makes no contribution to the saprophytic phase (Fransen, 1939; Gibbs & Brasier, 1980). If this is the case, then the pathogenic phase is a 'dead end', unable to contribute again to beetle inoculum.

This latter possibility is open to question on evolutionary grounds. Significant pathogenic variation exists within the *C. ulmi* population both at the sub-group level and in terms of individuals within the sub-groups (Brasier, 1982a, b). There is also evidence that the host tends to select more pathogenic genotypes from the population (Lea, 1977). However, unless the 'pathogenic' phase is able to contribute to beetle inoculum, and hence to feed back into the *C. ulmi* gene pool, it is difficult to see how pathogenic fitness can be maintained in the population: i.e. if only a bark to bark gene cycle occurs there is no point at which selection can operate, and maintenance of pathogenic fitness would have to be explained by other factors.

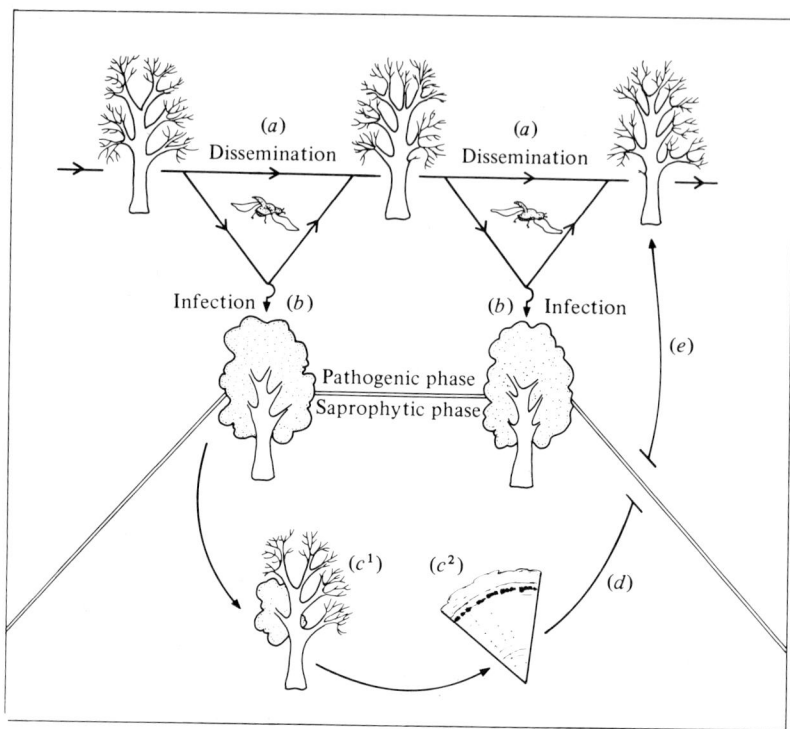

Fig. 10.4. (a) Emerging elm bark beetles contaminated with spores of *C. ulmi* fly to a new source of breeding material. Any twig crotch feeding takes place during this period. Some spores may be lost from the beetles during flight and feeding, but the remainder will be transported from the old breeding material to colonize the new in a simple bark to bark cycle. (b) During beetle feeding, infection of the host tree may occur through some of the feeding wounds. (c^1) External symptoms. Wilting and foliage loss occurs after infection. (c^2) Internal symptoms. Where *C. ulmi* is present in xylem vessels tyloses develop and are seen as dark spots in transverse section. (d) The *C. ulmi* in the xylem of a dying elm remains 'locked up' – the pathogenic phase is a dead end. (e) However, if *C. ulmi* is released from the xylem and contributes to elm bark colonization, *C. ulmi* from the pathogenic phase is fed back into the saprophytic gene pool. (Copyright Forestry Commission.)

Experimental investigation of inoculum origin. To resolve the question of saprophytic versus pathogenic inoculum source an experiment was devised which allowed us to monitor the fate of *C. ulmi* from the time of its infection of the xylem in summer, throughout its winter period as a bark saprophyte associated with beetle breeding, to the time of beetle

Table 10.1. *Percentage of* S. scolytus *beetles carrying wild-type or MBC-tolerant* C. ulmi *inoculum*

Year	Type of C. ulmi *inoculum carried*			
	All wild-type[b]	*Both wild-type and MBC-tolerant*	*All MBC-tolerant*[c]	*No inoculum detected*
1981	56.3%	37.5%	10.4%	2.1%
1982	8.3%	45.8%	35.4%	10.4%

[a] A total of 50 beetles were sampled each year. χ^2 analysis indicated a significant difference ($P < 0.001$) in the relative composition of *C. ulmi* inoculum carried by beetles when 1981 and 1982 date were compared.
[b] Derived from *C. ulmi* introduced by breeding beetles.
[c] Derived from *C. ulmi* in xylem.

emergence in the following spring. An isolate of the aggressive subgroup of *C. ulmi* tolerant to the fungicide MBC (a benomyl derivative) was developed using a simple laboratory procedure (Brasier & Gibbs, 1975). This genetically marked isolate was inoculated into the trunk of a mature elm in the summer of 1980, causing death within a few months and leaving the xylem well colonized by the marked isolate. Logs cut from this tree were then exposed to beetle colonization in the autumn, allowing 'wild-type' (fungicide-sensitive) *C. ulmi* to be introduced into the bark. Any movement of MBC-tolerant *C. ulmi* out of the xylem was traced by periodic sampling from the bark onto a selective agar medium containing the fungicide. Serial dilutions of surface washings from beetles emerging in the spring of 1981 were also tested on the selective medium, to search for evidence of *C. ulmi* inoculum of xylem origin on the beetles. A similar experiment was conducted in the season 1981–2 and some of the results obtained in each year are summarized in Table 10.1.

The periodic sampling of bark revealed that very soon after beetle breeding was initiated there was a steady outward movement of MBC-tolerant *C. ulmi* from the xylem into the bark. This continued during the overwintering period, resulting in the development of a complex mosaic of MBC-tolerant and wild-type individuals of *C. ulmi*. Moreover, the samples taken from emerging beetles showed that some carried inoculum derived entirely from the wild-type inoculum introduced by the breeding beetles, others carried MBC-tolerant inoculum derived from the xylem, and still others carried a combination of both

(Table 10.1). These results demonstrated a strong link or feedback between the pathogenic and saprophytic phases of the fungus.

When the MBC-tolerant isolates from bark and from beetles were analysed for differences in other genetic characters (vegetative incompatibility type and mating type) another interesting discovery was made. A proportion of the isolates were found to be of a different genotype from the MBC-tolerant isolate originally injected into the tree, and many were each of a unique genotype. This could only be as a result of genetic recombination (via the sexual stage of *C. ulmi*) occurring in the bark between MBC-tolerant xylem derived *C. ulmi* and wild-type *C. ulmi*. Therefore not only was there a feedback, but the pathogenic and saprophytic phases were genetically recombined, so that emerging beetles carried a proportion of recombinant inoculum.

There was also a significant, quantitative difference between the two experiments ($P<0.001$) in the contribution of the xylem derived *C. ulmi* to the inoculum; the majority in the 1980–1 experiment came from the wild-type *C. ulmi* introduced by beetles, whereas the majority in the 1981–2 experiment came from the xylem. The strength of the feedback from the xylem phase is therefore likely to vary from season to season, and probably also from tree to tree. One factor which may account for this variation is the ease with which *C. ulmi* escapes from the xylem to colonize adjacent bark. This may be a relatively slow process when *C. ulmi* has to grow out into the cambium through the pits of intact xylem vessels (Krause & Wilson, 1972; MacDonald, 1970; Scheffer & Elgersma, 1982). However during gallery excavation beetles and larvae sometimes groove the wood, especially if the overlying bark is thin, and this may well facilitate a more rapid physical release of the fungus from the xylem. If the inner bark is very thick, on the other hand, galleries may be cut entirely within the bark (Parker, 1939). In certain types and thicknesses of bark, therefore, the rate of release of *C. ulmi* from the xylem into the bark saprophytic phase could vary considerably.

Production of inoculum

In addition to its origin from the saprophytic or pathogenic phases, the quantity and quality of inoculum will also be determined by a number of other interlocking events preceding pupation during the long period of bark colonization by fungus and beetle which begins in autumn. Since this overwintering phase is a major part of the disease cycle lasting anything from eight to ten months (Fig. 10.2), it must have a profound influence on the eventual supply of inoculum to emerging

Production of inoculum

beetles; however its significance and essential features have received little attention during the first 50 years of research on Dutch elm disease.

Development of C. ulmi *in the beetle galleries*

Lea & Brasier (1983) have recently examined the role of fruiting structures of *C. ulmi* in beetle galleries within inner bark and recorded a distinct succession of fruiting structures between the time when beetles enter the bark in autumn and the time when their offspring emerge in spring (Fig. 10.5). Synnemata and mycelial conidia are abundant shortly after beetle gallery initiation whereas perithecial production peaks sharply during the winter. Only the mycelial conidia are abundant in the galleries at the time of beetle emergence. The following series of events therefore appears to take place:

(1) During early colonization of senescent bark by beetles, a primary inoculum of *C. ulmi* brought into the bark by beetles (or, also as we now know, released from the xylem) initiates the primary mycelia which produce abundant synnemata and conidia in the high nutrient environment. This results in a massive secondary inoculum of asexual spores which is dispersed by mites, nematodes and other arthropods which are by this time becoming increasingly abundant in the bark.

(2) Dispersal leads to secondary spread of the fungus in the galleries in bark and a complex mosaic of genotypes of *C. ulmi* is established (Lea, 1977; Brasier, 1984).

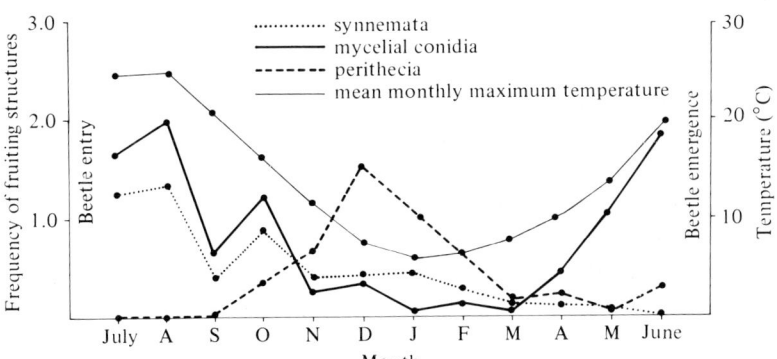

Fig. 10.5. Summary of changes in the frequency of the three spore stages of *C. ulmi* in beetle breeding galleries between beetle entry (left) and beetle emergence (right). (Copyright Forestry Commission.)

(3) Nutrient depletion and the onset of cold winter temperatures lead to a decline in vegetative growth and asexual reproduction. The mycelium and synnemata become degraded and predated by antagonistic microorganisms and arthropods.
(4) Abundant perithecia are formed and assist survival over the winter period. The perithecia are in turn destroyed by degradation and predation. The remaining ascospores and conidia surviving from the secondary inoculum contribute to a massive tertiary or resting inoculum. The beetle mother galleries and surrounding bark are saturated with these spores, which may facilitate further colonization of the bark by *C. ulmi* mycelium in the spring. This second phase of mycelial growth leads to a resurgence in production of mycelial conidia in the galleries.

What Lea & Brasier (1983) propose therefore is that a complex interplay of growth and fruiting precedes larval pupation, the dynamics of which is likely to have a major impact on the quality of inoculum received by emerging beetles. One of the most important aspects is the observation of the winter flush of perithecia. This is probably the event that is responsible for much of the genetic recombination of xylem origin and beetle origin *C. ulmi*.

Arthropod grazing pressure. The grazing of mycelium and spores and the predation of synnemata and perithecia is likely to reduce the quantity of *C. ulmi* in bark considerably. Indeed Lea & Brasier (1983) suggest that by the end of the winter period when most of the perithecia have disappeared (Fig. 10.5) much of the fungal biomass of *C. ulmi* and also that of competing saprophytes may have been diverted into bark microfauna. The melanization of the synnemata and perithecia of *C. ulmi* and the chaetae on the perithecia may serve to delay predation until spore formation has been secured (Brasier, 1978).

Grazing is probably carried out mostly by mites, and large numbers of both mycophagous and predatory mite species commonly inhabit beetle breeding galleries (Jacot, 1934, 1936; Fransen, 1939; Brasier, 1978; Doberski, 1980). The mite *Tyrophagus putrescentiae*, which is often plentiful in beetle galleries in Britain, has been observed to suck ascospores from the tips of perithecia by standing on its rear legs (Brasier, 1978). Other likely consumers of *C. ulmi* are nematodes, Collembola and various dipteran larvae. Nematodes are particularly abundant in breeding galleries, and some species also act as parasites of the scolytid vectors (Finney, 1974; Finney & Walker, 1977; Tomalak & Welch, 1982).

The effect of arthropods in reducing *C. ulmi* biomass is likely to be offset to some degree by their activity in smearing *C. ulmi* spores over the gallery and bark crevices. Fast running predaceous mites in particular are frequently covered in *C. ulmi* spores, including ascospores (Fig. 10.3*f*; Brasier, 1978), and must aid distribution and survival of both secondary and tertiary inoculum of *C. ulmi* and cause further intermixing of *C. ulmi* genotypes. In their latter role mites are probably important in the fertilization of *C. ulmi* protoperithecia and hence in the development of perithecia (Brasier, 1978).

Microbial antagonism. Apart from *C. ulmi* the bark is colonized by numerous other microorganisms. A factor likely to influence the above events and therefore ultimately the quantity of available inoculum is competition of these microorganisms with *C. ulmi*. Little is known about the effects of bacteria and actinomycetes, but Gibbs & Smith (1978) and Webber (1979) have considered the effects on *C. ulmi* of fungi from bark of *U. procera* and *U. glabra* respectively.

Webber (1979) found over 40 species of fungi in elm bark, categorized them according to their frequency at different stages of bark deterioration (healthy, senescing, long dead etc.), and investigated their potential to act as primary and secondary antagonists of *C. ulmi*. In log inoculation experiments four common bark inhabitants, *Phomopsis oblonga, Botryosphaeria stevensii, Nectria coccinea*, and *Fusarium solani*, showed as good, or even better, primary colonizing ability as *C. ulmi*. These fungi are therefore potentially strong competitors of *C. ulmi* during early colonization of bark, and, using the terminology of Gibbs & Smith (1978), are potential primary antagonists able to reduce the area of bark occupied by *C. ulmi*. *P. oblonga* in particular is a major primary colonizer of diseased elm bark (Webber & Gibbs, 1984), able not only to prevent establishment of *C. ulmi* but also to prevent successful beetle breeding (Webber, 1981, 1982).

All 40 bark inhabitants were screened for antagonism towards *C. ulmi in vitro*, and many exhibited volatile or non-volatile antibiotic activity against *C. ulmi*, whilst others showed direct parasitism and replacement. However, when tested on elm bark *in vivo* most fungi, including even potent antibiotic producers such as *Trichothecium roseum*, proved ineffective antagonists of *C. ulmi*. Only the three species which had shown physical parasitism and replacement of *C. ulmi in vitro* were effective antagonists *in vivo*. These three species, *Trichoderma viride*, *T. polysporum*, and *Gliocladium roseum*, are therefore potentially important secondary antagonists of *C. ulmi* and able to replace it in bark. This replacement may not mean elimination however, as *C. ulmi* can still be recovered from bark chips using a selective medium. It seems to suggest that a proportion of *C. ulmi* spores can survive secondary antagonism even though the mycelium may not.

Self-inflicted antagonism. Evidence is now emerging that competition and disturbance of *C. ulmi* in bark may come from a rather surprising source: from *C. ulmi* itself. Hyphal fusion between colonies of *C. ulmi* is regulated by vegetative (somatic) compatibility (v-c) genes, and there are a large number of different v-c groups of this fungus. Colonies of these different v-c groups have recently been shown to antagonize and replace each other in culture (Brasier, 1984). Such antagonism is also likely to operate during bark colonization by *C. ulmi* and to have a role in the establishment of the mosaic structure of the population.

In addition a cytoplasmically transmissible infectious particle, the 'd-factor', has recently been found, which markedly reduces the vigour of *C. ulmi* mycelium, and is spread between mycelia during the bark phase. Both the above phenomena are likely to influence the ability of different genotypes of *C. ulmi* to survive the bark phase to the time of larval pupation and to contribute to beetle inoculum (Brasier, 1983*b*).

Acquisition of inoculum

The type of inoculum received by beetles, whether of mycelial conidia, synnematospores, or ascospores, depends on the fruiting structures of *C. ulmi* available. However, it would be misleading to assume that fruiting structures formed in beetle breeding galleries contribute directly to the inoculum acquired by emerging beetles. This is because only the growing larvae are ever likely to come into contact with fruiting structures in galleries, and each larva, once it reaches its final instar, excavates a pupal chamber which is then physically cut off from the rest of the gallery system by a dense plug of frass.

Within the pupal chamber, larvae undergo their metamorphosis into adult beetles, losing their outer coat and intestinal lining in the process and hence losing any spore inoculum, internal or external (Fig. 10.6*a, b*), carried over from the larval stage (Fransen, 1939). After metamorphosis the adult beetles emerge by boring their way directly

Acquisition of inoculum 285

outwards from the pupal chamber to the exterior (Fig. 10.7). They do not, as is sometimes supposed, have to make their way through the old galleries containing fruiting *C. ulmi* to do so. The bulk of the inoculum acquired by beetles is therefore largely dependent on what *C. ulmi* fructifications, if any, are present in the pupal chambers. Colonization of these chambers is likely to result from spores being brought into the chambers by larvae and the larval coat that is shed during pupation (Fig. 10.6*b*, *c*), by mites and nematodes (Jacot, 1936; Brasier, 1978), and from mycelium growing into the chamber from the surrounding bark.

The freshly cut pupal chamber provides a good environment for further sporulation of *C. ulmi*. However, since the inner bark (phloem) is often of a higher moisture and nutrient content than the outer bark, the position in which pupation takes place within the bark is likely to influence the vigour of *C. ulmi* sporulation in the pupal chamber and hence the amount of inoculum available. The position chosen for pupation is itself likely to depend on bark thickness, on climate, and also vary from one *Scolytus* species to another. Thus with English elm in Britain, the thickness of outer bark on the main stem may vary from 8–25 mm according to age. In older elms many *Scolytus scolytus* larvae pupate entirely within this thicker outer bark layer, whereas in young trees pupal chambers are often cut in the inner bark or across the inner bark/outer bark interface (Fig. 10.7).

To examine the possible effect of pupal chamber position on inoculum, the proportion of occupied pupal chambers containing sporulating *C. ulmi* was compared in inner and outer bark, and in three different elm species (Table 10.2). The trees chosen were of a similar age and size, and bark was taken from the main stem. Most of the pupal chambers were found to lie either horizontally within the outer bark, or tangentially across the inner bark/outer bark interface (Fig. 10.7). Pupal chambers within the moister inner bark were scarce, but a much higher proportion of these contained fruiting *C. ulmi* irrespective of elm species. Where pupal chambers lay across the inner bark/outer bark interface, the part of the chamber in the inner bark was more likely to be carrying *C. ulmi* fructifications. Beetles emerging from pupal chambers in inner bark or across the inner bark/outer bark interface are therefore more likely to carry a significant *C. ulmi* inoculum load.

Some pupal chambers may even be cut in the underlying sapwood rather than in the bark. In this situation the chambers are termed pupal cells (Kirby, 1980; Kirby & Fairhurst, 1981). Pupal cells are only formed

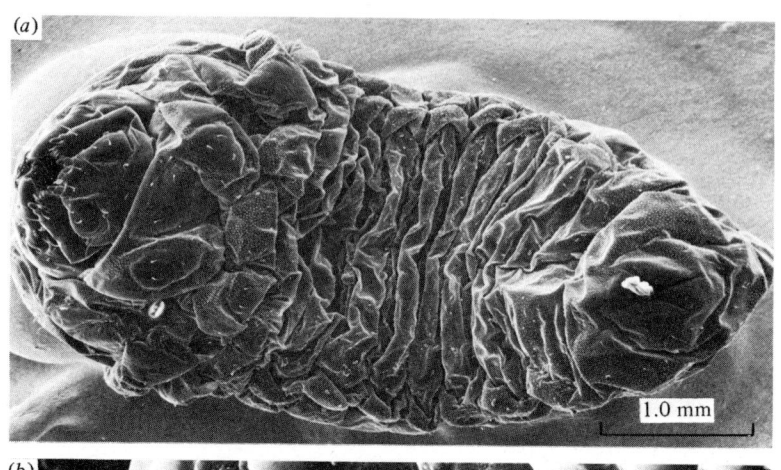

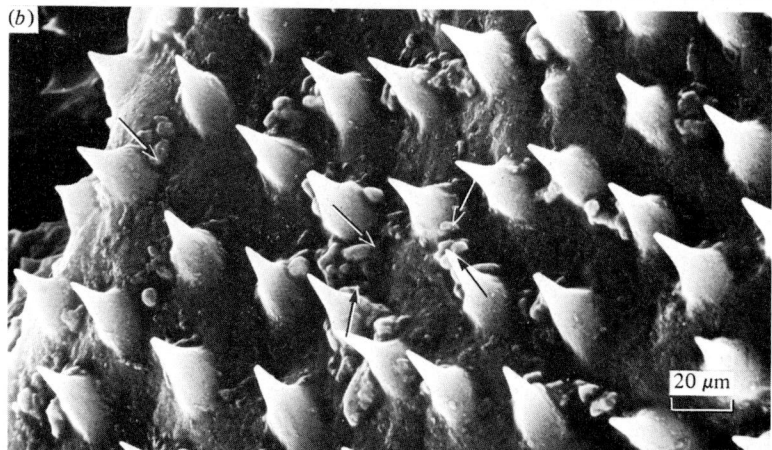

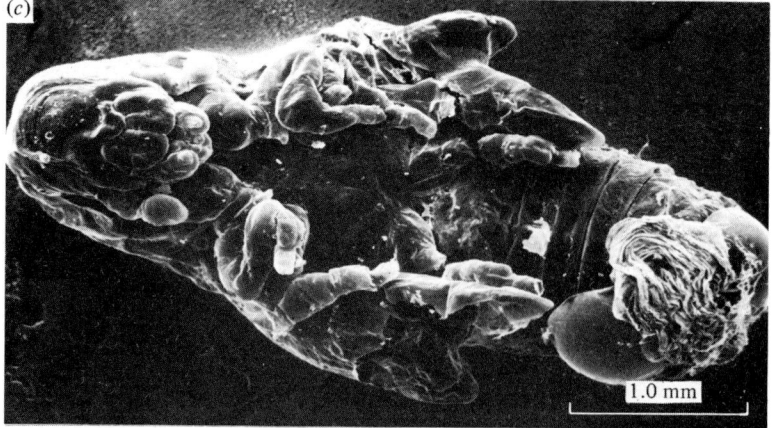

Table 10.2. *Effect of pupal chamber position on* C. ulmi *sporulation*

		Mean % of pupal chambers containing sporulating C. ulmi	
Elm species	Number of pupal chambers examined	Inner bark[a]	Outer bark
U. carpinifolia × sarniensis (Wheatley elm)	118	73.2 (7.5 mm)[b]	6.9 (6.60 mm)
U. procera (English elm)	60	77.8 (8.0 mm)	32.7 (14.0 mm)
U. carpinifolia (Smooth-leaved elm)	88	67.0 (7.5 mm)	8.0 (3.5 mm)

[a] Inner bark constitutes the phloem.
[b] Mean bark thickness given in parentheses.

Samples were taken in May and June 1980. Three trees were sampled of each elm species.

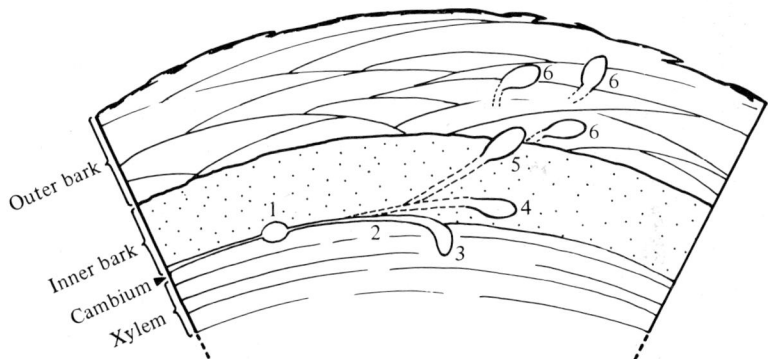

Fig. 10.7. Position of pupal chambers in elm bark. 1. Transverse section through mother gallery. 2. Longitudinal section through larval gallery. 3. Pupal cell excavated in the xylem. 4. Pupal chamber excavated wholly in the inner bark (phloem). 5. Pupal chamber excavated partly in inner and partly in outer bark. 6. Various positions observed of pupal chambers excavated in outer bark only. (Copyright Forestry Commission.)

Fig. 10.6. Developmental stages of *S. scolytus* beetle. (*a*) Ventral view of final instar larva. (*b*) *C. ulmi*-like conidia (arrowed) adhering to larval surface. (*c*) Ventral view of pupa, with larval coat and head capsule shed during pupation still visible. The latter may act as a source of *C. ulmi* inoculum which contributes to pupal chamber colonization.

occasionally by *S. scolytus*, and have not been observed in *S. multistriatus*. With the more northerly distributed *S. laevis*, the majority of pupal chambers are cut in the sapwood, suggesting that this behaviour is associated with climate (Kirby & Fairhurst, 1983). Whatever the cause, fruiting by *C. ulmi* is likely to be sparser in pupal cells because of the lower nutrient status (and possibly lower moisture content) of the sapwood compared with the bark. However, beetles boring outwards from pupal cells may come into contact with other fruit bodies of *C. ulmi* within the bark, in particular those which are sometimes formed in the space resulting from the loosening of the inner bark/outer bark interface during the spring and summer (J. F. Webber, unpublished observations).

Larvae generally excavate their pupal chambers from late winter (February/March) to early summer, and the time available for sporulation by *C. ulmi* in the pupal chamber may vary considerably depending on whether pupation and emergence is rapid or prolonged. To investigate this aspect, the frequency of fruiting structures of *C. ulmi* in occupied pupal chambers of *S. scolytus* was studied throughout a period of beetle emergence (Table 10.3).

In May, half the pupal chambers examined contained sporulating *C. ulmi*, most of these being mycelial conidia. By June, the number of pupal chambers carrying sporulating *C. ulmi* had increased to 64% and synnemata were now the major fruiting structure present. By July and August although the number of occupied chambers remaining was scarce, all contained sporulating *C. ulmi*, with both mycelial conidia and synnemata being plentiful; perithecia had also appeared in some of the chambers. These data suggest that the most frequent form of inoculum carried by beetles will be conidia, occasionally augmented by ascospores. They are in keeping with the classical observations of Wollenweber & Stapp (1928) and Fransen (1931*b*), who also concluded that synnemata were common in pupal chambers and that perithecia were only occasional, although these authors overlooked the occurrence of mycelial conidia.

These data also indicate that beetles emerging later in the season may carry a more abundant and varied spore inoculum. However, this may not be due only to a longer period of pupal chamber colonization by *C. ulmi*, since the later emerging beetles also tended to originate mainly from chambers in the inner bark where, as we have already indicated (Table 10.2), sporulation is likely to be more profuse. Nonetheless, the early dominance of mycelia conidia in the pupal chambers, followed

Table 10.3. Production of fruiting structures by C. ulmi in pupal chambers

Sampling time	Number of pupal chambers examined	% Pupal chambers in inner bark	% Pupal chambers containing sporulating C. ulmi	% Pupal chambers containing		
				Mycelial conidia	Synnemata	Perithecia
May	97	8.3	49.5	46.6	12.2	0
June	80	43.8	64.3	28.3	50.0	0
July/August	21	85.7	100.0	85.0	100.0	28.6

later by synnemata and finally perithecia (Table 10.3), also suggests that at least in inner bark a fruiting succession may occur in the pupal chambers similar to that occurring earlier in the breeding galleries (Fig. 10.5). The two successions are not unrelated events, since the earlier cycles of growth and sporulation in the bark around the galleries must determine both the quantity and genetic constitution of the *C. ulmi* colonizing and fruiting in the pupal chambers, and hence the quality of beetle inoculum.

Quantity of inoculum

The quantity of inoculum carried by beetles must be at least one of the factors that influence both the success of disease transmission during twig crotch feeding, and the colonization of bark used as breeding material. Many surveys have estimated the numbers of beetles acting as vectors in sampled populations in terms of the presence or absence of *C. ulmi* on individual beetles (Parker, 1939; Hoffman & Moses, 1940; Collins, 1941; Parker, Readio, Tyler & Collins, 1941; Parker *et al.*, 1947; Rankin, Parker & Collins, 1941; Gibbs, Brasier & Burdekin, 1973), but no information on the average spore load per vector insect has ever been published. The number of viable spores carried by individual beetles was therefore investigated over two successive seasons.

Logs with active beetle breeding in the bark were kept in nylon net bags, and collections of beetles were made twice daily during the major period of emergence. The beetles were then macerated individually in 5 ml sterile water. Serial dilutions of each beetle macerate were incorporated into dilution plates and the number of *C. ulmi* colonies present on the plates after incubation used to estimate the number of spores on each beetle. As this method does not distinguish between single spores or clumps of spores, it is likely to give an underestimate.

Fig. 10.8*a, b* compares the spore loads carried by emerging *S. scolytus* in each of the two seasons studied; Fig. 10.9*a, b* compares the spore load carried by *S. scolytus* and *S. multistriatus* emerging in the same season. The number of viable spores present on an individual *S. scolytus* beetle varied considerably, from 1 to well over 20000. An 'average' beetle carried a spore load of some 250–2500 spores, and roughly half of the *S. scolytus* carried 1000 spores or more. However, the numerical range of spores carried by *S. scolytus* was fairly constant from one year to the next, statistical analysis indicating no significant difference.

In contrast the smaller *S. multistriatus* was found to carry far fewer

Quantity of inoculum

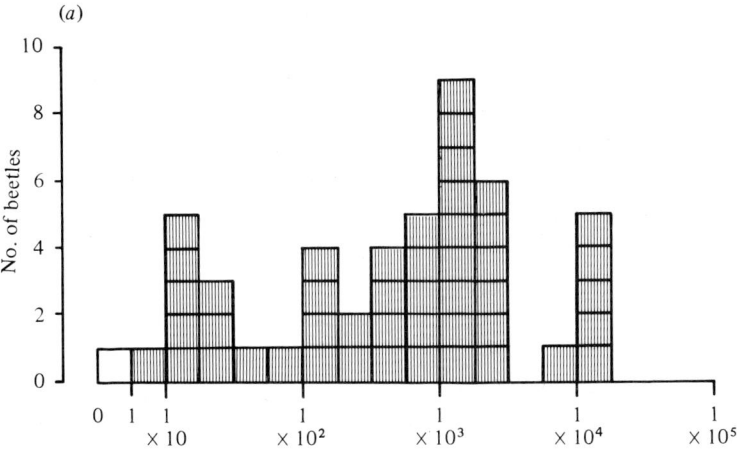

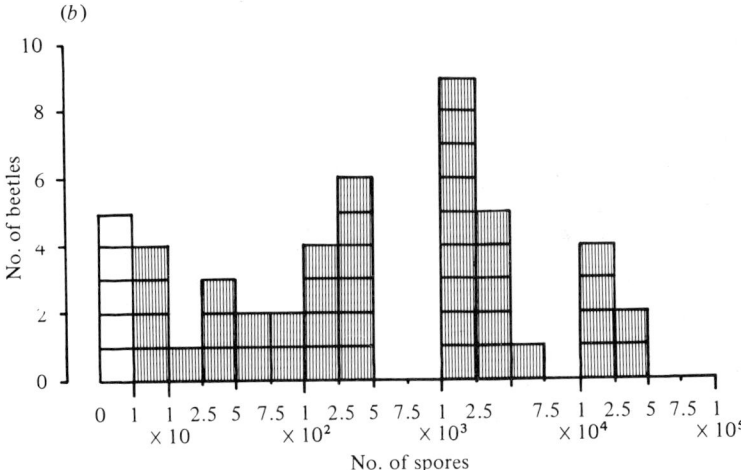

Fig. 10.8. Comparison of the number of spores carried by *S. scolytus* beetles. (*a*) Beetles emerging during the summer, 1981. (*b*) Beetles emerging during the summer, 1982. Each square denotes a single beetle; squares with no hatching indicate beetles with no detectable spore load, hatched squares indicate the beetles with detectable spore loads. (Copyright Forestry Commission.)

spores than *S. scolytus* and this difference was significant ($P < 0.01$) despite the small sample sizes (Fig. 10.9*a*, *b*). If these observations reflect a relationship between inoculum load and beetle size (in particular, beetle surface area), then it seems likely not only that most

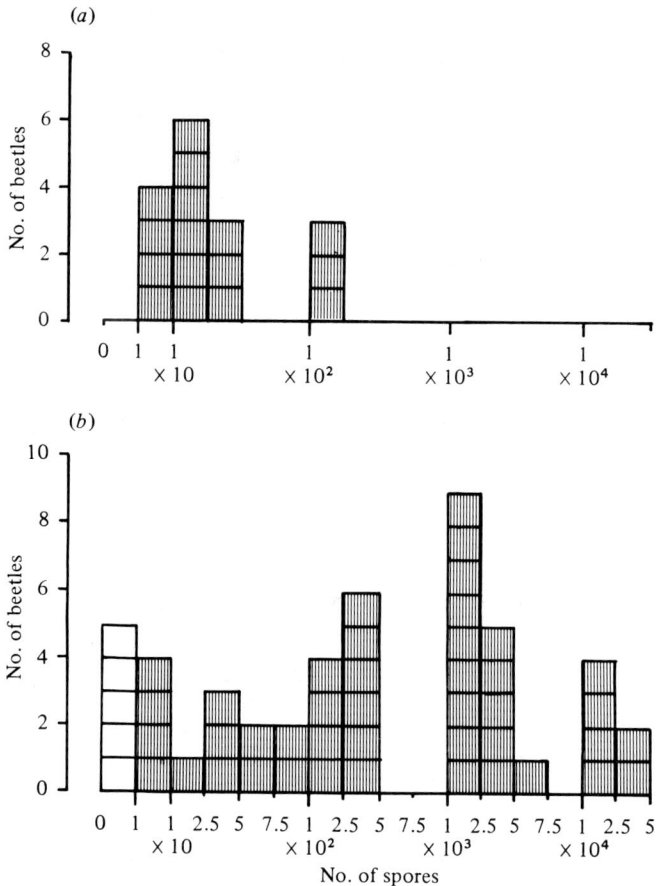

Fig. 10.9. Comparison of the number of spores carried by two species of elm bark beetle. (a) *S. multistriatus* emerging during summer, 1982. (b) *S. scolytus* emerging during summer, 1982. Both species of beetle emerged from the same breeding material. Further explanation as in Fig. 10.8. (Copyright Forestry Commission.)

individual *S. multistriatus* will be less efficient vectors than *S. scolytus* but that the even smaller scolytids such as *S. kirschi*, *S. pygmaeus*, and *S. ensifer* which are thought to be vectors in some European and Asian countries will be even less effective. If approximate body surface areas for *S. scolytus*, *S. multistriatus*, and *S. pygmaeus* are calculated from their average lengths of 4.5, 2.7 and 2.0 mm (Lanier & Peacock, 1981) (also using the assumption that body length equals four times body depth), the resulting surface areas are 22.5, 7.3 and 4.0 mm² respectively. On this basis *S. scolytus* would be three times more effective as a vector than *S. multistriatus* and 5.5 times more effective than *S. pygmaeus*. If one also assumes that a minimal or 'threshold' number of spores is usually required to initiate an infection (say 100 spores), then the relative effectiveness of *S. scolytus* is likely to be far greater. Such an inoculum threshold might sometimes be required to overcome competition with other microorganisms carried by the beetles (Stillwell, 1977).

Furthermore, it should not be assumed that the inoculum load remains constant after emergence. When Parker (1939) compared the number of beetles carrying *C. ulmi* just before emergence with a number carrying *C. ulmi* that had already emerged and were crawling on the bark surface, his results indicated a halving of the number of beetles acting as vectors after emergence (Table 10.4). We have also recorded at various sites a similar, if not greater, decline in the number of beetles acting as vectors. In two separate investigations (Table 10.4) when beetles were removed directly from pupal chambers, 80 and 98% were found to be contaminated with *C. ulmi*, whereas only 23 and 10% of beetles removed from twig crotches at the same sites were found to be contaminated. This represents a drop of 68 and 90% in the proportion of beetles carrying the fungus in each case.

Thus during flight and twig crotch feeding the number of beetles able to act as vectors is likely to fall dramatically. Presumably, the longer the emergent beetles take to begin crotch feeding or to enter fresh breeding material, the smaller the proportion of their original inoculum that remains viable and hence potentially transferable. Emerging beetles which convey very few spores may well lose much or all of their inoculum before the opportunity for transfer arises. Factors likely to reduce inoculum viability include desiccation, high temperature, and sunlight, and conidia and ascospores may vary in their ability to withstand such factors. Many spores may also simply be shaken or 'rubbed off' beetles during flight or feeding activity.

The infection process

The remaining step to be considered in the transmission of Dutch elm disease is the infection process via the beetle feeding groove. As the chink in the defensive armour of the elm through which xylem infection takes place, the feeding groove is the springboard of the pathogenic phase of the disease cycle. It may also be an important genetic sieve. Despite its significance however, the infection process has

Table 10.4. *Percentage of beetles contaminated with C. ulmi before and after emergence*

Year	Species	% Beetles contaminated with C. ulmi	
		Taken from pupal chambers	After emergence and flight
1938 (from Parker, 1939)	S. scolytus	56.5 (329)[a]	25.7 (412)
1980	S. scolytus	80.4 (46)	23.0 (74)
1981	S. scolytus	97.9 (48)	10.2 (69)

[a] Numbers in parentheses are those used to calculate the percentages.

received little critical research since the time of Fransen (1939) and Parker *et al.* (1941, 1947), and is frequently glossed over in both reviews and text-book accounts.

Feeding grooves vary in their structure and situation. Larger beetles such as *S. scolytus* may sometimes cut extensive and even branched horizontal grooves of 2–4 cm in length by tunnelling between the bark and the wood, particularly when several beetles feed in succession (Fransen, 1939). Smaller beetles often drill vertical holes. Grooves are commonly cut in the crotches of the junction between one- and two-year-old stems, but also occur on almost any roughened patch of bark so long as the bark is young and sappy (Webber & Kirby, 1983). The function of crotch feeding is not certain, but in some beetle species it may aid sexual maturation (Choudury, 1979). Beetles commonly aggregate for feeding in the tops of trees, and feeding grooves may be used as the site for mating (Svirha & Clark, 1980; Svirha, 1982). The grooves are usually occupied by beetles for up to several days, sometimes for more than a week.

To test the overall efficiency of the infection process studies were carried out at several disease sites in England between 1979 and 1982. In these investigations the proportion of beetles at a site carrying *C. ulmi* was assessed initially by removing beetles from feeding grooves and plating them on to a selective agar medium. The frequency of colonization of feeding grooves by *C. ulmi* was then determined by making isolations directly from the feeding groove tissues, and the frequency of resulting infection assessed by making isolations from the xylem a few cm below the feeding groove. The latter observations were supported by visual assessment of internal disease symptoms (xylem streaking). The results are summarized in Table 10.5.

Quite a high proportion of the feeding grooves, 30% or more, became detectably contaminated with *C. ulmi*. This was only slightly less than the number of beetles estimated to be carrying the fungus (Table 10.5). It therefore appears that most beetles which arrive at the feeding groove carrying viable *C. ulmi* are also likely to contaminate the feeding groove with the fungus. The discrepancy in the 1980 Crediton data in this respect may simply be a sampling error, although it could also be that more than one beetle had previously occupied and contaminated a feeding groove.

In contrast the level of resulting xylem infection was much lower, being only one-tenth of the rate of successful transfer of *C. ulmi* from beetle to host (feeding groove). In all four investigations carried out,

Table 10.5. Success of infection

Site and year	% Beetles in feeding grooves contaminated by C. ulmi	% Feeding grooves contaminated with C. ulmi	% Successful xylem infection
Devizes, Wiltshire 1979	47.5 (59)[a]	34.8 (250)	3.6 (165)
Crediton, Devon 1980	23.0 (74)	32.8 (204)	4.5 (155)
Crediton, Devon 1981	ND	30.6 (36)	2.8 (36)
Farnham, Surrey 1982	ND	ND	4.8 (167)

[a] Figures in parentheses are actual numbers examined. ND = No data.

The infection process

less than 5% of the feeding grooves actually led to disease. The greatest constraint on the infection process, therefore, appears to lie in the feeding groove itself. At present one can only speculate on the factors responsible. Much will depend on whether infection results from primary transfer of spores from the beetle surface directly into xylem vessels exposed by beetle feeding, or whether there is an intermediate mycelial phase, with spores of *C. ulmi* first colonizing the feeding groove tissues and then growing secondarily into functional xylem from their new food base. In either case the extent to which the xylem is exposed by feeding may be important.

Secondary infections of the xylem following mycelial colonization of the feeding grooves seems the more likely of the two processes, and this conclusion is supported by further evidence from the 1980 Crediton data that the genotype of the fungus isolated from the xylem beneath the feeding groove was frequently the same as that obtained from the feeding groove itself (J. F. Webber, unpublished data). Certainly the feeding groove is likely to be a good incubation chamber for growth and sporulation of *C. ulmi*, providing both nutrients and a high moisture level. The higher humidity level in the feeding groove may be particularly important for survival of *C. ulmi*, especially in hotter, dryer climates (Kais, Smalley & Riker, 1962). In this regard larger feeding grooves, and especially those that are partially covered-in 'tunnels', may produce the best conditions for colonization by *C. ulmi*.

As in the beetle breeding galleries, beetle feeding grooves often become colonized by other microfungi and by mycophagous animals such as mites. Following contamination, the ability of *C. ulmi* to colonize the feeding groove tissues may therefore again depend upon the extent to which it can compete with potential microbial antagonists. An initial inoculum threshold of *C. ulmi* spores (for example 100 spores) may sometimes be required to establish the fungus in the face of such microbial competition. Beetles with higher inoculum loads are therefore more likely to initiate feeding groove colonization by *C. ulmi* and hence to cause xylem infection.

The quality of inoculum is another factor likely to influence the success of infection from the feeding grooves. Clumps of mycelial conidia or synnemata spores carried into the feeding groove on the beetle surface are more likely to be of the same genotype, and hence to cooperate during colonization. In contrast, ascospores are more likely to be of different genotypes, and in particular of different vegetative-compatibility groups, and their derivative mycelia may well antagonize and compete with each other from the beginning of feeding groove colonization to the time of xylem infection. However, faster growing genotypes may be more likely to reach and infect the xylem in the face of competition both from other genotypes and from other microorganisms. When multiple xylem infections are initiated by different genotypes, the latter may also compete on the basis of their relative pathogenic ability. During the early stages of xylem infection, such competition may determine which genotype colonizes the larger portion of the tree, and hence is recycled back into the *C. ulmi* gene pool when the bark of the tree is colonized by beetles for breeding.

Finally it must be remembered that many if not all of the above processes may be opposed by the host's chemical or physical barriers such as growth inhibiting substances (for example polyphenols and mansonomes), thick bark, narrow xylem vessels, and tylose formation (Elgersma, 1982). In some circumstances these factors may prevent infection altogether.

Discussion

We have attempted to view the transmission of Dutch elm disease not as a spasmodic event centred on beetle twig crotch feeding, but as a continuum of interdependent processes, some of the more important of which occur long before the infection event in the feeding groove. The processes are those involved with the *C. ulmi* cycle of colonization, growth, reproduction and dispersal. They are in turn influenced by many biotic and abiotic factors including climate, competitive microorganisms, vector and host species and *C. ulmi* genotype.

During its long overwintering saprophytic phase associated with beetle breeding galleries in elm bark, *C. ulmi* must survive the depredations of climate, nutrient depletion, antagonistic microorganisms and animal predation. The fungus' strategy appears to be to produce such a large biomass that, although much of it is probably destroyed, enough remains to promote the vital process of beetle contamination in the spring. On present evidence the following broad sequence of events takes place.

Firstly, rapid mycelial colonization of large areas of bark around the breeding galleries, enhanced by the production of asexual spores (secondary inoculum) dispersed by mites and nematodes.

Secondly, saturation of the bark and galleries by a massive quantity of sexual and asexual spores, providing a resting or tertiary inoculum which assists survival over the winter period.

Discussion

Thirdly, recolonization of the bark, including newly exposed areas of inner and outer bark, at around the time of larval pupation in spring or early summer.

Lastly, provision of the bulk of the inoculum for emerging beetles from fruit-bodies produced in the pupal chambers.

Since a large proportion of pupal chambers contain sporulating *C. ulmi* (Table 10.3), this last process appears to present the fungus with few problems. However, once on the beetles, the biomass of *C. ulmi* is fixed and the fungus is evidently very vulnerable. During the flight period alone the number of beetles acting as vectors of *C. ulmi* may be effectively halved. Thus while 60–80% of beetles in pupal chambers may initially become contaminated with *C. ulmi*, only 10–50% of beetles may arrive at the feeding groove still contaminated (Tables 10.4 and 10.5). Some 30% of feeding grooves may also become contaminated with *C. ulmi*, but only 3–5% of all feeding grooves cut eventually lead to infection (Table 10.5).

Clearly a number of constraints combine to reduce the likelihood of *C. ulmi* from pupal chambers causing a xylem infection. The initial quantity of inoculum carried by a beetle is therefore likely to be crucial. Assuming that infection is more likely to result from beetles carrying the highest inoculum loads, and taking 5% as the average number of all beetles likely to initiate infection (Table 10.5) and the spore loads of beetles given in Fig. 10.8 as representative, then with *S. scolytus* initial spore loads of $> 100\,000$ may often be necessary for xylem infection to occur. Beetles carrying such large spore loads would be more likely to emerge from pupal chambers in the inner bark or inner bark/outer bark interface than from the outer bark (Table 10.2). If smaller beetles carry smaller inoculum loads, their chances of causing infection may be less. At the epidemic level this beetle size factor may be offset to some extent by other factors, for example by smaller beetles having larger populations or by their having other behavioural characteristics favouring disease transmission (Campana, 1978).

Environmental factors likely to cause a loss of inoculum during beetle flight and crotch feeding have been considered earlier. The flight period, the feeding groove and the infection court are all aspects of the problem which are in need of further research. With only limited information available, we have concluded, along with others (Campana, 1978), that xylem infection is more likely to follow a short period of saprophytic growth by the fungus than to occur directly, although in the absence of further evidence this conclusion must remain tentative. If mycelial growth of *C. ulmi* in the feeding groove does generally precede xylem infection, then the genetic characteristics of the mycelium – its growth-rate, temperature–growth optimum and pathogenicity – all of which vary enormously within and between the strains and races of *C. ulmi* (Brasier, 1982*a*, *b*) will be crucial to infection success. Where more than one genotype of the fungus is present, their vegetative-compatibility factors may also be important. The possibility of competitive interaction of *C. ulmi* genotypes in the feeding groove and xylem, and the possible effects of other microorganisms, humidity and host resistance on growth of *C. ulmi* in the feeding groove are outstanding problems central to the disease transmission process. The feeding groove could well be the weakest link in the fungal cycle, and might therefore be a good point at which to attack the fungus for disease control purposes.

Pathogenic feedback and the evolution of the pathogen

We have also considered here whether there are not in effect two distinct cycles of *C. ulmi* in Dutch elm disease, one involving the pathogenic phase of the fungus in the xylem and the other an entirely saprophytic bark to bark cycle (Figs. 10.2 and 10.4). Experiments with genetically marked isolates clearly demonstrate the existence of the two cycles. They also show that the *C. ulmi* from the pathogenic phase of the fungus, when fed back into the saprophytic gene pool, is genetically recombined with it before the next generation of beetles emerges. Probably most of this recombination occurs during the heavy flush of perithecial formation that occurs in bark in midwinter (Fig. 10.5). Some may also result from perithecia formed within pupal chambers. The net result is that emerging beetles carry a variable but significant proportion of inoculum derived from the fungus that killed the tree.

This feedback mechanism may well have played a central role in the evolution of this fungus–beetle association. In particular, assuming that crotch feeding was already established as an element of beetle behaviour, competitive survival of more pathogenic genotypes in the feeding groove and during xylem infection, coupled with the feedback mechanism, may have supplied an important ingredient of directional selection to the evolutionary process. Such selection may have fuelled the evolution of a once purely saprophytic *Ceratocystis*–scolytid association, similar to that of blue stain fungi such as *C. piceae*, towards that of being a complex pathogen–vector relationship. At some point the selective reinforcement of twig crotch feeding by the beetle may also have been involved.

Discussion

The *C. ulmi*–scolytid relationship may make a good blueprint for comparative evolutionary study of similar pathogen–vector systems on trees. Many systems have only been poorly investigated and further experimental information will be required before their relative extents of pathogenic selection or feedback can be evaluated. However, a reasonable amount is known about oak wilt (*C. fagacearum*) and this provides an interesting contrast to Dutch elm disease. Since the nitidulid beetle vectors of *C. fagacearum* do not breed in the bark of diseased trees (Gibbs & French, 1980) there is no comparable saprophytic (or bark to bark) cycle of the fungus. Spores on exposed fungus pressure cushions from which inoculum is taken by the beetles are derived either directly from the pathogen in the xylem where conidia are concerned, or from a combination of the xylem pathogen from one tree with that from another where ascospores (from perithecia) are concerned. When the inoculum is composed entirely of ascospores, all the inoculum will be genetically recombinant, but derived entirely from the xylem phase. There is also no equivalent of twig crotch feeding in oak wilt, infection taking place via wounds made by external agencies which are visited by the nitidulid beetles for sap feeding. A study of the pathogenic variation in *C. fagacearum* might indicate the intensity of selection which this more direct transmission system imposes on the fungus.

The fungus–vector relationship of oak wilt is therefore simpler and less specialized than in Dutch elm disease, and the evolutionary development of the two systems is likely to have been very different (cf. Dowding, Chapter 5). With its complex chain of transmission events, the *C. ulmi*–scolytid system could be considered rather sophisticated by comparison and likely to have evolved over a long period of time, certainly from before the first recorded epidemics of the disease occurred in Europe in the early years of this century. The species of *Scolytus* present in eastern Asia – the postulated centre of origin of the disease (Heybroek, 1966) – are unknown, and it remains a matter of speculation which *Scolytus* species may have been involved in the early evolution of the disease. Possibly at the centre of origin the host, pathogen and vector are in a better state of balance than they are in Europe and North America today.

Future of the disease: the role of the transmission process

Recent evidence suggests that the aggressive strain of *C. ulmi* responsible for the present epidemics in Europe and North America may well survive beyond the point at which most of the elms have been destroyed, and continue to attack the seedlings and suckers which arise to replace them. As a consequence, the elm in the medium term may no longer be a major landscape feature, but will largely be reduced to scrub or marginal populations (Brasier, 1983a). In the longer term, however, the greatest hope for the return of the field elm in Europe and North America must be the establishment of a more equal balance between host and pathogen.

Any readjustment in the balance must be effected through the disease transmission processes described here. An interesting feature of the transmission cycle is that it is probably extremely sensitive to small genetic changes in pathogen genotype, host resistance, and beetle behaviour. Looked at as a series of interlocking events, a slight change in the efficiency of one process at one point in the chain may have a compounding effect on many other points. It is in this way that the present elm–*C. ulmi* imbalance may eventually be improved.

In this context it is of interest to look at points in the transmission cycle where the fungus may be particularly sensitive to slight changes in the behaviour of the beetle. A critical factor is the position of the pupal chamber in the bark and its likely effect on the quantity of inoculum. Another critical point is the time of pupation, since delayed pupation and emergence may mean a qualitatively different inoculum, including more ascospores. Beetle behaviour during flight and in the feeding groove (for example length of flight time), and the depth and shape of the feeding groove produced probably all have a significant impact on the survival of *C. ulmi* inoculum. The proportion of the beetle population actually involved in twig crotch feeding will also influence the overall infection rate.

Similarly the fungus must ultimately influence the composition of the beetle population. In particular, beetle populations will be influenced by the strain or race of the fungus present. Indeed, the latter probably has by far the greatest influence on the whole transmission process, since, through killing more trees, a more pathogenic form of the fungus will provide much more beetle breeding material. This will result in a larger beetle population and eventually in an exploding disease situation as seen in present day epidemics. Since a more pathogenic form of the fungus may kill larger trees, rather thicker bark may be available for beetle breeding, leading to an expansion of the population size of larger vector species such as *S. scolytus*, and this in turn may increase the rate of disease transmission. A factor which can oppose this spiralling development is a higher level of host resistance to the fungus, and to the beetle. Thus the fungus and beetle are also sensitive to host genotype.

Acknowledgments. Our thanks are due to Mark Anderson for his help and advice in preparing this manuscript and also to George Gate for photographic assistance. We are grateful to the EEC for provision of a post-doctoral research grant to the senior author.

References

Atkins, P. M., O'Callaghan, D. P. & Kirby, S. G. (1981). *Scolytus laevis* (Chapuis) (Coleoptera: Scolytidae) new to Britain. *Entomologist's Gazette*, **32**, 280.

Brasier, C. M. (1978). Mites and reproduction in *Ceratocystis ulmi* and other fungi. *Transactions of the British Mycological Society*, **70**, 81–9.

Brasier, C. M. (1979). Dual origin of recent Dutch elm disease outbreaks in Europe. *Nature, London*, **281**, 78–80.

Brasier, C. M. (1982*a*). Genetics of pathogenicity in *Ceratocystis ulmi* and its significance for elm breeding. In *Resistance to Diseases and Pests in Forest Trees*, ed. H. M. Heybroek, B. R. Stephan & K. von Weissenberg, pp. 224–325. Proceedings of the Third International Workshop on the Genetics of Host–Parasite Interactions in Forestry, September 1980. Wageningen, Holland: Pudoc.

Brasier, C. M. (1982*b*). Occurrence of three sub-groups within *Ceratocystis ulmi*. Proceedings of the 1981 Dutch elm disease Symposium and Workshop, ed. E. S. Kondo, Y. Hiratsuka & W. B. G. Denyer, pp. 298–321. Winnipeg, Manitoba: Department of Natural Resources.

Brasier, C. M. (1983*a*). The future of Dutch elm disease in Europe. In *Research on Dutch elm disease in Europe*. Forestry Commission Bulletin no. 60, ed. D. A. Burdekin, pp. 96–104.

Brasier, C. M. (1983*b*). A cytoplasmically transmitted disease of *Ceratocystis ulmi*. *Nature, London*, **305**, 220–23.

Brasier, C. M. (1984). Intermycelial recognition systems in *Ceratocystis ulmi*: their physiological properties and ecological importance. In *The Ecology and Physiology of the Fungal Mycelium*, ed. D. Jennings & A. D. M. Rayner. Cambridge University Press (in press).

Brasier, C. M. & Gibbs, J. N. (1975). MBC tolerance in aggressive and non-aggressive isolates of *Ceratocystis ulmi*. *Annals of Applied Biology*, **80**, 231–5.

Campana, R. J. (1978). Inoculation and fungal invasion of the tree. In Dutch elm disease, perspectives after 60 years, ed. W. A. Sinclair & R. J. Campana, *Search (Agriculture)*, **8(5)**, 17–20.

Chapman, J. W. (1910). The introduction of a European scolytid, the smaller elm bark beetle, *Scolytus multistriatus* (Marsh) into Massachusetts. *Psyche*, **17**, 63–8.

Choudury, J. H. (1979). 'Flight activity, flight orientation and elm bolt infestation by *Scolytus multistriatus* (Marsh).' Unpublished Ph. D. thesis, Imperial College, University of London.

Clinton, G. P. & McCormick, F. A. (1936). Dutch elm disease. *Connecticut Experimental Station Bulletin*, **389**, 701–50.

Collins, C. W. (1941). Studies of elm insects associated with Dutch elm disease. *Journal of Economic Entomology*, **34**, 369–72.

Collins, C. W., Buchanan, W. D., Witten, R. R. & Hoffman, C. H. (1936). Bark beetles and other possible insect vectors of the Dutch elm disease. *Journal of Economic Entomology*, **29**, 169–76.

Doberski, J. W. (1980). Mite populations on elm logs infested by European elm bark beetles. *Zeitschrift für angewandte Entomologie*, **89**, 13–22.

Elgersma, D. M. (1982). Susceptibility and possible mechanisms of resistance to Dutch elm disease. Proceedings of the 1981 Dutch elm disease Symposium and Workshop, ed. E. S. Kondo, Y. Hiratsuka & W. B. G. Denyer, pp. 169–77. Winnipeg, Manitoba: Department of Natural Resources.

Finney, J. R. (1974). 'The physiological interactions between nematode parasites and their insect hosts.' Unpublished Ph.D. thesis, University of London.

Finney, J. R. & Walker, C. (1977). The DD-136 strain of *Neoaplectana* sp. as a potential biological control agent for the European elm bark beetle, *Scolytus scolytus*. *Journal of Invertebrate Pathology*, **29**, 7–9.

Fransen, J. J. (1931*a*). Enkele gegevens omtrent de verspreiding van de door *Graphium ulmi* Schwarz veroorzaakte iepenziekte door de iepenspintkevers, *Eccoptogaster (Scolytus) scolytus* F. en *Eccoptogaster (Scolytus) multistriatus* Marsh in verband met de bestrijding dezer ziekte. *Tijdschrift over plantenziekten*, **37**, 49–62.

Fransen, J. J. (1931*b*). Why do the coremia of *Ceratostomella ulmi* occur in the pupa beds of the elm sapwood beetles? Door welke oorzaak onstaande coremia van *Ceratostomella ulmi* in de poppenwiegen van de iepenspintkevers? *Fungus*, **7**, 39–42.

Fransen, J. J. (1939). Elm disease, elm beetles and their control. Iepenziekte Iepenspintkevers an Beider Bestrijding. Unpublished Dissertation, Wageningen.

Fransen, J. J. & Buisman, C. (1935). Infectieproeven op verschillende iepensoorten met behulp van iepen spintkevers. *Tijdschrift over plantenziekten*, **41**, 221–39.

Gibbs, J. N. & Brasier, C. M. (1973). Correlation between cultural characters and pathogenicity in *Ceratocystis ulmi* from Britain, Europe and America. *Nature, London*, **241**, 381–4.

Gibbs, J. N. & Brasier, C. M. (1980). Further studies on carbendazim tolerance in *Ceratocystis ulmi*. *Annals of Applied Biology*, **94**, 273–310.

Gibbs, J. N., Brasier, C. M. & Burdekin, D. A. (1973). Dutch elm disease. *Forestry Commission Report on Forest Research*, 1973, 94–7.

Gibbs, J. N. & French, D. W. (1980). The transmission of oak wilt. *USDA Forest Service Research Paper* NC-185, 1–17.

Gibbs, J. N. & Smith, M. (1978). Antagonism during the saprophytic phase of the life cycle of two pathogens of woody hosts – *Heterobasidion annosum* and *Ceratocystis ulmi*. *Annals of Applied Biology*, **89**, 125–8.

Guyot, M. (1921). Notes de pathologie végétale. *Bulletin de la Société de pathologie végétale de France*, **8**, 132–6.

Heybroek, H. M. (1966). Dutch elm disease abroad. *American Forests*, **72**, 26–9.

Heybroek, H. M. (1981). Elm cultivation. In *Compendium of Elm Disease*, ed. R. J. Stipes & R. J. Campana, pp. 3–5. American Phytopathological Society.

Hoffman, C. H. & Moses, C. S. (1940). Mating habits of *S. multistriatus* and the dissemination of *Ceratostomella ulmi*. *Journal of Economic Entomology*, **33**, 818–19.

Jacot, A. P. (1934). Acarina as possible vectors of the Dutch elm disease. *Journal of Economic Entomology*, **27**, 858–9.

Jacot, A. P. (1936). Three possible mite vectors of Dutch elm disease. *Entomological Society of America*, **29**, 627–35.

Kais, A. G., Smalley, E. B. & Riker, A. J. (1962). Environment and development of Dutch elm disease. *Phytopathology*, **52**, 1191–6.

Kirby, S. G. (1980). Biology of scolytid beetles in relation to Dutch elm disease in northern England. *Forestry Commission Report on Forest Research*, 1980, 60–1.

Kirby, S. G. & Fairhurst, C. P. (1981). Towards an understanding of the biology and ecology of elm bark beetles in northern England. *Arboricultural Journal*, **5**, 243–9.

Kirby, S. G. & Fairhurst, C. P. (1983). Ecology of elm bark beetles in northern Britain. In *Research on Dutch elm disease in Europe*. Forestry Commission Bulletin, no. 60, ed. D. A. Burdekin, pp. 28–38.

Krause, C. R. & Wilson, C. L. (1972). Fine structure of *Ceratocystis ulmi* in elm wood. *Phytopathology*, **62**, 1253–6.

Lanier, G. N. & Peacock, J. W. (1981). Vectors of the pathogen. In *Compendium of Elm Diseases*, ed. R. J. Stipes & R. J. Campana, pp. 14–16. American Phytopathological Society.

Lea, J. (1977). A comparison of the saprophytic and parasitic stages of *Ceratocystis ulmi*. Unpublished Ph.D. thesis, Queen Mary College, University of London.

Lea, J. & Brasier, C. M. (1983). A fruiting succession in *Ceratocystis ulmi* and its significance in Dutch elm disease. *Transactions of the British Mycological Society*, **80** (in press).

Lekander, B., Bejer-Peterson, B., Kangas, E. & Bakke, A. (1977). The distribution of bark beetles in Nordic countries. *Acta Entomological Fennica*, **32**, 1–100.

MacDonald, W. L. (1970). Electron microscopy of elm infected with *Ceratocystis ulmi* (Buism.) C. Moreau. Unpublished Ph.D. thesis, Iowa State University, Ames, Iowa.

Marchal, E. (1927). Rapport sur les résultats des recherches effectuées à la station de Phytopatologie de l'Etait à Gembloux sur la maladie l'orme. *Forestière de Belgique*, **35**, 162–4.

Parker, D. E. (1939). Investigations on the relation of elm insects to the Dutch elm disease in Great Britain and other European countries during the years 1935–38. US Department of Agriculture, Bureau of Entomology and Plant Quarantine, Forest Insect Investigations Report, pp. 1–37. Morristown, New Jersey: USDA.

Parker, K. G., Collins, D. L., Tyler, L. J. Connola, D. P., Ozard, W. E. & Dietrich, H. (1947). The Dutch elm disease. *Cornell Experimental Station Memoir*, **275**, 5–44.

Parker, K. G., Readio, P. A., Tyler, L. J. & Collins, D. L. (1941). The transmission of the Dutch elm disease by *Scolytus multistriatus* and the development of infection. *Phytopathology*, **31**, 548–51.

Peace, T. R. (1962). *Pathology of Trees and Shrubs with special reference to Britain*. Oxford University Press.

Rankin, W. H., Parker, K. G. & Collins, D. L. (1941). Dutch elm disease fungus prevalent in bark beetle infested wood. *Journal of Economic Entomology*, **34**, 548–51.

Scheffer, R. J. & Elgersma, D. M. (1982). A scanning electron microscope study of cell wall degradation in elm wood by aggressive and non-aggressive isolates of *Ophiostoma ulmi*. *European Journal of Forest Pathology*, **12**, 25–8.

Schwarz, M. B. (1922). The twig and vascular disease of the elm. Das Zweigsterben der Ulmen, Trauerweiden und Pfirsichbäume. *Mededelingen uit het Phytopathologisch laboratorium 'Willie Commelin Scholten'*, **5**, 1–73.

Smucker, S. J. (1935). Air currents as possible carriers of *Ceratostomella ulmi*. *Phytopathology*, **25**, 442–3.

Stillwell, M. A. (1977). Microflora associated with elm bark beetle feeding niches suggest biological control of Dutch elm disease. *Canadian Forestry Service Bi-monthly Research Notes*, **33 (3)**, 20.

Svirha, P. (1982). The behaviour of *Scolytus multistriatus* in California. Proceedings of the 1981 Dutch elm disease Symposium and Workshop, ed. E. S. Kondo, Y.

Hiratsuka & W. B. G. Denyer, pp. 395–405. Winnipeg, Manitoba: Department of Natural Resources.

Svirha, P. & Clark, J. K. (1980). The courtship of the elm bark beetle. *California Agriculture*, **34**, 7–9.

Tomalak, M. & Welch, H. E. (1982). *Neoplectana carpocapsae* Weiser (Rhabditoidea, Nematoda) DD-136 as a potential biocontrol agent against *Hylurgopinus rufipes* Eichhoff (Scolytidae, Coleoptera). Proceedings of the 1981 Dutch elm disease Symposium and Workshop, ed. E. S. Kondo, Y. Hiratsuka & W. B. G. Denyer, pp. 14–23. Winnipeg, Manitoba: Department of Natural Resources.

Webber, J. F. (1979). Interactions between bark saprophytes and the Dutch elm disease pathogen *Ceratocystis ulmi*. Unpublished Ph.D. thesis, University College of Wales, Aberystwyth.

Webber, J. F. (1981). A natural biological control of Dutch elm disease. *Nature, London*, **292**, 449–51.

Webber, J. F. (1982). Natural biological control of Dutch elm disease by *Phomopsis oblonga*. Proceedings of the 1981 Dutch elm disease Symposium and Workshop, ed. E. S. Kondo, Y. Hiratsuka & W. B. G. Denyer, pp. 24–35. Winnipeg, Manitoba: Department of Natural Resources.

Webber, J. F. & Gibbs, J. N. (1984). Colonisation of elm bark by *Phomopsis oblonga*. *Transactions of the British Mycological Society*, **82** (in press).

Webber, J. F. & Kirby, S. G. (1983). Host feeding preference of *Scolytus scolytus*. In *Research on Dutch elm disease in Europe*. Forestry Commission Bulletin, no. 60, ed. D. A. Burdekin, pp. 47–9.

Westerdijk, J. & Buisman, C. (1929). De iepenziekte. *Rapport over het ondeszoek verricht op verzoek van de Nederlandsche Heidemaatschappij. Nederlandsche Maatschappij.* Arnham.

Wollenweber, H. W. (1928). Elm blight and its cause, *Graphium ulmi* Schwarz. *Bartlett Research Laboratories Bulletin*, **1**, 26–31.

Wollenweber, H. W. & Stapp, C. (1928). Untersuchungen überdie als Ulmensterben bekannte Baumkrankheit. *Arbeiten aus der biologischen Bundesanstalt für Land- u. Forstwirtschaft*, **16**, 282–324.

11
The interrelationships between microbial entomopathogens and insect hosts: a system study approach with particular reference to the Entomophthorales and the eastern spruce budworm

D. F. PERRY AND G. H. WHITFIELD*

Environment Canada, Canadian Forestry Service, Forest Pest Management Institute, Sault Ste. Marie, Ontario P6A 5M7, Canada

Key words: entomopathogenesis; system modelling; pest management; Spruce Budworm; temperate forest; *Choristoneura fumiferana*; *Zoophthora radicans*; *Nosema fumiferanae*

Introduction

The basis for the management of pest species is derived from an ecological approach to the study of the numerous factors that interact in the system under investigation (Clark, Geir, Hughes & Morris, 1967; Watt, 1968; Clark, 1970; Tummala, Ruesink & Haynes, 1975; Haynes & Tummala, 1976; Barfield & Stimac, 1980; Haynes, Tummala & Ellis, 1980). Modern pest management attempts to integrate the judicial use of chemical and biological agents, with cultural practices, in an effort to decrease damage to a given resource.

In the agrocoenosis the relative simplicity of the plant communities and their cultural and harvesting procedures make integrated management programmes easier to employ than in the forest ecosystem with its heterogeneous stand composition, long-term silvicultural practices and physiographical diversity. Nonetheless, in forestry as in agriculture, successful management depends upon a thorough understanding of both global and specific interactions.

* Agriculture Canada Research Station, Lethbridge, Alberta T1J 4B1, Canada

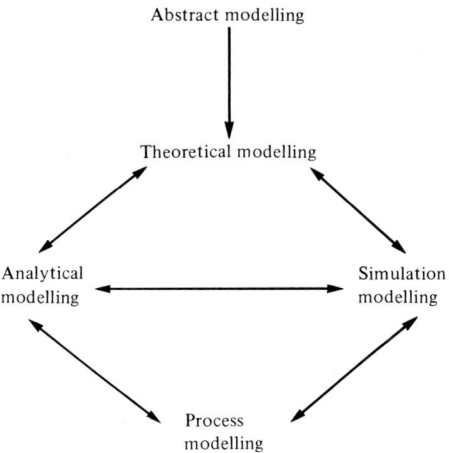

Fig. 11.1. Modelling methods used in systems science.

The systems approach

The broad complexity inherent in any ecosystem necessitates the formation of a conceptual model before significant analysis can be undertaken (Patten, 1972). The combination of investigative research with systems techniques (Pielou, 1969, 1981) should allow a theoretical framework to be drawn up to study host–pathogen interactions, so that an ecosystem can be defined in terms of a large number of biological entities (states) and the accompanying influences of various exogenous factors. This approach has been successfully used to address economic, industrial, social, and biological problems (Tummala, 1974; Koenig, Haynes & Tummala, 1975; Gutierrez & Wang, 1976; Ruesink, 1976; Levins & Wilson, 1980; Haynes et al., 1980). The application of these techniques in the examination of microbial pathogen–insect relations will be discussed in this paper.

Systems science may involve several different methods (Fig. 11.1). Abstract modelling produces a qualitative framework, into which theoretical modelling can incorporate functional elements based on known interactions. Analytical modelling concerns the mathematical description of interrelationships within the system, and simulation modelling provides an output which can be tested against observed biological phenomena. Simulation models often have important predictive capabilities. Process modelling requires increased resolution in terms of biotic and abiotic events so that causal relationships can be established. While in some instances the different forms of modelling

The systems approach

Fungal development

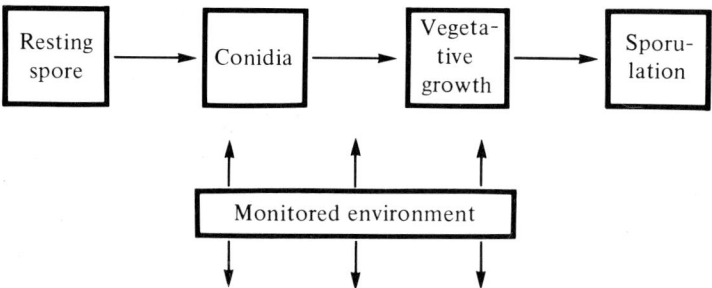

Spruce budworm development

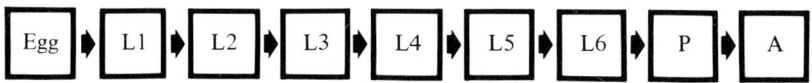

Fig. 11.2. Block model for fungus–budworm development sequence. L1 to L6 larval stages; P, pupa; A, adult.

are distinct units of a problem solving exercise, e.g. pesticide treatment strategies, or environmental impact, the modelling of ecological interactions is more of a feedback process where all stages of modelling are constantly being up-dated and modified as knowledge from each step is obtained.

Abstract and theoretical frameworks are used to delineate components of the ecosystem to be modelled (host–pathogen) and their environment. This provides a system–environment dichotomy relating separated model components to common abiotic factors (Fig. 11.2). Successful simulation of biological processes requires a high degree of understanding and an adequate data base to accurately reflect structure/ function relationships (Getz & Gutierrez, 1982).

Utilization of the modelling approach to the study of the dynamic relationships between organisms has been widespread, and proven useful in describing the interactions between host and animal disease-causing organisms (e.g. Crofton, 1971; Lanciani, 1975; Anderson & May, 1979a, b, 1982b; Perrin & Powers, 1980; Levin & Pimentel, 1981; May, 1983), invertebrate species (Hardman, 1976; Royama, 1977, 1981; Anderson & May, 1980; Régnière, 1982), entomogenous parasites and predators (Miller, 1959; Holling, 1964, 1966; Gage & Haynes, 1975) and

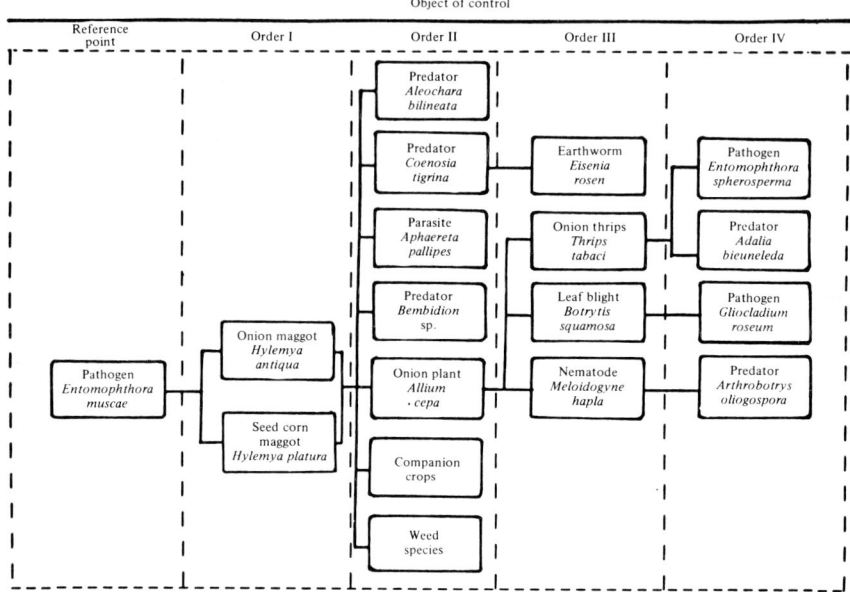

Fig. 11.3. Conceptual model of the object of control for an onion agroecosystem using *Entomopthora muscae* as the reference point.

phytopathogenic fungi (see Kranz, 1974). Such modelling has primarily used a combination of analytical and simulation techniques, where data yield formulae that express basic trends and simulation represents these mathematical equations in a structural sense. In contrast, the less prevalent use of process-oriented models stemming from a detailed knowledge of functional responses to causal agents has offered enhanced insights into microbial–animal relations.

For entomogenous fungi, a simulation model based on the biology and ecology of *Nomuraea rileyi* was developed to predict the incidence of the fungus on the velvet bean caterpillar *Anticarsia gemmatalis* in soybean (Kish & Allen, 1978). An epizootic–epidemiological analysis of *Zoophthora* spp. on the alfalfa weevil *Hypera postica* has also been constructed (Brown & Nordin, 1980, in press; Millstein, Nordin & Brown, 1982). The development of process-oriented simulation models has been undertaken for the study of *Entomophthora muscae* on the onion maggot, *Hylemya antiqua*, and on the seed corn maggot, *Hylemya platura* (Carruthers, 1981). The conceptual model of this study system (Fig. 11.3) draws upon a reference fungal pathogen and describes, in terms of order, the biological interactions. This same approach has been taken in an attempt to model the discrete life stages of the entomophthoralean species, *Zoophthora radicans* (= *Entomophthora sphaerosperma*) and *Entomophthora egressa* (Zygomycotina, Entomophthorales) and the eastern spruce budworm, *Choristoneura fumiferana* (Lepidoptera, Tortricidae) (Perry, unpublished). (*C. fumiferana* has been the subject of an extensive ecological modelling effort – see Cuff & Baskerville, 1982.)

Microbial–insect interactions

Microorganisms influence insect population levels and as such are being studied for integration into pest management schemes. Aside from causing direct mortality, these microbes can reduce fecundity and vigour, curtail feeding, retard development and influence susceptibility to other diseases. To investigate the host–pathogen interactions it is necessary to obtain and analyse biological data on host occurrence and behaviour, and pathogen frequency, and relate these to environmental conditions.

The terminology used to describe microbes that debilitate their supportive hosts is varied: parasite, pathogen and disease are often utilized. Recently two adjunct terms have been employed:

Microparasite, broadly including viruses, rickettsiae, bacteria, fungi, and protozoa. These organisms may induce host immunity in recovered individuals and are further distinguished by diminutive size, rapid regeneration and high reproductivity within the host. *Macroparasites*, notably the parasitic worms (helminths) differ from the above in relative size and do not multiply within their host (Mims, 1976, 1981; May, 1983; Anderson & May, 1981, 1982c).

Entomopathogenic fungi, as microparasites, are distinguished in that little evidence exists for sub-lethality or acquired invertebrate host immunity, although a cellular response has been observed. Furthermore, unlike some of the other microbial pathogens, vertical transmission between generations is unknown and reproductive propagules are formed outside as well as within the host. The microparasites, including the fungi, do, however possess a unified disease pattern based on four developmental phases:

(1) *initial infection*; host–pathogen encounter;
(2) *incubation*; the provision of a suitable habitat for growth of the obligate parasite;

(3) *transmission*; vertical or horizontal;
(4) *persistence*; within and external to the host, in the presence and in the absence of the host, or whenever environmental conditions for the above three phenomena permit occurrence.

The inclusion of environmental parameters, behavioural and immunological host responses, nutritional and allelopathic factors and genetics as components of host susceptibility add to the complexity of the system.

Biological observations – spruce budworm and pathogens

Choristoneura fumiferana is a univoltine diapausing species that occurs throughout north-eastern North America causing defoliation and floral bud damage, primarily to balsam fir (*Abies balsamea*) and spruce (*Picea glauca, P. rubens* and *P. mariana*).

Oviposition of 10–20 egg masses per female (*c.* 200 eggs) takes place in mid-July (Fig. 11.4). Eclosion 10 days later yields first instar larvae (1) which search out crevices to spin up hibernacula, without feeding, soon after moulting to overwinter in the second instar until early May (2–4). Activity in the spring begins with larval dispersal; feeding on current year shoots is begun by third instar larvae and continues into early June (sixth instar), on older foliage if necessary (5). Pupation in mid-July is followed by adult emergence 8–12 days later (6–7).

The spruce budworm is subject to infection by a number of disease organisms, notably *Nosema fumiferanae* (Protozoa, Microsporida), a

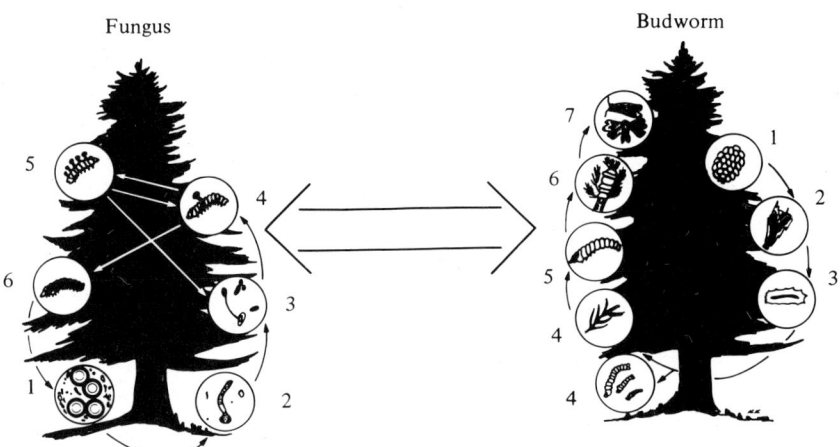

Fig. 11.4. Host–pathogen life cycles.

complex of at least four viruses – entomopox virus (EV), cytoplasmic virus (CV), granulosis virus (GV), and nuclear polyhedrosis virus (NPV) – and several fungal species: *Beauveria*, *Metarhizium*, *Isaria*, *Hirsutella* (Fungi Imperfecti), *Conidiobolus* spp., *Zoophthora radicans* and *Entomophthora egressa* (Zygomycotina, Entomophthorales). Two other microsporidia are reported: *Pleistophora schubergi* and *Thelohania* spp., and other microbial pathogens may also exist.

Nosema is found in all life stages of *C. fumiferana*; infection is manifest in reducing larval feeding and curtailing adult fecundity. Lethal effects are observed when spore levels per larva are high (10^8 spores /5th instar; Wilson, 1974).

Viral infection may often cause slower development, supernumerary moults and mortality. There is no published evidence that spruce budworm virus may be transmitted vertically from adult to offspring, as is the case with the microsporidia *N. fumiferanae*. However, infection may occur during feeding or through oviposition by parasitic insects. Contamination of healthy larvae and reinfection of diseased individuals may occur through ingestion of microspores of *Nosema* or viral propagules.

In contrast, fungi are not transmitted vertically; penetration of the exocuticle during germination phases is the most common mode of infection, although some fungi such as *Metarhizium anisopliae* have been shown to cause infection *per os*, on other hosts. Imperfect fungi occur in far lesser numbers than Entomophthorales species and appear to predominate on pupae, with the exception of *Hirsutella gigantea*. While *Conidiobolus* spp. are found occasionally, occurrence of *Z. radicans* and *E. egressa* on fifth and sixth instar larvae is often spectacular in causing high mortality. Of the above mentioned pathogens, high densities occur sporadically of *N. fumiferanae*, *Z. radicans*, and *E. egressa*. Epizootics of virus seem to be less frequent.

This paper will focus upon the interactions between the spruce budworm and the entomophthoralean fungi.

Entomophthoraceous fungi attacking *Choristoneura fumiferana*

The classical entomophthoralean life cycle consists of the formation of resting spores, hyphal bodies and conidia (Fig. 11.4). *Zoophthora radicans* resting spores germinate in the spring producing primary germ conidia (1–3), which in turn can infect the host through penetration of the cuticle during germ tube formation (4). Conidial germination may also result in the development of different forms of secondary conidia(3). The conidial stage represents the infective propagule, although the respective role of each type of spore in determining infection is unclear.

After penetration of a susceptible host (second to sixth instar) vegetative growth occurs, killing the larva in 2–5 days through rampant invasion of tissues. Sporulation begins 1–2 days after death, resulting in conidiophore production on the surface of cadavers, or resting spore formation within (5–6). Infected insects can be found attached to foliage, or when wind and rainfall cause them to fall from the needles, they may either lodge in lower branches or reach the soil surface.

The resting spores are presumed to overwinter most successfully in the soil although survival of the fungus in the canopy has been observed (Perry, unpublished). Conidial persistence in the soil is also possible. This winter period is necessary to break the dormancy of *Z. radicans* resting spores (Perry, DeLyzer & Tyrrell, 1982).

Entomophthora egressa also produces resting spores, but germination requirements are not yet known. This species also differs from *Z. radicans* in that vegetative growth within the host involves protoplast development, whereas with *Z. radicans*, as in the majority of Entomophthorales species, *in vivo* growth is more commonly through the formation of hyphal bodies.

Although isolated from several insect species, *Z. radicans* and *E. egressa* exist overwhelmingly on *C. fumiferana* where it occurs in relative abundance. Nonetheless, alternate hosts are present within the forest.

Experimentation and microclimate

The determination of development rates, survivorship and attrition (in the sense of mortality and/or inactivation of development at stages of an organism) requires a complex series of experiments if laboratory data are to be extrapolated into the field. This is a necessary step for validation of a process model. The analysis of the effect of microclimate on host–pathogen synchrony must also consider the abiotic parameters important to the interactions.

Microenvironment can be referred to as that portion of the biosphere represented by a soil–plant–atmosphere interface where physical conditions influence and are modified by physiological processes. *Microclimate* is considered in terms of the interrelationships between environment and life-related phenomena using the principles of energy exchange, where continuous energy transfer occurs in the form of radiant energy, sensible heat (conduction and convection), and latent heat (evaporation, transpiration and condensation) (Precht, Christophersen, Hensel & Larcher, 1973). A primary consideration is the fluctuation in conditions that determine the direction, amplitude and rates of energy transfer.

Processes are usually described in terms of the development times for an organism to change from one stage to the next, e.g. moulting, morphogenesis, etc., at a given temperature. Most poikilotherms respond to changes in temperature as chemical reactions, that is, rates generally increase up to a maximum before falling off steeply (Precht *et al.*, 1973). As a result both maximum and minimum thresholds exist. That organisms differ in their response to temperature (both under constant and fluctuating conditions) and not all responses are linear leads to subtle influences on the associations between host and pathogen.

Investigating the effects of temperature can follow a stepwise process:

(1) Determination of rates at constant experimental temperatures while varying the acclimation temperature at which the organism is held during precedent stages.
(2) Determination of rates at fluctuating temperatures with a cycle of known mean, amplitude and frequency.

In both cases, monitoring of field conditions for the determination of experimental limits can aid in designing the experiment.

Constant temperature rates for spruce budworm and *Z. radicans* development have been determined yielding maxima and mimima (Perry, unpublished; Régnière, unpublished; van Roermund, Perry & Tyrrell, 1983). The effect of fluctuating temperatures is being studied. The rates of development determined from the above experiments were used as a basic framework for a process model, as described below.

Choristoneura fumiferana – Zoophthora radicans: **a process model**
Distributed delays

There are many methods used to approximate biological development, each acknowledging that inherent individual variability may be described by a statistical distribution for development rates, even at single constant temperatures (Stinner, Gutierrez & Butler, 1974; Stinner, Butler, Bacheler & Tuttle, 1975; Logan, Wollkind, Hoyt & Tanigoshi, 1976; Sharpe & DeMichele, 1977). This variability can be incorporated into a model through the use of Monte Carlo techniques, binomial expansions and probability density functions (Hardman, 1976; Ruesink, 1976; Whitfield, Drummond & Haynes, 1981).

The processes leading to infection of *C. fumiferana* by *Z. radicans* (Fig. 11.2) were modelled through the use of time varying distributed delays, linked together in series (Manetsch, 1976; Manetsch & Park, 1977, 1981), which has the following advantages:

(1) continuous description of instantaneous as well as cumulative development from one life stage to the next;
(2) accurate approximation of aggregative processes through the incorporation of variability;
(3) precision with processes exhibiting a lag response.

This technique permits aggregates of entities to enter a stage at a point in time according to a rate function based on temperature (Fig. 11.5). Aggregates flow from one stage to the next after a lag period, specific to each entity, so that while individuals may enter the delay at the same time they do not necessarily exit together; departure from the delay is in accordance with a probability function based on inherent variability. Transit times for the movement of individual entities through the delay process are calculated using an Erlang density function. In this way, attrition in the form of inactivation or mortality could be approximated instantaneously for any stage throughout the model (Manetsch, 1976). For each stage, a proportion of individuals per unit time is removed according to a defined attrition function. Through integration over time, stage mortality is assessed.

The linked compartments (Fig. 11.2) represent free-body models of the host, *C. fumiferana*, and the fungus *Z. radicans*. Interconnecting the two organisms necessitated the constraint that the host must be present for infection, incubation and transmission to occur. The density levels of each life stage represent an instantaneous state which is up-dated at each time step. Data for temperature-dependent development for budworm (Bean, 1961; Eidt and Cameron, unpublished) in pre-emergence and post-emergence stages were used (Régnière, 1982) to determine development rate parameters for use in matched asymptote equations (Logan *et al.*, 1976). From these equations, nonlinear temperature dependent development curves for male and female larvae were determined (Fig. 11.5). The relation between total larval development and the proportion of individual instar development time at a constant temperature was used to determine rates for the development of

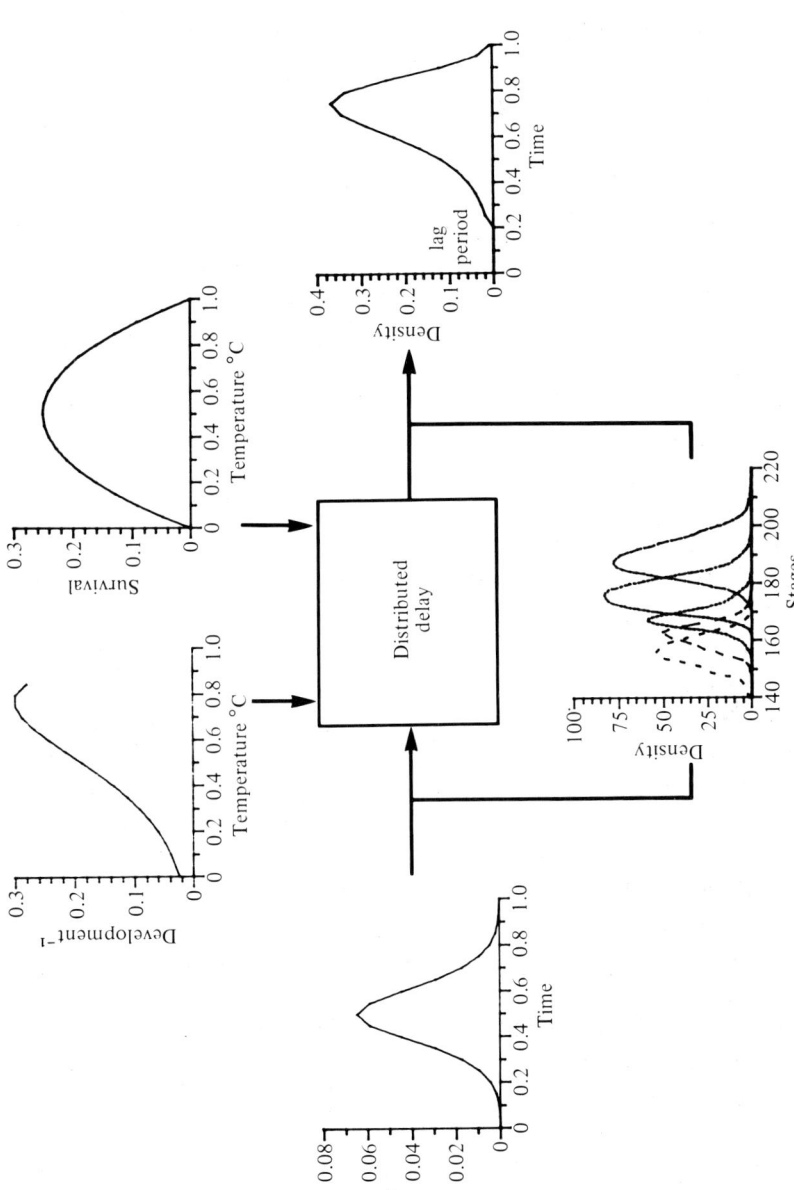

Fig. 11.5. The relationships between environmental variables and biological development in a distributed delay system.

post-emergent stages (second to sixth instar) (Régnière, 1982). For each time increment instantaneous temperature was employed to move budworm aggregates from one larval stage to the next. Transit through the distributed delays is based on a determined mean lag period and an associated variance derived from the Erlang density function. A normal distribution and variance was assumed for the lag period for each biological process.

Heat accumulation – physiological time

A sub-routine for degree-day calculations was used for the purposes of model output presentation. Several methods for this have been used successfully in phenological studies (Eckenrode & Chapman, 1972; Gage & Haynes, 1975; Gutierrez et al., 1975; Tummala, 1974). Most techniques involve the utilization of modified sine wave approximations and base temperatures for the simulation of continuous daily temperature from maxima and minima and the accumulation of heat units (Baskerville & Emin, 1969; Allen, 1976). This technique is extremely precise in the linear portion of the temperature curve, but is subject to some bias when temperatures occur outside this range. In this case two half cosine functions were used to simulate continuous diurnal temperatures from daily maxima and minima, assumed to occur 12 hours apart. A modified sine function was used to calculate degree–day accumulation. Development rates at a constant temperature were taken from the experimental data.

Mycosis

Fungal life stages were modelled in a manner similar to that used for the host. As necessary, table look-up techniques allowed interpolation and extrapolation from the developmental curves. Abolition of resting spore dormancy in *Zoophthora radicans* is dependent upon a period of cold. Germination rates and maxima vary with the duration of storage at $4 \pm 2\,°C$. No germination is observed after a storage period of less than 4 months and germination does not attain 100%.

Attrition, or the proportion of spores that would not germinate, needed to be calculated on a continuous basis. Survival, as an exponential function of temperature, enabled the determination of the proportion of resting spores which remained inactive per unit time during the delay.

Temperature also determined the type of conidial germination that

Discussion

would take place. Experiments with nutritive agar suggested that true germ tubes would be formed between 24 and 32 °C, capilloconidia from 4–28 °C. The delay was designed so as to incorporate the germination of secondary conidia by recycling this aggregate twice. All other stages pass through a delay once.

Output

Results from the model are represented graphically for a weather set from Gargantua, Lake Superior Provincial Park, Ontario, 1982 and Black Sturgeon Lake, Ontario, 1971 (Figs. 11.6 and 11.7). Field observations for the Gargantua plot in 1982 are also provided (Fig. 11.8), showing the relation between simulated occurrence and recorded host stages or mycosis. The host section utilizing the model based on a series of distributed delays corresponds with that obtained using other techniques (Régnière, 1982) and correlates well with field data.

In the case of Gargantua, second instar emergence was simulated to begin on day 125 and subsequent peak stage occurrence was on days 140, 150, 160, 170, 180, and 195 for third, fourth, fifth, and sixth instars and pupae. The observed dates were 130, 140, 150, 155, 171, and 187 for larval stages, and pupation was delayed until day 200. Fungal development beginning with resting spore germination prior to day 145 gave rise to conidia from day 115 to day 155 and a simulated vegetative growth period between days 120 and 200. Secondary sporulation of cadavers began near day 150. Although mycosis of the population at no time exceeded 3%, fungal disease was detected from day 160 up to day 200 through intensive sampling. It would thus appear that widespread mycosis might be dependent on a threshold level of initial infection prior to cadaver sporulation. Notably, the fifth instar stage begins near day 155, both simulated and observed, corresponding with initial field detection (160) and sporulation (150). The simulated results for Black Sturgeon Lake weather data are included to demonstrate the sensitivity of the model to different temperatures, and potential variation in seasonal life stage development times.

Discussion

The model output represents discrete budworm and fungal development for a theoretical system where temperature is a single motivating force. Testing of each section has demonstrated the extent of realism inherent in the model. The fungal model accurately depicts the sequential life stage occurrence, but only for the single driving variable of temperature. From this simulation the effect of temperature, separate from other microclimatic influences, can be analysed.

The effects of relative humidity, leaf wetness, soil moisture, solar radiation, and wind on developmental rates may not be as significant as

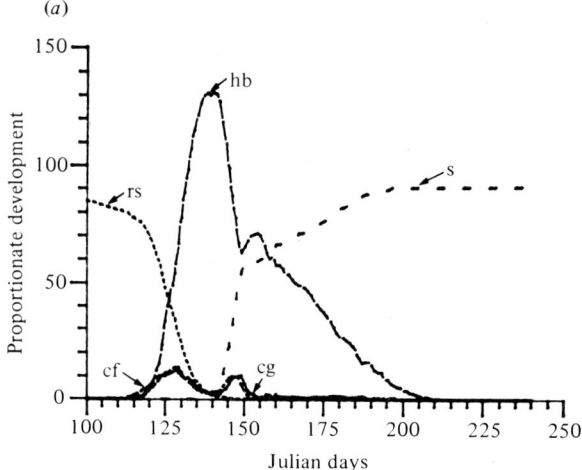

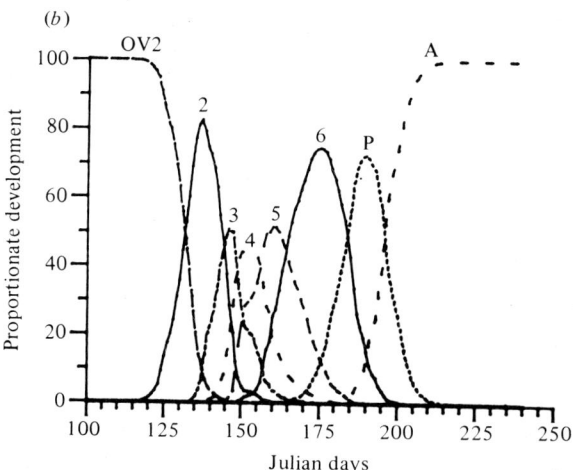

Fig. 11.6. Simulated fungal (a) and host (b) development for Gargantua, 1982. Abbreviations: cf, conidial formation; cg, conidial germination; hb, hyphal body growth; rs, resting spore germination; s, resting spore formation; OV2, overwintering second instar; 2, 3, 4, 5, and 6, second, third, fourth fifth, and sixth instar; A, adults; P, pupae.

Discussion

that of temperature, although these factors may strongly influence periodicity, density dependent infection responses, and key developmental processes such as the formation of conidia or resting spores. These physical factors may readily be incorporated into the process

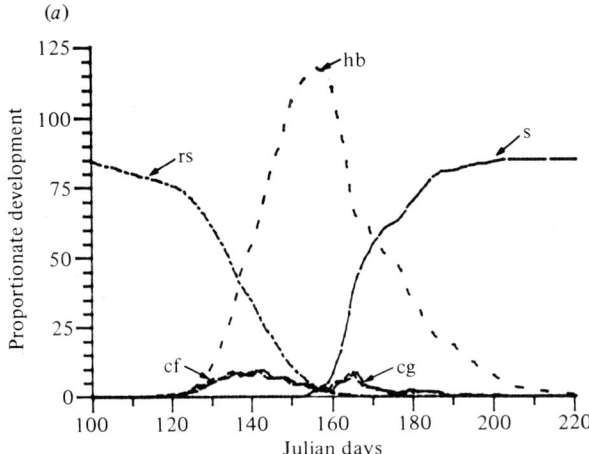

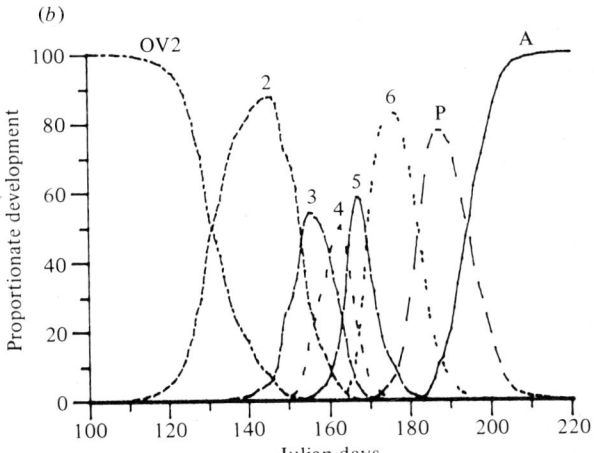

Fig. 11.7. Simulated fungal (*a*) and host (*b*) development for Black Sturgeon Lake, 1971. Abbreviations: cf, conidial formation; cg, conidial germination; hb, hyphal body growth; rs, resting spore germination; s, resting spore formation; OV2, overwintering second instar; 2, 3, 4, 5, and 6, second, third, fourth, fifth, and sixth instar; A, adults; P, pupae.

model. This is a natural beginning point for the development of phenological models for the examination of population dynamics for use in pest management programs (Welch, Croft, Brunner & Michels, 1978). The experimental designs have attempted to parallel the naturally occurring influences on development and as such, a satisfactory

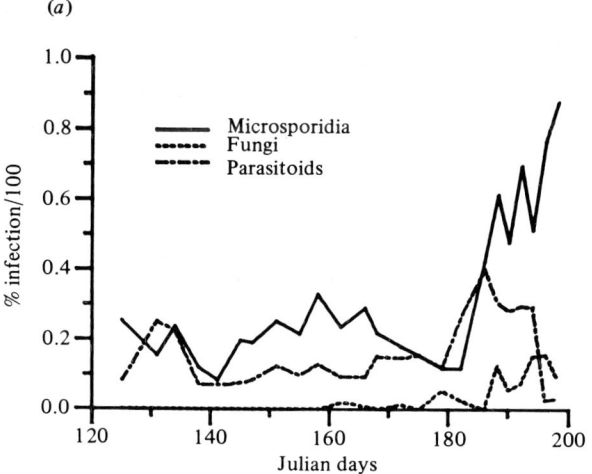

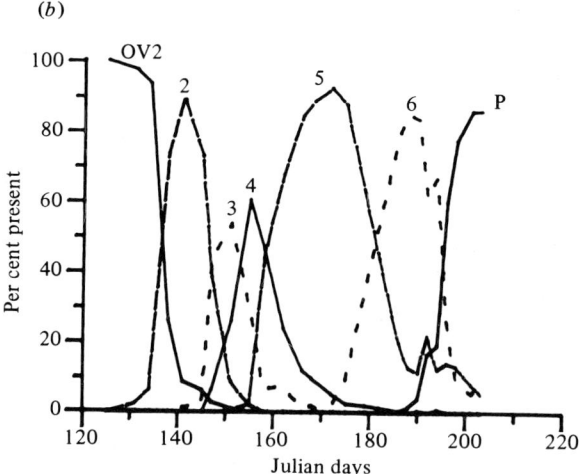

Fig. 11.8. Field observations for stage occurrence for selected natural enemies (*a*) and host (*b*) at Gargantua, 1982. OV2, overwintering second instar; 2, 3, 4, 5, and 6, second, third, fourth, fifth, and sixth instars; P, pupae.

Discussion

degree of biological realism has been attained. However, increased confidence in the ability of the model to emulate detailed interactions between environmental parameters and the host–pathogen relationship requires further experimentation, field observations and analyses.

The above approach has involved the use of a systematic problem-solving methodology in an effort to clarify some basic observations on spruce budworm–microbe interactions. Two points of discussion come to mind:

(1) As seen in the field data presented in Fig. 11.8 as well as in the graphic simulations (Figs. 11.6 and 11.7), mycosis in spruce budworm larvae due to Entomophthorales species is observed only in the fifth and sixth instars, while laboratory and field experiments show that the second to sixth instar larvae are susceptible. Furthermore, when overwintering larvae are forced out of the branches in the laboratory, mycosis occurs prior to fifth instar.

(2) Life table studies of this nature, designed to look at interactions between host and pathogen or one pathogenic organism and another, and the overriding biotic influences, are complex labour-intensive researches. How can the systems approach be applied to other animal–microbial interactions?

No data exist to demonstrate conclusively the most probable milieu for initial infection. In the case of *Zoophthora radicans*, survival of resting spores is possible in the canopy, within or without diseased individuals, as well as in the soil. As mentioned earlier, conidial persistence might also offer a source of inoculum for initial spring infection. In each of these ways, a susceptible individual can come in contact with the fungus either through airborne conidia originating in the canopy or on the soil surface, or soil-borne conidia, during larval dispersal throughout the life cycle. Because winter temperatures would be more moderate on the soil surface than in the air below snow-line, this interface might be the site where conidia would be most prevalent in early spring because of increased survival. Snow cover might conceivably delay a host–pathogen encounter for one or two months either through inhibition of resting spore germination (temperature cooling) or as a physical barrier. This alone could account quite nicely for the occurrence of infection later in the season. An analysis of the effect of temperature on conidial and resting spore survival and germination is indicated.

As an example, excising the resting spore section of the model permits the study of the effect of temperature on activation, survival, germination, and conidial production.

Resting spore dormancy is broken at a constant $4 \pm 2\,°C$ over a four-month period (Fig. 11.9). Between four and ten months, germination and development rates are linearly proportional to the length of treatment at $4\,°C$; 10 months' storage induces a maximum germination rate of 5% per day. Although this rate remains constant for periods

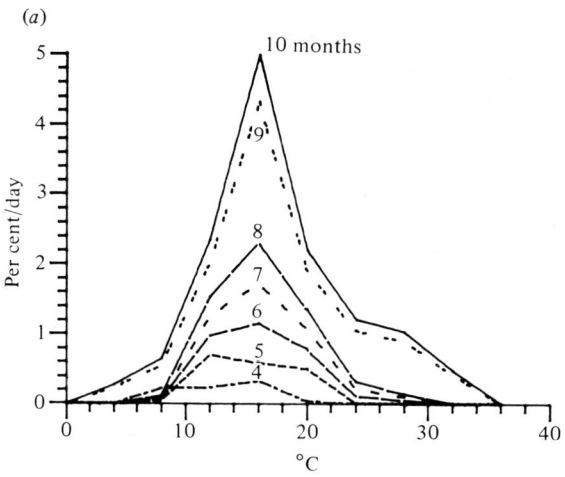

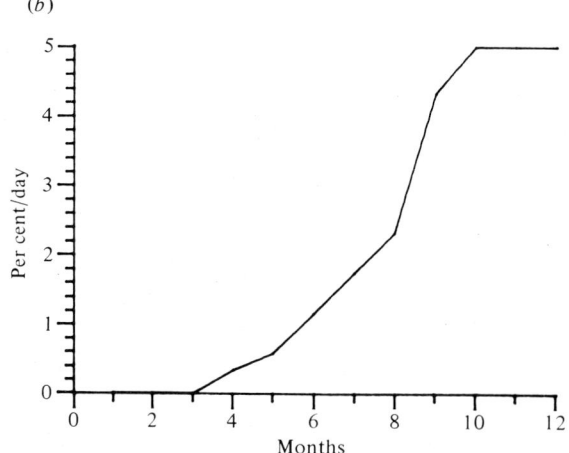

Fig. 11.9. Resting spore germination. (a) Temperature–rate response and (b) maximum rate v. storage period.

Discussion

exceeding 10 months, resting spores die off rapidly so that less than 10% germination is possible after 12 months in the cold.

The data demonstrates through simulation that depending upon the duration of cold received, resting spore germination at any given date may vary by as much as 40% or more (Fig. 11.10). This again emphasizes the possibility of a temperature-induced synchrony of mycosis in the fifth instar.

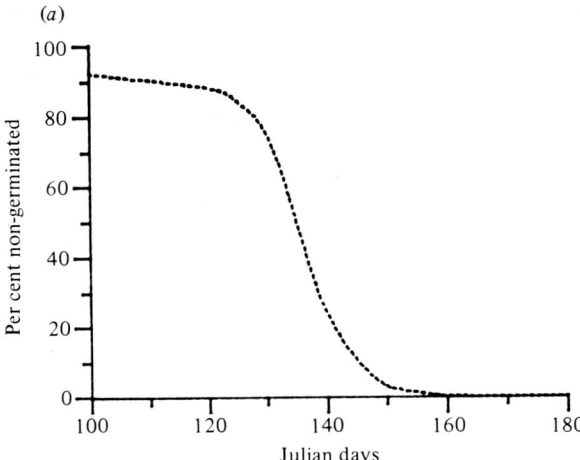

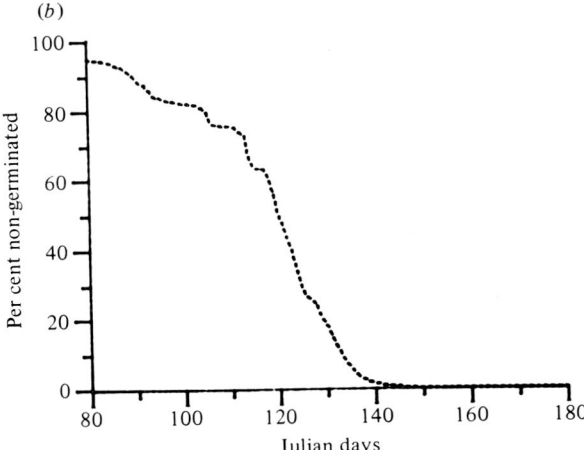

Fig. 11.10. Simulated resting spore germination for Gargantua, 1982, utilizing storage rates of (a) 4 months and (b) 10 months.

It should be noted that there is a major problem as regards the above data and their simulation. The accumulation of physiological time based on the breaking of dormancy at $4 \pm 2\,°C$ is difficult at best. In other words, how can 4–10 months at $4\,°C$ be related to natural varying temperatures?

This in itself represents the lack of a limit on the number of distinct alternative hypotheses consistent with the presently available data, emphasizing the absolute necessity for experimental data and field observations. Several experiments are proposed, involving the determination of decay rates of activated resting spores and the subsequent use of their inverses in determining a distribution for the abolition of dormancy at winter soil temperatures. This coupled with actual field observations in spore development should enable the definition of realistic limits in order to delineate experimental, varying temperature regimes to emulate natural conditions in the laboratory.

Although this paper has dealt primarily with entomophthoralean fungi, the systems approach can be used to study other microbial pathogens. The techniques are the same although the processes can differ. Consider the protozoan *Nosema fumiferanae*, a microsporidian parasite of the spruce budworm. The development of *N. fumiferanae* on spruce budworm differs from that of *Z. radicans* in several interesting ways: (1) sub-lethal doses are prevalent, with the budworm population response being perhaps most evident in terms of reduction in fecundity and increased susceptibility to other mortality factors than in terms of direct death; (2) vertical disease transmission occurs between generations; and (3) infection occurs *per os*.

Thus, initial infection of a spruce budworm could occur through the ingestion of spores present on a needle. Combined infections of *Nosema* and fungi or parasitoids may also be common, and it is possible that the microsporidian could be transmitted during oviposition penetration by a contaminated (surface or internally-infected) insect. Reproduction within the host is dependent upon spore germination and multiplication within infected cells, spores either being regurgitated or excreted into the environment. Transmission of the disease can occur in this way or in a transovum or transovarial fashion from adult female to egg. Persistence of the spores outside the host may occur from season to season, but is assured within first and second instar larvae.

The above is a description of the budworm *Nosema* interaction which can be used to define a model for development (Fig. 11.11), emphasizing inherent dfferences between the biology of *N. fumiferanae* and

Conclusion

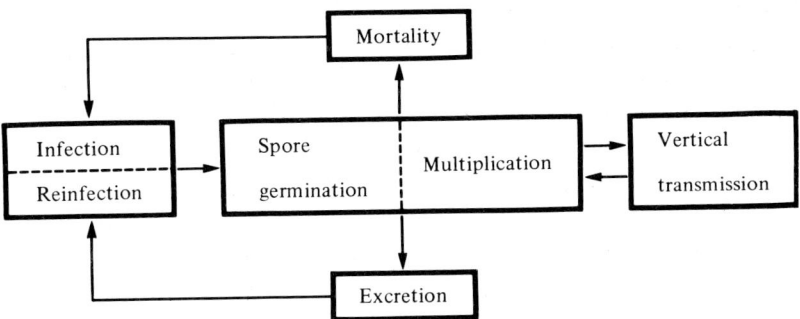

Fig. 11.11. A model for the development of *Nosema fumiferanae*.

Zoophthora radicans. In many respects the relationship between *Nosema* development and temperature is considerably more complex, due to the passive nature of the microsporidian life cycle, where spore germination and multiplication within the infected cell are active temperature-dependent processes governed by the microparasite, and infection and reinfection are based on the activity rates of the host (and to some extent parasitoid activity).

Most feeding in spruce budworm occurs in the fourth, fifth and sixth instars. Although burrowing and spinning may account for the ingestion of small numbers of spores, it would be expected that increases in intensity within the host would be most dramatic in the larger instars, which is in fact true. This is due to several factors other than feeding rates, including larval size, spore germination and multiplication; again, through the above-mentioned approach to a microbe–insect interaction, the system is broken down into specific factors of importance.

From the field observations (Fig. 11.8) it seems that an increase in infected individuals occurred after day 180, but this is misleading as the values provided are based on dead insects. In fact, what is indicated is that the actual proportion of insects dying (regardless of cause) and being infected with *Nosema* increases. Returning to the model (Fig. 11.10) enables the deduction that increased spore loads, perhaps due to reinfection through ingestion, enhanced germination and multiplication, may yield higher mortality. The same trend was observed amongst the other natural enemy groupings, implying that there exists some interaction based on stress of the host by the pathogens.

Conclusion

An effort has been made to explore the use of systems science in the study of microbial–insect interactions. Despite the use of some textbook terminology, it is hoped that this case history offers a clear step-wise progression of the application of systems techniques in biology. The particular examples used were limited, but the references cited provide access to an extensive base for the application of systems methodology in a multitude of ways. Perhaps the formalization of the science of systems represents one of the most recent steps in the evolution of experimental design.

References

Allen, J. C. (1976). A modified sine curve method for calculating degree days. *Environmental Entomology*, **5**, 388–96.
Anderson, R. M. & May, R. M. (1979a). Population biology of infectious diseases. Part I. *Nature, London*, **280**, 361–7.
Anderson, R. M. & May, R. M. (1979b). Population biology of infectious diseases. Part II. *Nature, London*, **280**, 455–61.
Anderson, R. M. & May, R. M. (1980). Infectious diseases and population cycles of forest insects. *Science*, **210**, 658–61.
Anderson, R. M. & May, R. M. (1981). The population dynamics of microparasites and their invertebrate hosts. *Philosophical Transactions of the Royal Society of London, Series B*, **291**, 451–524.
Anderson, R. M. & May, R. M. (1982a). Directly transmitted infectious diseases: control by vaccination. *Science*, **215**, 1053–60.
Anderson, R. M. & May, R. M. (1982b). Coevolution of hosts and parasites. *Parasitology*, **85**, 411–26.
Anderson, R. M. & May, R. M. (eds.) (1982c). *Population Biology of Infectious Diseases*. New York: Springer-Verlag.
Barfield, C. S. & Stimac, J. L. (1980). Pest management: an entomological perspective. *Bioscience*, **30**, 638–9.
Baskerville, G. L. & Emin, P. (1969). Rapid estimation of heat accumulation from maximum and minimum temperatures. *Ecology*, **50**, 514–17.
Bean, J. I. (1961). Predicting emergence of second-instar spruce budworm larvae from hibernation under field conditions. *Annals of the Entomological Society of America*, **54**, 175–7.
Brown, G. C. & Nordin, G. L. (1980). An epidemiological model of *Entomophthora phytonomi – Hypera postica* populations. *Proceedings of the Society for Invertebrate Pathology*, 00–00.
Brown, G. E. & Nordin, G. L. (1983). An epizootological model of an insect–fungal pathogen system. *Bulletin of Mathematical Biology* (in press).
Carruthers, R. I. (1981). 'The biology and ecology of *Entomophthora muscae* (Cohn) in the onion agroecosystem.' Unpublished Ph.D. thesis, Michigan State University, East Lansing, Michigan.
Clark, L. R. (1970). Analysis of pest situations through the life systems approach. In *Concepts of Pest Management*, ed. R. L. Rabb & F. E. Guthrie, pp. 45–57. Raleigh: North Carolina State University.
Clark, L. R., Geir, P. W., Hughes, R. D. & Morris, R. F. (1967). *The Ecology of Insect Populations in Theory and Practice*. London: Methuen & Co.
Crofton, H. D. (1971). A quantitative approach to parasitism. *Parasitology*, **62**, 179–93.
Cuff, W. & Baskerville, G. (1982). Ecological modelling and management of spruce

budworm infested fir-spruce forests of New Brunswick, Canada. *Third International Conference on State-of-the-Art in Ecological Modelling.* Colorado State University.

Eckenrode, C. J. & Chapman, R. K. (1972). Seasonal adult cabbage maggot populations in the field in relation to thermal-unit accumulation. *Annals of the Entomological Society of America,* **65,** 151–6.

Gage, S. H. & Haynes, D. L. (1975). Emergence under natural and manipulated conditions of *Tetrastidus julis*, an introduced larval parasite of the cereal leaf beetle, with reference to regional population management. *Environmental Entomology,* **4(3),** 425–34.

Getz, W. M. & Gutierrez, A. P. (1982). A perspective on systems analysis in crop production and insect pest management. *Annual Review of Entomology,* **27,** 447–66.

Gutierrez, A. P., Falcon, L. A., Lowe, W., Leipzig, P. A. & van den Bosch, R. (1975). An analysis of cotton production in California: a model for Acala cotton and the effects of defoliators on its yields. *Environmental Entomology,* **4,** 125–36.

Gutierrez, A. P. & Wang, Y. H. (1976). Applied population ecology: models for crop production and pest management. In *Pest Management, International Institute for Applied Systems Analysis Proceedings Series*, ed. G. A. Norton & C. S. Holling, pp. 255–80. Oxford: Pergamon Press.

Hardman, J. M. (1976). Deterministic and stochastic models simulating the growth of insect populations over a range of temperatures under Malthusean conditions. *Canadian Entomologist,* **108,** 907–24.

Haynes, D. L. & Tummala, R. L. (1976). Development and use of predictive models in the management of cereal leaf beetle populations. In *Modelling for Pest Management, Concepts Techniques and Applications; USA/USSR*, ed. R. L. Tummala, D. L. Haynes & D. A. Croft, pp. 53–60. East Lansing: Michigan State University.

Haynes, D. L., Tummala, R. L. & Ellis, T. (1980). Ecosystem management for pest control. *Bioscience,* **30,** 690–6.

Holling, C. S. (1964). The analysis of complex population processes. *Canadian Entomologist,* **96,** 335–47.

Holling, C. S. (1966). The functional response of invertebrate predators to prey density. *Memoirs of the Entomological Society of Canada,* **48,** 1–86.

Kish, L. P. & Allen, G. E. (1978). The biology and ecology of *Nomuraea rileyi* and a program for predicting its incidence on *Anticarsia gemmatalis* in soybean. *Florida Agriculture Experimental Station Bulletin,* **795,** 1–47.

Koenig, H. E., Haynes, D. L. & Tummala, R. L. (1975). Systems engineering – prospects for technological development of an approach to systems in biometeorology. In *Biometeorology. Proceedings of 7th International Biometeorological Congress,* vol. 6, ed. H. E. Landberg, pp. 73–81.

Kranz, J. (ed.) (1974). *Epidemics of Plant Diseases. Mathematical Analysis and Modelling.* New York: Springer-Verlag.

Lanciani, C. A. (1975). Parasite-induced alterations in host reproduction and survival. *Ecology,* **56,** 689–95.

Levin, S. & Pimentel, D. (1981). Selection of intermediate rates of increase in parasite–host systems. *American Naturalist,* **117,** 308–15.

Levins, R. & Wilson, M. (1980). Ecological theory and pest management. *Annual Review of Entomology,* **25,** 287–308.

Logan, J. A., Wollkind, D. J., Hoyt, S. C. & Tanigoshi, L. K. (1976). An analytic model for the description of temperature-dependent rate phenomena in arthropods. *Environmental Entomology,* **8,** 114–16.

Manetsch, T. J. (1976). Time varying distributed delays and their use in aggregative models of large systems. *Transactions on Systems, Man and Cybernetics*, vol. SMC-6, No. 8, August 547–553.

Manetsch, T. J. & Park, G. L. (1977). *Systems Analysis and Simulation with Application to Economic and Social Systems.* Part II. (3rd ed.) East Lansing: Michigan State University.

Manetsch, T. J. & Park, G. L. (1981). *Systems Analysis and Simulation with Application to Economic and Social Systems.* Part I. (3rd ed.) East Lansing: Michigan State University.

May, R. M. (1983). Parasitic infections as regulators of animal populations. *American Scientist*, **71**, 36–44.

Miller, C. A. (1959). The interaction of the spruce budworm, *Choristoneura fumiferana* (Clem.), and the parasite *Apanteles fumiferanae* Vier. *Canadian Entomologist*, **8**, 457–77.

Millstein, J. A., Nordin, G. L. & Brown, G. C. (1982). Microclimatic humidity influence on conidial discharge in *Erynia* sp. (Entomophthorales: Entomophthoraceae), an entomopathogenic fungus of the alfalfa weevil (Coleoptera: Curculionidae). *Environmental Entomology*, **11**, 1166–9.

Mims, C. A. (1976). *The Pathogenesis of Infectious Diseases*. London: Academic Press.

Mims, C. A. (1981). Vertical transmission of viruses. *Microbiology Reviews*, **45**, 267–86.

Patten, B. C. (1972). *Systems Analysis and Simulation Ecology.* New York: Academic Press.

Perrin, W. F. & Powers, J. E. (1980). Role of a nematode in natural mortality of spotted dolphins. *Journal of Wildlife Management*, **44**, 960–3.

Perry, D. F., DeLyzer, A. J. & Tyrrell, D. (1982). The mode of germination of *Zoophthora radicans* zygospores. *Mycologia*, **74**, 549–54.

Pielou, E. C. (1969). *An Introduction to Mathematical Ecology.* New York: John Wiley.

Pielou, E. C. (1981). The usefulness of ecological models: a stock taking. *Quarterly Review of Biology*, **56**, 17–31.

Precht, H., Christophersen, J., Hensel, H. & Larcher, W. (1973). *Temperature and Life.* New York: Springer-Verlag.

Régnière, J. (1982). A process-oriented model of spruce budworm phenology (Lepidoptera: Tortricidae). *Canadian Entomologist*, **114**, 811–25.

Royama, T. (1977). Population persistence and density dependence. *Ecological Monographs*, **47**, 1–35.

Royama, T. (1981). Fundamental concepts and methodology for the analysis of animal population dynamics, with particular reference to univoltine species. *Ecological Monographs*, **51**, 473–93.

Ruesink, W. G. (1976). Status of the systems approach to pest management. *Annual Review of Entomology*, **21**, 27–44.

Sharpe, P. J. H. & DeMichele, D. W. (1977). Reaction kinetics of poikilotherm development. *Journal of Theoretical Biology*, **64**, 649–70.

Stinner, R. E., Butler, G. D., Bacheler, J. S. & Tuttle, C. (1975). Simulation of temperature-dependent development in population dynamic models. *Canadian Entomologist*, **107**, 1167–74.

Stinner, R. E., Gutierrez, A. P. & Butler, G. D. (1974). An algorithm for temperature dependent growth rate simulations. *Canadian Entomologist*, **106**, 519–24.

Tummala, R. L. (1974). General principles of systems modelling and control for pest ecosystems. *Proceedings of the Entomological Society of America*, **5(29)**, 74–82.

Tummala, R. L., Ruesink, W. G. & Haynes, D. L. (1975). A discrete component approach to the management of the cereal leaf beetle ecosystem. *Environmental Entomology*, **4**, 175–86.

van Roermund, H. J. W., Perry, D. F. & Tyrrell, D. (1983). The influence of temperature, light, nutrients and pH in the determination of the mode of conidial germination in *Zoophthora radicans*. *Transactions of the British Mycological Society* (in press).
Watt, K. E. F. (1968). *Ecology and Resource Management.* New York: McGraw-Hill.
Welch, S. M., Croft, B. A., Brunner, J. F. & Michels, M. F. (1978). PETE: an extension phenology modelling system for management of multi-species pest complex. *Environmental Entomology*, **5(7)**, 487–94.
Whitfield, G. H., Drummond, F. A. & Haynes, D. L. (1981). A simulation model for the survival and development of the onion maggot. *Proceedings of the Entomological Society of Ontario*, **8**, 39–55.
Wilson, G. G. (1974). Effects of larval age at inoculation and dosage of microsporidian (*Nosema fumiferana*) spores on mortality of spruce budworm (*Choristoneura fumiferana*). *Canadian Journal of Zoology*, **52**, 993–6.

Index of specific names

Abies balsamea, 312
Acanthamoeba sp., 40
Acanthamoeba polyphaga, 20, 42
Acanthocinus sp., 139
Acanthoscellides obsoletus, 167
Acer pseudoplatanus, 109
Acer rubrum, 15
Acer saccharum, 99
Acetobacterium sp., 184
Acheta domestica, 208, 211
Achorutes muscorum, 23
Acremonium sp. (= *Nectria candicans*), 21, 107, 140
Acremonium fungicol (= *Nectria violacea*), 21
Acrostalagmus aphidum, 232
Acrostalagmus cinnabarinus (= *Verticillium tenerum*), 8
Acrostalagmus fragrans, 8
Actinomyces spp., 10
(Aerobacter) = *Klebsiella aerogenes*, 4, 6, 26
Agathidium species, 24
Allium porrum, 52
Alternaria spp., 13, 17
Alternaria brassicicola, 17
Amanita phalloides, 8
Amanita rubescens, 17, 18
Amaurochaete atra, 29
Amaurochaete tubulina, 29
Amitermes sp., 190
Amitermes minimus, 187
Amphicyllis globiformis, 24
Amphicyllis globus, 24
Amylostereum sp., 145, 146
Amylostereum areolatum, 111
Amylostereum chailletii, 160–3
Ancistrotermes spp., 111
Androniscus dentiger, 28
Anisotoma spp., 23–5

Anisotoma humeralis, 23, 24
Anisotoma plasmodiophaga, 23
Anobium punctatum, 117
Anopheles quadrimaculatus, 259
Anticarsia gemmatalis, 310
Aphanocladium album, 21
Aphanomyces astaci, 231, 235, 236, 238, 250, 254, 257–9
Aphelenchus avenae, 47
Aphelenchus bicaudatus, 51
Aphis fabae, 235
Aphis pisum, 233, 235, 259
Apis mellifera, 252
Arcyria denudata, 9
Arcyria incarnata, 9, 25, 28
Arcyria occidentalis, 13
Armillaria sp., 96
Arthrobotrys sp., 21
Aspergillus sp., 8, 13, 21, 118
Aspergillus flavus, 232, 236, 239, 240, 252, 254, 256, 258, 259
Aspergillus flavus-oryzae, 248
Aspergillus glaucus, 13
Aspergillus nidulans, 247
Aspergillus niger, 8, 258, 259
Aspergillus parasiticus, 235, 248
Aspidophorus lareyniei, 24
Aspidophorus orbiculatus, 24
Astacus astacus, 235, 238, 254
Astacus fluviatilis, 215
Atta colombica, 165
Atta texana, 164
Athelia epiphylla, 134
Atheta species, 24
Aureobasidium pullulans, 140, 147
Auricularia mesenterica, 16
Autranella congolensis, 100

Bacanius rhombophorus, 24

Index of specific names

Bacillus sp., 178
Bacillus mycoides, 10
Bacillus subtilis, 11, 181
Bacteroides sp., 178, 182, 183
Bacteroides termitides, 193
Badhamia curtisii (= *B. obovata*), 4
Badhamia foliicola, 10, 17
Badhamia lilacina, 9
Badhamia magna, 9, 10, 13, 17
Badhamia rubiginosa, 13, 17
Badhamia utricularis, 10, 12–21, 26, 27
Baeocera species, 24
Beauveria spp., 231, 313
Beauveria bassiana, 231–40, 245, 248, 251–9
Beauveria brogniartii (= *B. tenella*), 248
Betula sp., 16
Betula pendula + *pubescens*, 20, 100
Bjerkandera adjusta, 17, 97
Bjerkandera fumosa, 103
Bombyx mori, 232, 233, 237, 239, 248, 251, 252, 255, 258
Botryosphaeria stevensii, 283
Botrytis sp., 147
Botrytis cinerea, 241
Bouteloua gracilis, 48, 50–2
Bradysia sp., 22, 25, 97
Brefeldia maxima, 13, 17
Bremia lactucae, 249
Brittenia fraxinicola, 25
Bulgaria polymorpha, 14
(*Byssostilbe stilbigera*) = *Stilbella tomentosa*, 21

Calanus finmarchicus, 215
Cancer pagurus, 215
Candida brevis, 8
Candida scottii, 12
Candida tropicalis (= *Monilia candida*), 8
Carpophilus spp., 145
Cephalosporium sp., 142
Ceratocystis spp., 110, 113, 139–49
Ceratocystis coerulescens, 141, 142
Ceratocystis fagacearum, 112, 141–4, 301
Ceratocystis fimbriata, 144, 145
Ceratocystis minor, 141, 142, 146
Ceratocystis piceae, 147, 300
Ceratocystis pilifera, 141, 142
Ceratocystis ulmi (= *Ophiostoma ulmi*), 112, 143, 272–302
Cetonia auratia, 260
Chaetomium cochliodes, 8
Chaetomium globosum, 8
Chilo simplex, 232
Chironomus sp., 259
Chlorella spp., 18

Chlorococcum sp., 18
Choristoneura fumiferana, 255, 259, 311–16
Circinella simplex, 8
Circinella spinosa, 8
Cis boleti, 25
Citrobacter sp., 193
Citrobacter freundii, 188
Cladonia spp., 19
Cladosporium sp., 107, 113, 147
Cladosporium herbarum, 72, 79
Clambus spp., 24
Clostridium sp., 181
Coelomyces sp., 256
Coelomyces psorophorae, 230, 257
Coelomyces punctatus, 259
Colletotrichum graminicola, 235
Colletotrichum lindemuthianum, 243, 258
(*Collybia*) = *Flammulina velutipes*, 8
Colloderma oculatum, 19
Comatricha nigra, 14, 21
Comatricha typhoides, 9, 23
Conidiobolus spp., 313
Coniophora sp., 18
Coniophora puteana, 103
Coprinus micaceus, 18
Coptotermes sp., 96
Coptotermes formosanus, 96, 177, 183, 189, 190, 192, 196
Coptotermes lacteus, 188, 195
Coptotermes niger, 99
Coptotermes sjöstedti, 100
Cordana pauciseptata, 107, 115
Cordyceps militaris, 240, 250
Coriolus versicolor, 16, 72, 73, 96, 97, 103, 105
Cornitermes sp., 189, 190
Cortinarius semisanguineus, 8
Corylus avellana, 95, 100
Coryne sarcoides, 14
Cossus cossus, 231
Cribraria piriformis, 23
Crotalaria juncea, 145
Cryptocercus punctulatus, 155, 216
Cryptococcus fagisuga, 112
Cryptococcus (= *Torulopsis*) *laurentii*, 12, 27
Cryptococcus (= *Torulopsis*) *neoformans*, 27
Cryptostroma corticale, 109
Cryptotermes brevis, 190
Cryptovalsa sparsa, 8
Cubitermes sp., 189, 190
Cubitermes severus, 177, 215–22
Culex fatigans, 229, 252
Culex pipiens, 215, 230, 233, 252
Culex pipiens quinquefasciatus, 256
Culicinomyces clavosporus, 230, 252

Index of specific names

Culiseta inornata, 257
Cyathus stercoreus, 13
Cylindroiulus sp., 209, 211
Cylindroiulus punctatus, 94
Cylindronotus sp., 94
Cypha longicornis, 24

Daedalea dickinsii, 96
Daedalea quercina, 8
Daedalea unicolor, 111
Daldinia concentrica, 17
Daphnia magna, 215
Dasyhelea obscura, 252
(Dematium chodatii) = Endomycopsis burtonii, 8
Dendrobaena rubida, 124
Dendroctonus sp., 137
Dendroctonus frontalis, 110, 139
Dendryphiella sp., 21
Diacheopsis insessa, 19
Diaporthe oxyspora, 8
Dictydiaethalium plumbeum, 7, 8, 9
Dictyostelium discoideum, 6, 26
Diderma sp., 21
Didymium sp., 21, 28
Didymium aquatile, 18
Didymium difforme, 5, 11, 12, 18
Didymium iridis, 2, 11, 28
Didymium nigripes, 11
Didymium nigripes var. xanthopus, 9
Didymium squamosum, 7
Dichaena rugosa, 134
Dilophospera sp., 109
Dinothrombium giganteum, 236, 239
Dipodascus macrosporus, 20, 21
Diprion similis, 259
Drapetis nigritella, 22
Drosophila melanogaster, 145
Drosophila repleta, 25

Echinostelium minutum, 4, 5
Endoconidiophora spp, 140, 142
Endomycopsis burtonii (= Dematium chodatii), 8
Endothia parasitica, 8
Enerthenema papillatum, 28
Enicmus species, 24
Enterobacter aerogenes, 6
Enterobacter agglomerans, 189
Entomophthora spp., 230, 236, 241, 256
Entomophthora aphides, 233, 235, 259
Entomophthora apiculata, 235, 239, 240, 255
Entomophthora coronata, 237, 239, 252, 254–6
Entomophthora egressa, 255, 259, 311, 313, 314
Entomophthora fresenii, 235

Entomophthora grylli, 259
Entomophthora muscae, 310
(Entomophthora sphaerosperma)
  = Zoophthora radicans, 311
Entomophthora tenthredinis, 259
Epicypta testata, 22
Erysiphe graminis, 241
Escherichia coli, 4, 6, 11, 27, 181, 189
Eucinetus morio, 24
Euglena sp., 18
Eurygaster integriceps, 232
Eurysphindus spp., 24
Eurysphindus hirtus, 24
Eutypa acharii, 109
Eutypella scoparia, 8
Exeristes comstokii, 259
Exidia glandulosa, 13

Fagus grandifolia, 116
Ficus sp., 145
Fistulina hepatica, 103
Flammulina (= Collybia) velutipes, 8, 72, 73
Flavobacterium sp., 47
Folsomia candida, 52, 66, 67
(Fomes applanatus) = Ganoderma applanatum, 8
(Fomes connatus) = Oxyporus populinus, 99
Fomes fomentarius, 16
Fraxinus excelsior, 95, 100
Fuligo sp., 26
Fuligo cinerea, 18
Fuligo septica, 13, 17, 18, 22, 23, 26
Fusarium spp., 110, 113
Fusarium oxysporum, 47
Fusarium solani, 249, 283

Galleria mellonella, 237, 239, 242, 248, 252, 255, 256, 259
Gammarus fossorum, 168
Ganoderma applanatum (= Fomes applanatus), 8, 96, 97, 109
Gigaspora rosea, 52
Gliocladium sp. (= Nectria sporangiicola), 21
Gliocladium roseum, 284
Gloeophyllum trabeum (= Lenzites trabea), 96
Glomeris marginata, 63, 70–82, 216, 218
Glomus fasciculatus, 52, 53
Gnathomitermes perplexus, 183
Graphium sp., 140, 142
Guepinia spathularia, 13

Harposporium sp., 21
Helicostylum piriforme, 8
Heliothis zea, 231–42, 248, 257

Index of specific names 336

Hemitrichia clavata, 9, 13, 17
Hemitrichia vesparium, 9, 13
Heterobasidion annosum, 136, 142, 146
Heterotermes sp., 96, 190
Heterotermes indicola, 105
Hevea sp., 145
Hevea brasiliensis, 29
Hirneola hispida, 17
Hirsutella spp., 313
Hirsutella gigantea, 313
Holomastigotoides hartmanni, 196
Hordeum vulgare, 241
Humicola sp., 107
Hyalophora (= *Platysamia*) *cecropia*, 254, 258
Hydnum septentrionale, 8
Hylastes ater, 137, 141, 147, 148
Hylemyia antiqua, 310
Hylemyia cilicrura, 22, 25
Hylemyia platura, 310
Hylobius pales, 233, 235, 252, 253
Hylurgopinus rufipes, 273
Hypera postica, 310
Hyphoderma setigerum, 97, 108, 116
Hypolithus bicolor, 239

Incisitermes sp., 192
Incisitermes minor, 189, 190
Incisitermes = (*Kalotermes*) *schwarzi*, 179, 180
Ipomoea spp., 145
Ips spp., 148
Ips acuminatus, 137
Ips grandicollis, 137
Ips typographus, 137
Isaria spp., 313
Isaria felina, 8
(*Isaria*) = *Paecilomyces fumosorosea*, 258
Iulus scandinavius, 70

Kalotermes sp., 96, 192
Kalotermes flavicollis, 105
(*Kalotermes*) = *Incisitermes schwarzi*, 179, 180
Klebsiella (= *Aerobacter*) *aerogenes*, 4, 6

Labiotermes sp., 189, 190
Laetiporus sulphureus, 103
Lagenidium giganteum, 256
Lambdina fiscellaria fiscellaria, 259
Lamproderma sp., 17
Lamproderma scintillans, 18
Lathridius consimilis, 24
Lathridius nodifer, 24
Lentinus pallidus, 100
Lenzites betulinus, 13
Lenzites trabeus (= *Gloeophyllum trabeum*), 96

Leocarpus fragilis, 9, 13, 16, 17
Leptinotarsa decemlineata, 235, 237
Leptocerca fontinalis, 22, 25
Leptographium sp., 140, 142
Licea flexuosa, 12
Ligia oceanica, 215
Limonius californicus, 239
Lindbladia effusa, 13
Liriodendron tulipifera, 120
Listerella paradoxa, 19
Lithobius variegatus, 94
Locusta migratoria, 216
Locusta migratoria manilensis, 256
Lonchaea vaginalis, 22, 25
Lucilia cuprina, 215
Lumbricus rubellus, 61, 70, 94
Lumbricus terrestris, 94
Lycogala epidendrum, 9, 13, 17, 25, 27
Lycogala flavofuscum, 27

Macrotermes sp., 189
Macrotermes bellicosus, 104, 111
Macrotermes natalensis, 158, 159, 161
Macrotermes subhyalinus, 111, 161
Macrotermes ukuzii, 190, 195
Macrosporium spp., 113
Malacosoma alpicola, 239
Manduca sexta, 242
Marasmius androsaceus, 65, 72
Marginitermes hubbardi, 179, 183
Mastotermes darwiniensis, 179, 185, 187, 188, 190
Megasternum obscurum, 24
Melanophis bivittatus, 215
Melolontha melolontha, 248, 258
Merulius americanus, 13
Mesodiplogaster lheritieri, 20, 41
Metarhizium spp., 313
Metarhizium anisopliae, 230–60, 313
Methanospirillum sp., 182
Methanospirillum hungatii, 183
Microcerotermes edentatum, 95, 188
Micromonospora propionici, 187
Microsporum canis, 247
Microtermes spp., 111
Mixotricha paradoxa, 179
Mollisia sp., 8
(*Monilia candida*) = *Candida tropicalis*, 8
Monochamus sp., 139
Monochamus notatus, 113
Monochamus scutellatus, 113
Monosporella unicuspidata, 252
Mortierella isabellina, 107
Mortierella rammaniana, 107
Mucilago spongiosa, 27
Mucor sp., 27
Mucor hiemalis, 107, 259
Mucor javanicus, 8

Index of specific names 337

Mucor mucedo, 20, 21
Mucor plumbeus, 66, 67
Mucor ramannianus, 8, 72
Mycena galopus, 65
Mycetophilus vittipes, 25
Mycobacterium sp., 10
Myelophilus minor, 137, 148
Myelophilus piniperda, 137, 138, 141, 147

Narcissus sp., 145
Nasutitermes corniger, 187–90
Nasutitermes ephratae, 191
Nasutitermes exitiosus, 179, 188, 190, 195, 196
Nectria spp., 21, 145
Nectria species, 21
Nectria coccinea, 112, 283
Nemopodia nitidula, 22
Neocapritermes sp., 191
Neurospora crassa, 247
Nidularia pulvinata, 13
Nocardia (= *Actinomyces*) *asteroides*, 7, 10
Nomuraea rileyi, 241, 253, 310
Nosema fumiferanae, 312, 313, 326, 327
Nummularia pithodes, 17

Odontosphindus clavicornis, 24
Odontosphindus denticollis, 24
Odontosphindus grandis, 24
Odontotermes spp., 111
Odontotermes obesus, 237
Oedocephalum sp., 142
Oidium sp., 142
(*Oidium*) = *Trichosporon cutaneum*, 8
Oniscus asellus, 94, 167, 208, 211, 221, 223
Onychiurus encarpatus, 53
Oospora humi, 8
(*Ophiostoma*) = *Ceratocystis ulmi*, 112
Orchesella villosa, 70
Orconectes sanborni, 251
Orthosoma brunneum, 116
Oryctes spp., 251
Oryctes monoceros, 251
Oryctes nasicornis, 155, 215
Oryctes rhinoceros, 253, 255, 260
Oscillatoria rubescens, 11
Ostrinia (= *Pyrausta*) *nubilalis*, 236
Oxyporus populinus (= *Fomes connatus*), 99
Oxytelus tetracarinatus, 24

Paecilomyces farinosus, 235
Paecilomyces (= *Isaria*) *fumosorosea*, 258
Pelodera sp., 47, 52
Penicillium spp., 8, 13, 17, 21, 107, 113, 115, 147
Penicillium camembertii, 8
Penicillium roquefortii, 9

Periplaneta americana, 155, 195, 215, 216, 218, 220, 221, 251
Phallus impudicus, 72
Phellinus cryptarum, 103, 105
Philonthus fimetarius, 24
Philoscia muscorum, 168
Phlebia gigantea, 113, 142, 143
Phlebia merismoides, 16
Phlebia radiata, 97
Phomopsis oblonga, 283
Phronia sp., 25
Phthiracarus sp., 215
Physarella oblonga, 27
Physarum cinereum, 11, 13
Physarum confertum, 27
Physarum didermoides, 18
Physarum flavicomum, 5, 13, 17, 22
Physarum gryrosum, 11, 18
Physarum nutans, 14, 18
Physarum polycephalum, 2, 5, 11, 12, 13, 17, 18, 26, 27
Physarum pusillum, 11
Physarum reniforme, 29
Physarum rigidum, 26
Physarum tenerum, 13, 17
Physarum virescens, 13, 17
Physarum viride, 9, 13, 26
Physarum viride var. *rigidum*, 17
Picea sp., 139
Picea abies, 113
Picea glauca, 312
Picea mariana, 312
Picea rubens, 312
Pinus banksiana, 136
Pinus caribaea, 100
Pinus radiata, 145
Pinus silvestris, 49, 62, 117
Pityogenes spp., 148
Pityogenes chalcographus, 137
Pityogenes quadridens, 137
Platanus spp., 145
(*Platysamia*) = *Hyalophora cecropia*, 254
Platurocypta punctum, 25
Platurocypta testata, 25
Pleistophora schubergi, 313
Pleurotus cornucopiae, 17
Pleurotus ostreatus, 97
Polydesmus sp., 209
Polydesmus angustus, 70, 94
Proactinomyces spp., 10
Procubitermes aburiensis, 177, 211, 221–3
Protosphindus chilensis, 24
Prunus spp., 145
Pseudomonas sp., 40, 42
Pseudomonas fluorescens, 11
Pseudomonas maltophilia, 181
Pseudopityophthorus spp., 112
Pteronarcys proteus, 168

Index of specific names

Pterotermes occidentis, 179
Pullularia nigrans, 8
Pyrausta (= Ostrinia) nubilalis, 236
Pyrsonympha verteus, 177

Quedius cruentus, 24
Quedius mesomelinus, 24
Quercus spp., 92, 103, 134
Quercus petraea, 95, 100
Quercus robur, 68, 95, 109

Reticularia jurana, 23, 29
Reticularia lycoperdon, 16, 22, 23, 25, 27, 29
Reticulitermes spp., 96, 97, 192
Reticulitermes flavipes, 175–84, 189–95, 252, 254, 256
Reticulitermes santonensis, 97, 196
Reticulitermes tibialis, 183, 184
Revelieria californica, 24
Rhizopus oryzae, 8
Rhodotorula spp., 107
Rhodotorula minuta, 12
(Rhodotorula mucilaginosa) = Torula aclotiana, 12
Rhynchonectria longispora, 21
Rhynchotermes perarmatus, 189, 190
Rozites caperata, 16

Sarcina lutea, 11
Scatopse fuscipes, 22
Schistocerca gregaria, 208, 210, 219, 232, 252–4
Schizophyllus commune, 17
Schizopora paradoxa, 97, 116
Scolytus sp., 301
Scolytus ensifer, 273, 292
Scolytus kirschi, 273, 292
Scolytus laevis, 273, 288
Scolytus multistriatus, 137, 138, 273, 288–93
Scolytus pygmaeus, 273, 292, 293
Scolytus scolytus, 137, 138, 273, 279, 285–95, 299, 302
Scopulariopsis spp., 21
Serpula lacrimans, 96, 123
Sesquicillium microsporum, 21
Sirex cyaneus, 160–3
Sirex gigas, 162
Sirex noctilio, 145, 146
Sirex phantoma, 162
Solenopsis richteri, 252
Sphaerosoma piliferum, 24
Sphindus species, 24
Sphyrapicus spp., 144
Spicaria sp., 239
Spirotrichonympha leidyi, 196
Spodoptera frugiterda, 232

Spongipellis pachyodon, 99
Sporothrix spp., 21, 140–2, 276
Sporotrichum sp., 107, 113, 115
Staphylococcus sp., 178
Staphylococcus aureus, 27
Stemonitis axifera, 28
Stemonitis flavogenita, 28
Stemonitis fusca, 5, 9, 11, 13, 14, 27
Stemonitis herbatica, 5, 6, 7, 19
Stemonitis splendens var. flaccida, 9
Stereum sp., 96
Stereum chailletii, 111
Stereum hirsutum, 14, 17, 26, 96, 102
Stereum sanguinolentum, 111, 147
Sterigmatocystis sp., 13
Stilbella species, 21
Streptococcus sp., 178, 193
Stysanus sp., 13
Stysanus stemonitis, 8
Symbiotes latus, 24
Symphytocarpus flaccidus, 29

Tachypodiulus spp., 209, 211, 223
Tenebrio sp., 167
Tenebrio molitor, 232, 235, 252
Termitomyces spp., 103, 104, 111, 117, 118, 156–61, 195
Thelohania spp., 313
Thyridopteryx ephemeraeformis, 235
Thyronectria denigrata, 8
Tipula abdominalis, 215
Tipula flavolineata, 93–6, 100, 102, 106–23
Tomocerus minor, 70
Torula sp., 107
Torula aclotiana (= Rhodotorula mucilaginosa), 12
Torula nigra, 259
(Torulopsis) = Cryptococcus laurentii var. laurentii, 12, 27
(Torulopsis) = Cryptococcus neoformans, 27
Tremella sp., 13
Tremella mesenterica, 13
Trichamphora pezizoidea, 16, 27
Trichia decipiens, 13, 17, 21
Trichia floriformis, 9, 21
Trichia fragilis, 5
Trichia persimilis, 13
Trichia scabra, 13
Trichoderma spp., 21, 107, 115, 117, 143
Trichoderma alba, 107
Trichoderma konigii, 107
Trichoderma lignorum, 8
Trichoderma polysporum, 284
Trichoderma viride, 107, 243, 284
Trichomitopsis termopsidis, 181–3
Trichoniscus sp., 28

Index of specific names

Trichonympha sphaerica, 181, 182
Trichophisia ni, 235, 240
Trichophyton rubrum, 247
Trichosporon sp., 107
Trichosporon (= Oidium) cutaneum, 8
Trichothecium roseum, 283
Trinervitermes trinervoides, 189, 191
Tubifera ferruginosa, 22, 27
Tyromyces sp., 96
Tyromyces palustris, 96
Tyrophagus putrescentiae, 28, 282

Ustulina zonata, 17
Ulmus spp., 272
Ulmus carpinifolia, 287
Ulmus carpinifolia × *sarniensis*, 287
Ulmus glabra, 282
Ulmus procera, 283, 287

Verticicladiella sp., 140, 142

Verticillium spp., 230
Verticillium species, 21
Verticillum albo-atrum, 244
Verticillium tenerum (= Acrostalagmus cinnabarinus), 8
Vicia faba, 241
Vuilleminia comedens, 110

Xantholinus punctulatus, 24
Xestobium rufovillosum, 103, 105
Xylaria hypoxylon, 14, 15
Xyleborus sp., 29
Xyleborus ferrugineus, 145

Zoophthora sp., 310
Zoophthora radicans (= Entomophthora sphaerosperma), 313–18, 323, 326
Zootermopsis sp., 186, 189, 191, 192
Zootermopsis angusticollis, 180, 181
Zootermopsis nevadensis, 184, 215, 216

Subject index

Acari, see mites
acetate, 179–96
N-acetyldopamine, 230
N-acetylglucosamine, 12, 230–42, 245–6, 249
Actinomycetes: consumption of 3, 7, 10–11; in termite guts, 178, 187, 221–2; nitrifying, 75; nutrient immobilization, 35, 37; on bark, 283; pathogenic, 7
agricultural ecosystem, see ecosystems
algae, 3, 7, 11, 18–19, 134
allelopathic substances, 89, 94, 99–100, 125, 167, 312
ambrosia beetles, 110–12, 125, 139, 163
amino acids, 222; as nutrient source, 40, 164, 245–7, 257; in attractants, 11; in gut secretions, 78, 212; in food, 78, 186, 196; in root decomposition, 40, 41
ammonia, 194, 217
ammonium: excretion by animals, 79, 217; in nitrogen mineralization, 43–9, 63, 66, 70–82; uptake by mycorrhizae, 49, 82
ammonification: by bacteria, 43, 47; by fauna, 47, 79; by fungi, 47, 63, 66, 72; through interactions, 43, 45, 47, 63, 66, 69–80
amoebae, 40–4; see also myxamoebae
amylose, 160, 162
anaerobes, see bacteria
anaerobic conditions, see guts
antimicrobial agents: acquired, 167; fungitoxic, 232–5, 253–7, 283; in experimental techniques, 79, 115, 179, 181, 188, 221, 279–80; in myxomycetes, 27; of resin, 135
ants, see Hymenoptera
aphids, 232, 236
arthropods: and entomopathogenic fungi, 229–60; bacterivores, 60, 62, 67;
detritivores, 89, 167–8; fungivore, 60, 67, 71, 82, 109, 136–9, 142–5, 155–68, 188, 297; grazing activity, 22–5, 37, 65–82; guts, 205–24; myxomyceticolous, 22–5; secondary colonizers, 114; exoskeleton decomposition, 40; soil, 37, 49, 64; see also Collembola, Insects and arthropod groups, mesofauna, mites, termites and vectors
ATP, 179
axenic cultures: of insects, 253–4; of myxomycetes, 2–5, 10–11, 18; of protozoa, 181, 185

bacteria: acetogenic, 184–5; adherence, 5–7, 177–9, 205–11; ammonifiers, 78; anaerobes, 177–8, 193–4; biomass, 62–3; cellulolytic, 173–87; chitinolytic, 48, 195; consumption by invertebrates, 42–53, 60, 62, 65, 77, 196; consumption by myxomycetes, 2–7, 10–12, 26; consumption by nematodes, 40–53, 65; consumption by protozoa, 40–8, 65, 185; fermentive, 185–8; in gnotobiotic cultures, 40–3; in faeces, 114–15; in termite guts, 173–9, 182–5; interactions with fungi, 39–54, 215; methanogenic, 177, 182–5; nitrifying, 75; nitrogen fixing, 188–92; nutrient quality of, 182, 194, 196; plant growth influence, 38–9; populations, 44, 46, 51–2, 178; uricolytic, 193–5; see also microorganisms
bark beetles, 110–11, 125, 134–49, 271–303; see also ambrosia beetles
beetles (Coleoptera): acquired enzymes of, 166; fungivores, 103, 109, 144, 149; galleries (tunnels), 113–16, 137–42, 146–8, 274–93, 298; myxomyceticolous, 22–5;

Subject index

beetles (Coleoptera)—*contd.*
pupal chambers, 138–42, 147–8, 284–93, 299; vectors, 109–12, 144–9; wood-boring, 90, 93–4, 103, 105, 113, 125; *see also* ambrosia beetles and bark beetles
biomass: of higher plants, 50, 89; of invertebrates, 42, 61–2, 69, 71, 76–7, 80, 174; of microbes, 44, 62–3, 77, 282–3; of myxomycetes, 2
birds, 99, 112–13, 134, 144
butyrate, 180, 185

calcium: in fungi, 102, 104; in guts, 181, 214; in litter, 70; mobilization, 70, 75–6, 82, 91, 101, 120–3; sources, 123
canker of apple, 145
capillitium, 21
caprylic acid, 232
carbon: bidirectional flow, 54; cellulolysis fermentation product, 179–86; limitation, 78; mineralization, 44–7; nutrient ratio, 100–3, 120; source, 40–1, 49, 100–2, 125, 185, 196, 231, 244–7; uricolysis, 193–4; utilization from lignin, 125; *see also* carbon–nitrogen ratio and respiration
carbon dioxide: effects on fungi, 140; *see also* respiration
carbon–nitrogen ratio: in humus, 49; in litter, 82; in soils, 60; of fungi, 101, 104; of insects, 101–2; of wood, 100–2, 120, 188; theory, 60, 78
carboxymethylcellulose, 181, 184, 187
catabolism: catabolite repression (CR), 244–7; fungal, 73; microbial, 64
cellobiose, 156–7, 181, 186
cellulytic activity (cellulolysis): acquired, 155–68; fermentative, 179–86; in animals, 97–8, 155–63, 166–8, 179–88, 197; in detritus, 167–8; of actinomycetes, 187; of bacteria, 184, 187; of fungi, 99, 103, 155–68; of protozoa, 156, 173, 175, 179–85, 197; of termites, 98, 106, 156, 187, 197; *see also* enzymes
cellulose: decomposition by fungi, 99, 103; depolymerization, 179–88; effect on decomposition rate, 46–8; nutrient source, 40–8, 174, 185; *see also* cellulolytic activity, in animals and enzymes
chemotaxis: of fungal spores, 231; of invertebrates, 94–7, 144; of myxomycetes, 6, 9, 11, 26
Chilopoda, 94
chitin: and nutrient mobilization, 48–51; as mineral source, 46–52, 195, 244–5; bacterial decomposers of, 48; decomposition rate, 46, 47; digestion of, 242; fungal decomposers of, 48; in fungal walls, 40, 195; in invertebrate cuticle, 40, 206–7, 230, 239–46, 249; in microbial faeces, 117; in myxomycetes, 22; structure of, 230; *see also* enzymes, chitinase, chitobiase
chlamydospores, 52
climate, 48, 53, 275, 285, 288, 298
Coleoptera, *see* beetles
Collembola: effects on microorganisms and nutrient mobilization, 37, 52, 62, 65–71; food and feeding, 23, 62, 65; vectors, 109, 114
colonization (= invasion): after microbial 'conditioning' 94–106; animal facilitation of, 106–8, 111–15; by mesofauna and effects on decomposition, 93–126; by termites, 92–3, 96–100, 103–6; by the cranefly and effects on decomposition, 92–6, 102, 106–7, 112, 114, 117, 120–3; by wood fungi, 92–7, 109–15; fungal facilitation of, 93–4, 98–9; gut, 177–9, 205–24; indirect animal effects on, 115–18; interactions in, 103, 111, 117–18, 144–9; primary animal colonists, 93–4, 134–9; primary fungal colonists, 20–2, 28–9, 139–43, 146; secondary colonists, 114, 136, 147; *see also* beetles, infection, vectors and beetles, wood boring
comminution, effects on decomposer community, 61, 64, 67, 71; effects on decomposition, 64, 77, 117; effects on nutrient mobilization, 64, 123–5
communities: animal, 93; decomposer, 44, 60–61, 69, 90, 118–20, 126, 134; function, 35, 37; in wood, 93–4, 106–7; microbial, 44, 65, 106–8, 116, 173; plant, 307
competition, 143, 166, 253, 282–4, 293, 297–300
conidia, 156, 158, 230, 249–50; as food, 8, 258; capilloconidia, 319; viability and germination of, 142, 231, 253–4, 259, 293, 313–14, 318–24; *see also* disease transmission, infection, spores and vectors
coprophagy, 29, 156, 194
coprophily, 28, 29
Crustacea, 205, 211–17, 229, 231
cutin, 230, 249
Cyanobacteria, 3, 11
cyclic AMP, 6, 11

death watch beetle, 103, 105
decomposer communities, *see* communities
decomposition: and resource quality, 61, 69, 94, 100–6, 120–3, 192; and wood softening, 94, 98–9; energy flow during,

Subject index

37; fast-slow cycle, 40–1; fungal effects on, 90–100, 139–40; mesofauna effects on, 53, 59–82, 90–3, 106–18; microfauna effects on, 37, 46, 49, 59–82; microorganism effects on, 59–82, 89–126; myxomycete effects on, 13–15; of dung, 29; pathway, 64, 90–2, 124; physical environment conditions and, 67, 75–7, 115–20; process of, 64, 69, 77–82, 90–3; rate of, 46, 49, 60, 80, 115–20; systems, 41, 49, 59–62; *see also* cellulolysis, communities, communition, litter, roots and wood
destruxin, 255–6
detritus, 89, 167–8
dictyostelids, 6
diet: algae, 3, 7, 19; bacterial, 2–7, 42–53, 65, 77; cellulose, 40, 173–4, 180–1, 188; decayed wood, 99–105, 111, 156, 174; decay fungi, 97, 102–8; detritus, 167–8; dung, 156, 174; fungi, 2–3, 7–10, 40–53, 65, 72, 97, 100–2, 139, 142, 144, 149, 156–68, 188, 195; grasses, 156, 167, 174; gut bacteria, 182, 194, 196; hemicellulolose, 174; humus, 174; insect exoskeletons, 195; leaf litter, 59–82, 192; leaves, 91, 174, 312; lignin, 174; plants, 40, 164; roots, 174; sap, 112, 144; soil, 174, 209, 217, 222, 253; sound wood, 97–105, 110–11, 117, 138, 149, 155–8, 161, 166, 173, 180, 188; wood litter, 192; *see also* resources
digestion: anaerobic (fermentation), 175, 178–88, 216; assimilation efficiency, 62, 78, 80, 102, 125, 179, 196; attraction products of, 26; by acquired enzymes, 98, 110–12, 125, 155–68; by arthropods, 173–88, 207–19; by symbionts, 155, 173–88; extracellular, 2, 12–13, 20, 135, 244, 247, 250; of bacteria, 7–10; of cuticle, 236–52; of fungi, 7–9, 12–18, 28, 252; of gut symbionts, 182, 194, 196; of recalcitrant materials, 89–126, 155, 164, 168, 174–88, 196–7, 216–17; susceptibility to, 7–8, 13, 17; *see also* cellulolysis, enzymes, lignin, spore viability and wood
Diplopoda, *see* millipedes
Diptera: and decomposition, 90, 94, 125; and nutrient cycling, 64, 68; larvae and gut symbionts, 125; microbial grazing by, 64, 68, 136, 139, 142, 144, 147, 149, 168; myxomycete feeding by, 22–5; pathogenesis of, 236; predatory, 136, 139; vectors, 109, 145, 148; *see also* grazing, infection and vectors
disaccharides, 140, 157, 185–6
disease transmission, 136, 139, 142–9, 271–302, 316, 326; *see also* infection and vectors
dispersal by wind, 113, 136, 140, 143, 146–7, 323
dormancy and overwintering, 137–8, 143–4, 149, 278–82, 314, 318–26
dung, 28; *see also* excretory products
Dutch elm disease, 133, 144, 271–302

earthworms, 29, 61, 67, 69, 80, 94, 109, 124
eelworms, *see* nematodes
ecosystems: agricultural, 39, 60, 80; coniferous forest, 9, 62, 76, 81, 90, 136; deserts, 48, 53; grassland, 48, 61, 80; savanna, 90; temperate deciduous forest, 15, 59–62, 73–6, 80, 82, 89–93; tropical forest, 1, 15, 25, 27, 29, 89–93
encapsulation, of invading fungi, 258–60
enchytraeids, 49, 64, 69, 70
energy flow: contribution by fauna and microflora, 37; in soil, 37; of myxomycetes, 2
energy source, 49, 179, 193, 231, 244
entomopathogenesis, *see* pathogenic interactions
enzyme activity: acquired, 155–68; and pathogenicity, 248–9; cuticle degrading, 231, 235–52; fungal-wall degrading, 253; induction of, 244–51; inhibition of, 125, 167, 250; in insect gut, 164, 211–14; repression, 245–51; termite microbial symbionts, 173–97; termite, 187; in wood, 90, 94, 97–100; *see also* digestion, extracellular
enzymes: acetylglucosaminidase, 243; aminopeptidase, 243; amylase, 160, 167; arabinase, 244; carbohydrase, 182, 187, 251; carboxymethylcellulase = C_x cellulase; C_1-cellulase, 156–63, 184, 187; C_x-cellulase, 156–65, 184, 187; cellobiase, 156–63, 181, 184, 187; cellobiohydrolase = C_1-cellulase; cellulase, 166, 167, 181, 182, 243; CMCase = C_x-cellulase; chitinase, 12, 164, 195, 239–51, 258; chitobiase, 242; collagenase, 251; cutinase, 244, 249; EC 3.2.1.4 = C_x-cellulase; EC 3.2.1.21 = cellobiase; EC 3.2.1.91 = C_1-cellulase; endoglucanase = C_x-cellulase; esterase, 243; ferrodoxin oxidoreductase, 181; β-glucosidase = cellobiase; glutamine synthetase, 194; hydrogenase, 181; laminarinase, 160; lipase, 242, 248–9; lipolytic, 231, 237, 248; nitrogenase, 189; panprotease, 251; papain, 251, pectinase, 160, 166, 243; pepsin, 251; phosphatase, 250; polyphenoloxidase,

enzymes—contd.
135, 258; polysaccharase, 125, 244, 249; pronase, 251; protease, 242–247, 250, 256, 258; proteolytic, 164, 167, 248; proteinase, 164, 239, 244, 250; purine nucleoside phosphorylase, 193–4; pyruvate, 181; trypsin, 251; uricase, 193, xanthine dehydrogenase, 193–4; xylanase, 160–3, 244
excretory products, 93, 137–8, 209–217, 224, 284; as food source, 29, 108, 156, 174, 194; decomposition, 120; enzymes in, 164, 166; microbes in, 78, 108–10, 114–15, 120; nutrients in, 29, 60–4, 77–9, 103, 117, 135, 176; termite, 103, 193; see also urine
exudate: cuticle, 235; gut, 215; resin, 100, 135, 146, 149; root, 38–40; sap, 29, 114; toxic, 100, 135, 146, 149, 215
evolution: of interactions, 111, 133, 136, 272, 300–1; of molecular mimicry, 260; selection process, 9, 125, 136, 277, 300–1; of wood exploitation, 125–6, 133–6

faeces, see excretory products
feeding: grooves, 293–300; inhibitors, 10, 27; preference, 192; specialization, 22, 25–6, 217; specificity, 10, 65, 71–3;
 see also bacteria, diet, fungi, grazing, myxomycetes, phagocytosis, predation and saprophagy
fermentation: aerobic, 179, 185–6; anaerobic, 179, 184, 186; by gut symbionts, 175, 178–88, 216
ferric iron, 40
food chains (webs): anaerobic, 182; detritus, 89; length and nutrients, 53; of myxomycetes, 2–3, 22; soil, 40
forestry management practices, 59, 81, 133–40, 143, 145, 307
forests, see ecosystems
formate, 185–6
fulvic acid, 124
fungal fruiting bodies: 123, 148, 285; and nutrient status, 281, 288, 289; ascocarps, 142; basidiocarps, 16–17, 109; conidiophores, 17, 142, 147, 156, 158, 275–6, 281–4, 288–9, 297, 314; mycotêtes, 103; perithecia, 8, 141–3, 252, 275–6, 281–3, 288–90, 297, 300–2; synnemata, 275–6, 281–2, 288–9
fungal groups: Ascomycetes, 90–1, 97, 109–10, 113, 125, 139–40, 272; Basidiomycetes, 13, 72, 90–126, 136, 139–43, 147; Discomycetes, 14; Deuteromycetes, 72, 107, 110, 124, 139, 140, 255, 260, 313; Mastigomycotina, 259; Moniliales, 106; Mucorales, 13, 106–7, 117; Polyporaceae, 17, 109; Pyrenomycetes, 17; Thelephoraceae, 17; Xylariaceae, 95, 97, 125, 158; Zygomycetes, 109, 259
fungi: and nutrients, 65–7, 78–80, 100–6, 123, 166, 192, 281, 288–9; as food, 2, 12–17, 25, 46–54, 60–82, 109, 136, 139–49, 297; and termites, 103–4, 109, 111, 117, 155–9, 174; brown-rot, 96, 105, 125, 145; chitinolytic, 48, 164, 239; decay, 13, 14, 26, 89–126, 139–40; dermatophytic, 247; dry-rot, 123; entomopathogenic, 21, 229–60, 307–27; growth and grazing, 65–7; host penetration by, 233–60, 313–14, 326; mycelial antogonists of, 97, 283–4, 298, 300; non-cellulolytic, 140, 147; non-decay, 140–9; phytopathogenic, 39, 229, 233–5, 249–50, 271–302; saprophytic, 244, 275–82, 290, 299–300; soft-rot, 125; walls of, 12–13, 17, 40, 195; white rot, 96–103, 109, 124–5; see also digestion, enzyme activity, grazing, hyphal bodies, infection, interactions, microorganisms, mycorrhiza, nutrients and spores
fungus combs (gardens): of ants, 155, 163–5; of termites, 103, 104, 109, 111, 117, 156–9, 174

D-galactosamine, 22
galacturonic acid, 230, 243
genes: and 'd'-factors, 284; and host genotype, 302, 312; and pathogenic genotype, 230, 233, 248, 277, 290, 297–302; mutant, 5, 248, 279; pools, 300; recombination, 280–2; regulatory, 5, 97, 248; selection of, 277–293, 298–301; vector, 277; vegetative compatibility, 284, 300; see also evolution
germ tube, 234–7, 250, 313, 319; and appressoria, 234–9
β-glucans, 163, 242, 258
D-glucosamine, 12, 231, 245–6, 249
glutamine, 194
glutathione, 216
glycogen, 181
glucose: and nitrogen immobilization, 42, 78–9; as carbon source, 41, 196, 246; as energy source, 231, 244; fermentation of, 186; in cellulose, 156–7, 181, 230; in fungal walls, 12
grassland, see ecosystems
grazing: and biomass, 44, 65; and decomposition, 37, 40, 49, 64–5, 107–8, 126; and nutrient flow, 37, 41–9, 62–82; by Collembola on microbes, 37, 52, 65–82; by insects, 22, 23; by millipedes, 63–82; by nematodes on microbes, 41–3, 47–52, 142; effect on fungal species, 65,

Subject index

71–3; effect on microbes, 44, 51, 53, 64–5, 107–8; intensity, 42, 46–7, 64–5; on fungi, 12–18, 62–82, 126; species-selective, 65–7, 71–3, 95–7; *see also* predation

growth: dynamics, 65–7; efficiency, 42, 44; inhibitors, 39, 298; promotors, 38–9, 181

guts: anatomy, 175–9, 205–24; anaerobic conditions in, 177–87, 193–4, 197, 205, 216; microsites in, 216; millipede, 78, 209, 222; pathogenic microbes in, 209, 252–7; physicochemical conditions, 114–15, 159–62, 176–7, 211–16, 252–3; symbionts, 98, 111, 173–97, 205–24; termite, 173–81, 210, 215–23; transit time in, 176, 209–12, 217; *see also* digestion and enzymes

gypsy moth, 91

hemicellulose, 162, 174, 179–80, 186, 235

heterotrophic interactions: and physical environment, 115; animal–microbial, 37, 40–54, 60–82, 89–126, 144–9, 155–68, 173–97; effect on nutrient flux, 62–3, 68, 77; effect on decomposition, 49, 118–26; fungal–bacterial, 39, 42–3, 50; microbial attractants in, 10, 26, 94–7; nitrification 75; respiration, 62; *see also* colonization, feeding, symbionts and vectors

hormones, 38, 257

host response: against colonization, 140, 300, 302; against infection, 167, 232–3, 248–9, 257–60, 298, 311; against predation, 7, 24, 282; immunity, 251, 258–9, 311–12

host susceptibility: of invertebrates to infection, 251–7, 311–14, 323; of plants to colonization, 92, 146; of plants to infection, 109, 272

humic acid, 124

humidity, *see* moisture

humification, 124

humus: as termite food, 174; layers, 62, 64, 73–5, 81; soil type, 68

hydrogenase, 181

Hymenoptera: ants, 23–6, 155, 163–5; fungus-growers, 155, 163–5; parasitic, 137; wasps, 99, 109–12, 138, 145, 167; *see also* Siricidae (woodwasps)

hyphal bodies, 254–9, 313–14, 320–1

infection: and surface nutrients, 231–2, 235, 297; at oviposition, 110–11, 146, 161, 313, 326; by ingestion, 20, 229, 252, 326; infection structure, 234–5; of insect cuticle, 229–60, 311–19, 323, 326–7; of wood by fungi, 112–14, 143–9, 271–302; *see also* colonization, canker, disease transmission, Dutch elm disease, host susceptibility, oak wilt and vectors

ingestion: of bacteria, 4–7, 11; of fungal enzymes, 155–68; susceptibility to, 7–10, 20; *see also* feeding

Insecta, 3, 22–3, 29, 59, 92, 107, 109, 112, 114, 133, 136, 139, 143–9, 155, 166–7, 205–6, 212–19, 223, 229–43, 249, 258, 326–7; *see also insect groups*

insect cuticle, 206–14, 217–24, 230; invasion of, 229–53, 313; nutrient source, 40, 231–5, 257, 297; properties of, 232–3, 257

insecticides, 27, 53, 135, 252, 256; *see also* toxic compounds

inulin, 181

Isopoda, *see* woodlice

lactate, 185–6
laminarin, 160–3
lanolin, 231, 242
leaf-litter, 60–82, 89, 163–6, 192; *see also* litter
Lepidoptera, 215, 230, 232
lichens, 3, 19, 134
life cycle: and colonization synchronization, 111–12; and degree of fungal decay, 103–5; and pathogen synchronization, 325; barkbeetle, 137–8, 274; Dutch elm disease fungus, 274, 278, 281–2; pathogen stages, 278, 312; strategy type, 117, 136
lignin: degradation, 40, 97, 99, 124–5, 179, 196–7; inhibition of polysaccharase, 125; in termite diet, 174; in wood, 89, 125–6; soil content, 60
lignocellulose, 182, 216
lipid, 22, 231–3, 237, 240, 242, 249
litter: decomposition, 53, 59–82, 92, 96; fungal conditioning of, 71; in interactions and nutrient flow, 60–9; leachates of, 62–3, 66, 69–82, 123; nutrient quality of, 60, 80, 192; resource, 60–2, 66, 69–77, 89, 163–6, 192; *see also* detritus, leaf-litter, roots and wood
locust, 232, 253
lysis: autolysis, 3, 244, 250; cuticle, 237, 240, 251, 254, 256; hyphal, 12, 244, 250; of bacteria, 196; of protozoa, 196; root cell wall, 38–9; yeast, 27
lysates, 38–9

macrofauna, 40, 64, 69, 81, *see also* earthworms, millipedes, molluscs and woodlice
magnesium, 63, 91, 101–4, 121–2, 213–14
mammals, 109–12, 134

mannan, 179
mannose, 12
melanin, 257–9, 282
mesofauna, 35, 40, 62, 64, 67, 69, 71, 81, 93, 94; *see also* collembola, enchytraeids, mites and termites
methane, 177, 182–5
microclimate, 115–17, 314, 320
microcosms, 40–52, 63–82
microcysts of myxomycetes, 7, 19–22
microfauna, 2, 35–6, 40, 62, 71, 91; *see also* nematodes and Protozoa
microflora, 37–40, 71, 93, 107, 112, 114, 133; *see also* Actinomycetes, bacteria, fungi, microorganisms and yeasts
microorganisms (= microbes): as food, 2, 7, 68, 168; catabolism of, 64; effect on hormones of, 38, 115–117; in arthropod gut, 114–15, 173–97, 205, 209–12, 216–24; inoculation of, 109–15; interactions in decomposition, 90–115; in wood, 94, 97–9, 106–7, 283; myxomycete interactions, 2–4; nutrient immobilization by, 35, 42–6, 53, 59, 62–5, 76–80, 102; rhizosphere-associated, 38–40; soil, 62–82; termite hindgut, 186, 196–97; *see also* Actinomycetes, bacteria and fungi
microsites: in nutrient fluxes, 49, 77–8; insect gut, 216; rhizosphere 35, 49, 54, 65; soil, 68
millipedes (Diplopoda): and nutrient cycling, 62–4, 70–82; grazing by, 3, 23, 62–4, 70–82; guts of, 209, 212, 214, 216, 222–3; in wood, 94, 124
mites (Acari): fungivorons, 142, 149, 282, 297; grazing by, 3, 62–4, 68, 282; guts of, 214, 215: in wood, 114, 134, 136, 142; pathogenesis in, 229, 239; predatory, 53, 282–3; vectors, 145, 271, 281, 285, 298
models, 36, 272, 308–10; of interactions, 3, 68, 71, 77, 309–10; pathogens and vectors, 272, 303–28; phenological, 322; process, 68, 71, 308, 310, 314–15; simulation of fungal and host development, 308–10, 320–1, 325–6; soil nutrient cycles in, 60, 68, 77, 82; of spruce budworm infection, 316, 318; systems, 308, 326–8
moisture (= water, humidity), 8, 9, 48, 76, 112, 115–16, 120; status and tree infection, 134–5, 140–5, 285, 288, 293, 297, 300; conditions and insect infection, 232–3, 236, 256, 320
molluscs, 3, 22, 62, 109
moulting (= ecdysis), 242–3, 312–15; fluid, 239, 242–3
mucigels, 38–9

mucilage: adhesive, 230, 235, 241; mucus in, 146, 209, 213; secretion of, 9, 38–9, 235, 275; toxic mucus in, 146
mycetangia, 100
mycorrhiza: biomass, 52; grazers, 41, 51, 52; growth of, 51–4; in nutrient transport, 40–1, 49, 82, 123, 192; in soil, 36, 82
mycosis, 152, 256–9, 318–19, 323, 325
mycotêtes, 103
Myriapoda, 205, 212, 215
myxomycetes: and nutrient flow, 2, 13, 21; as fungal food, 20–2; as invertebrate food, 22–5; biomass, 2; chemotactic interactions, 6, 9, 11, 26; coprophilous, 28–9; dispersal of, 22, 27–8; ecological role of, 1; energy flow and, 2; fungal feeding of, 12–17; in forests, 1, 9; in wood decomposition, 14–15; microbial ingestion by, 1–12, 17–20; myxamoebae, 1, 3–10; parasitic associations, 17, 20–2; plasmodia, 1, 3, 10–20; osmotrophic nutrition of, 4–5, 10; spores and sporangia, 1, 3, 7, 16, 21, 23–8; swarm cells, 1, 3–10; trophic interactions of, 2–26; walls of, 22

naphthalene, 258
nematodes: bacterial feeders, 40–54; fungivorous, 46–51, 142, 149, 282; grazing and nutrient flow, 62, 65, 69, 71; in wood, 91, 113, 124, 134, 136, 142; myxomyceticolous, 3, 20, 21, 23; vectors, 109, 145, 281, 285, 298
nitrate-N, 36, 70, 74–6, 82, 245, 247
nitrogen: assimilation efficiency, 62; capital, 59; conservation, 192–6; content of fungi, 102, 103, 168; content of plants, 48–50, 168; content of wood, 100–3, 125, 174, 192; effect of grazers on mobilization, 40–53, 59–82; fixation, 38, 188–9, 192; fixation in termites, 188–97; flow, 60, 62, 81, 91, 120–3, 174, 193; in chitin, 47–8, 52, 195, 244–5; in soil, 41; mineralization, 40–53, 59–82, 188–97; mineralization and excretion, 60–2, 77–9, 120; nitrification, 75, 77; nitrifying bacteria, 75; Relative Nitrification Index (RNI), 75; source, 123, 164, 231, 244–7; *see also* carbon–nitrogen ratio, and nutrient fluxes
nutrient: availability, 53, 59, 78, 81, 100; budgets, 60–2; immobilization, 35, 42–6, 53, 59, 60, 62–5, 71, 76; pools, 60, 68, 79–81; translocation by fungi, 40–1, 49, 82, 123, 192
nutrient fluxes: and grazing, 38–53, 61–82; and effect of predator–prey interactions,

Subject index 347

37, 53, 61; and soil properties, 48–9, 60, 68, 73–7, 80–2; and species complexity, 42–6; conservation in, 59, 195; direct animal effects in, 38–51, 60–3, 77, 123; fast-slow cycle in, 40–1; in decomposition, 90, 100–6, 120–3; indirect animal effects in, 51–4, 60–9, 77–82; in litter, 59–82; input, 38–9, 61, 91; in rhizosphere, 35–54; in wood, 120–3; leaching loss and, 66–80, 123; microbial regulation of, 37; mobilization in, 37–53, 59–82, 282; myxomycetes in, 2; pathways, 2, 61; physical variables of, 75–7; proceeses of, 53–4, 59–61, 69–73; rate of, 40–1, 48, 81, 123; relative contributions of organisms in, 37, 49, 61–2; release of nutrients and, 40, 46, 60, 62, 65, 77, 123; seasonality of, 62
nutrient quality: and fungal growth, 281–2, 288; and decomposition, 69, 94, 120–3; of bacteria, 168, 182, 194, 196; of decayed wood, 100–6, 192; of detritus, 167–8; of fungi, 53, 66–7, 103, 168, 195; of humus, 124; of litter, 49, 60–1, 80, 192; of plant material, 48–9; of wood, 100–6, 120–5, 192

oak wilt, 112, 143–5, 301
oidia, 99, 111
Oligochaeta, *see* earthworms and enchytraeids
osmotrophic nutrition (absorptive), 4–5, 10, 14
oxalic acid, 251

paramecia, 19
pathogenic and parasitic interactions: and decreased host resistance, 311, 326–7; and enzymes, 244; and genotype, 248, 277, 290, 297–302; and nutritional state, 312; animal–Actinomycete, 7; balance, 272, 301–2; between fungi, 283; between myxomycetes and fungi, 17, 20; cuticle-degrading enzymes in, 236–51; density-dependent infection in, 319, 321; evolution of, 136, 272, 300–2; inhibition by microbes, 39; insect (entomo-pathogenic), 21, 229, 260, 307–26; life cycles of, 111–12, 137–8, 274, 278, 281–2, 312, 325; models of, 272, 303–27; nematode–insect, 282; of microbes in guts, 209, 252–7; plant (phytopathogenic), 39, 134, 143–9, 229, 233–5, 249–50, 271–302, 310; strain specificity in, 230–1, 234, 248–9, 260, 298, 302; sublethal doses in, 311, 326; synchrony in, 111–12, 314, 325; *see also* fungi, infection, vectors and virulence

pectin, 140, 160–4, 230
penetration: by infection peg, 235–55; host response to, 257–60; gut protection from 209, 217, 219; of gut, 209–11, 236, 252–4, 313; of haemocoel, 254–7; of insect cuticle, 255–60, 313–14; of plant cuticle, 244, 249; *see also* infection
perithecia, *see* fungal fruiting bodies
pests, and their management, 133–6, 145, 300, 302, 307, 310–11, 322; *see also* pathogenic interactions
pesticides, 135, 256, 307, 309
pH: enzyme activity and, 161–4, 182, 211, 244–5, 251; in guts, 177, 211, 215–16, 221–4, 253; in soil, 49, 68, 70, 75, 80–1; of myxomycete cultures, 18; of protozoan cultures, 182
phagocytosis, 6–12, 17, 258
phenols: in insect cuticle, 257–8, 298; in wood, 99, 135
phloem, 110, 134, 136, 138, 141–2, 146–7, 287
phytoalexin, 258
phytopathogenesis, *see* pathogenic interactions
plants (vascular), 48–50, 168, 174; *see also* detritus, leaf-litter, litter, roots and wood
polysaccharides: dissimilation in wood, 106, 125, 179, 182–6, 192; in protective slime, 27; in plant cell walls, 38, 125, 155, 243–4, 249; microbial, 38, 258; myxomycete, 22; types, 160, 164, 197; *see also individual polysaccharides*, decomposition, digestion and enzymes
phosphorus: cycle in soil, 40, 53, 62, 65, 81; cycle in wood, 91, 101–4, 121–3, 214; effect of grazing on mobilization, 41–44, 53, 65; microbial immobilization of, 42–6, 63; mineralization of, 37, 41–6, 53, 65; nutritional interaction and, 18; stability of, 37
populations: bacteria, 44, 46, 51, 178; grazer, 65–7, 71, 81; growth and nutrients, 65–7; gut symbionts, 174–5, 178, 187, 216, 221–2; host insect, 319, 326; host tree, 272, 302; interactions and, 51–3, 65–7, 71, 81, 272, 322; models of, 322; pathogen, 272, 277; predator-prey, 37, 61; Protozoa, 44, 175, 178; termite, 183, 189; vector, 282, 290, 299, 302, 311; *see also* biomass and standing crop
potassium: in gut, 212–13, 221; in wood, 101–4, 121–3; mobilization and grazing, 62–3, 70, 73, 76, 82, 91
predation: by myxomycetes, 2–19; in interactions, 37, 61, 136, 139, 282–3;

Subject index 348

predation—*contd.*
  nutrient flow and, 53; on myxomycetes, 7–29; *see also* grazing
production: efficiency, 42; primary, 51, 53, 59, 62; secondary, 53, 60, 69, 71
proprionate, 180, 185–6
protein: enzyme digestion of, 164, 166, 240–7; in insect cuticle, 230, 240–5, 249, 251; in myxomycetes, 22; nutrient source, 245; *see also* digestion and enzymes
protoplasts, 255, 259, 314
Protozoa: anaerobic fermentation of cellulose, 179–86, 197; as food, 19–20; biomass, 42, 178; grazing and nutrient flow, 37–8, 44–6, 49, 54, 62, 65, 69, 71; gut symbiont, 125, 155–6, 173–88, 196–7; Microsporidia, 312–13, 322, 326; *see also* amoebae and bacteria

quinone, 258
quinolone, 217

resources: *see individual resources* (bacteria, cuticle, detritus, excretory products, fungi, litter, plants, roots and wood), diet, nutrient quality and nutrient source
respiration, 37, 42–7, 62–5, 76, 116
reproduction, 4, 20, 71, 103–5, 117, 149, 192, 274–5, 282, 298, 311, 326; *see also* genes and life cycle
resin, 100, 135, 146, 149
rhizosphere, 35–58; *see also* root(s) and soil
root(s): associated microbes and plant growth, 38–41; as termite food, 174; exudate, 36–40; grazers, 36–7, 41–9, 53; growth, 48, 54, 81; infection route, 113, 136, 146–7; nutrient flow and, 36–7, 40–54, 59–60, 89; *see also* mycorrhiza and rhizosphere
rumen bacteria, 178–83

saprotrophy, 126, 134
saprophagy, 25, 60, 71, 138–9, 275–80; associations, 146–9
saprophytic, 244, 275–82, 290, 299–300
sapstain, 113, 125, 133, 143
sclerotia, 141, 144, 146, 301
season, 62, 136, 138, 143, 145, 274–5, 278–82, 288–90, 298–9, 312, 314, 319–26; *see also* dormancy
secretion: adhesive, 235; in gut, 78, 193, 212; of attractants, 6;;of enzymes, 166, 185, 237–51; root, 38–9
selection, *see* evolution
shipworms, 113
siderophores, 40

Siricidae (woodwasps), 155, 160–4
slugs and snails, *see* molluscs
sodium: in guts, 213–14, 219, 221; in soil, 70, 73, 82
soil: acid, 49, 68, 70, 75, 81; agricultural, 39–40, 60, 80; effect of particle size, 35, 49, 54, 64–8, 71, 116–17; forest, 60–1, 73, 75, 81; grassland, 61, 80; interactions in, 59–82, 107, 222; moder, 64, 68; myxomycetes in, 1–2, 9, 11; nutrient flow in, 35–54, 59–82; spores in, 314, 320, 323; *see also* leachates, rhizosphere and roots
solar radiation, 48, 142, 146, 293
Spirochaetes, 178–81, 197, 219
species: complexity and nutrient flow, 42–5, 49; dominance and interactions, 65, 67, 107, 117; selection by fungivores, 71–3, 118, 125; specific host response, 249, 260
spores; and insect pathogenic infection, 231–6, 249–54; ascospores, 8, 142–4, 252, 275–6, 282–4, 288, 293, 297, 301; asexual, 142, 149, 281; as food, 7–10, 17–18, 139, 142, 144, 149, 282–3; basidiospores, 8; blastospores, 255; germination, 26, 117, 143, 231–6, 250, 253–4, 313–14, 318, 323; resting, 313–14, 318–26; sexual, 149; sporangiospores, 8; toxin production by, 27, 252; viability, 7, 17–18, 28, 109, 118, 142, 253–4, 284, 293, 318, 323; wall composition of, 8–9, 17, 26; zoospores, 230–1, 238; *see also* conidia, disease transmission, infection and vectors
spruce budworm, 309–27
standing crop: bacterial, 81; faunal, 61; fungal, 65–7, 81
starch, 125, 140, 164, 167, 181, 221, 244
succession, 14, 107, 112–17, 139, 147, 288–90
sugars, 22, 26, 89, 125, 134, 221, 242, 246, 257
symbiosis: evolution of, 111, 125–26; fungal exosymbionts, 98, 103–6, 110–12, 117–18, 155–68; gut microflora in, 105–6, 125, 155–61, 173–97, 210–11, 219–23; insect-decay fungus, 103–4, 109–12, 125, 146, 149; of myxomycetes, 11, 18–20; wood-boring beetles and microflora, 112, 125, 139; *see also* ambrosia, bark and wood-boring beetles, guts and termites

tannins and tannin complexes, 135, 216, 230, 236, 240
temperate forests, *see* ecosystems

Subject index

temperature: and enzyme activity, 221; and fungal pathogens, 281–2, 293, 315–27; and beetle activity, 275, 315–27; soil metabolic activity, 75–7, and wood-decomposer organisms, 115

termites: attractants of, 96–7; diet of, 173–4, 181–3, 188, 192–6; fungus-growing, 97–8, 103–4, 111, 117, 155–61; gut symbionts in, 105–6, 125, 173–97, 210–11, 215–23; in wood decomposition, 90–3, 99–100; pathogens of, 237, 252, 254, 256; *see also* digestion

toxic compounds: anti-haemolytic, 260; effects of pathogen, 256–7; enzymes, 253, 256; fungal products, 96–7, 252–6; gut products, 215, 217; insect products, 232, 246; wood products, 135; *see also* allelopathic substances, antimicrobial agents, insecticides and pesticides

trophallaxis, 176, 194, 196

trophic interactions: between arthropods and fungi, 104–6, 155–68; in soil, 35–54, 59–82; in wood, 89–126, 133–49, 313; nutritional, 98, 103–25, 178–97, 216; of myxomycetes, 2–28; *see also* interactions

tropical rain forest, *see* ecosystems

tryptophan, 6, 217

turpentine, 100

tyloses, 135, 278, 298

urea, 78, 217, 246–7

uric acid, 78, 117, 193–5, 217; uricolytic bacteria and, 193–5

urine, 176, 212, 217

vectors, 109–11, 139, 142–9, 271–93, 298–303; bird, 144; mite, 145; nematode, 109; of myxomycete spores, 22, 27–9; opportunist, 111, 113

virulence: and aggressive strains, 133, 272, 279, 301; in entomopathogenesis, 230, 233, 244, 248, 255, 259; in phytopathogenesis, 244, 248; nutrition and, 20

virus, 311, 313

wasps, *see* Hymenoptera, and Siciricidae (woodwasps)

water, *see* moisture

wood: as resource, 89–126, 192, 285, 288; bark resistence to invasion, 89, 111–16, 134–40; biomass, 89; canopy, 90–2, 97, 99, 103, 109–12, 134–6, 273–4, 314, 323; decay process, 89–126; digestion, 155–68, 173–97, 216–17; heartwood, 99–100, 105, 110–13; litter, 90–9, 106–7, 115–26, 192; living, 134–8, 146, 271–4, 277; microbial conditioning of, 94–106; microclimate in, 115–17, 120, 140–3; nematodes in, 91, 109; nutrient source, 89, 94, 100–6, 111, 120–5, 174–98, 285, 288; relative density, 91–101, 118–19; sapwood, 103, 110, 117, 138, 140, 144–5; stumps, 90, 97, 111, 123, 136–9, 149, 156; wound invasion, 91–2, 112, 135, 142–7, 271, 273, 278; *see also* decomposition, diet, digestion, ecosystems, enzymes, infection, pathogens, phloem, termites, wood-boring beetles and xylem

woodlice (Isopoda), 3, 22–3, 28, 64, 68, 94, 209, 217

xylan, 160, 162, 164, 179

xylem, 134–41, 274–81, 293–9

yeasts, 3, 10, 12, 20–1, 27, 107, 114–15, 134